Multivariate Statistical Analysis for Geographers

MULTIVARIATE STATISTICAL ANALYSIS FOR GEOGRAPHERS

Daniel A. Griffith
Syracuse University

Carl G. Amrhein
University of Toronto

PRENTICE HALL, *Upper Saddle River, New Jersey 07458*

Library of Congress Cataloging-in-Publication Data

Griffith, Daniel A.
Multivariate statistical analysis for geographers / Daniel A. Griffith, Carl G. Amrhein.
p. cm.
Includes bibliographical references and index.
ISBN 0-13-605692-X
1. Geography--Statistical methods. 2. Multivariate analysis.
I. Amrhein, Carl G. II. Title
G70.3.G746 1997
910'.07'27--dc21 96-53264
CIP

Acquisition Editor: ***Daniel Kaveney***
Production Editor: ***Joanne E. Jimenez***
Cover Designer: ***Karen Salzbach***
Manufacturing Manager: ***Trudy Pisciotti***

Printed in the United States of America

10 9 8 7 6 5 4 3 2 1

ISBN 0-13-605692-X

PRENTICE-HALL INTERNATIONAL (UK) LIMITED, LONDON
PRENTICE-HALL OF AUSTRALIA PTY. LIMITED, SYDNEY
PRENTICE-HALL CANADA INC. TORONTO
PRENTICE-HALL HISPANOAMERICANA, S.A., MEXICO
PRENTICE-HALL OF INDIA PRIVATE LIMITED, NEW DELHI
PRENTICE-HALL OF JAPAN, INC., TOKYO
SIMON & SCHUSTER ASIA PTE. LTD., SINGAPORE
EDITORA PRENTICE-HALL DO BRASIL, LTDA., RIO DE JANEIRO

This volume is dedicated to Edwin and Patricia Amrhein, and Edward and Mary Calabrese, my parents and parents-in-law. Ellen and I are forever grateful for their continued encouragement and support to excel in all we do.

Carl G. Amrhein

✪

This volume also is dedicated to the memory of Donald S. Griffith, Sr. (1921-1995).

Daniel A. Griffith

CONTENTS

PREFACE

The classic geography text concerning multivariate statistical analysis is by Johnston (1978). Since its publication, there has been no comprehensive effort to produce an updated text that incorporates recent developments in spatial statistics, either those statistics developed from first principles for spatial analysis (e.g., trend surface analysis, the treatment of aggregation effects) or adjustments to classical statistics that incorporate the reality of spatial relations (such as spatial autoregressive models).

Our first book, *Statistical Analysis for Geographers* (Griffith and Amrhein, 1991), introduces spatial statistical concerns in univariate and bivariate statistical analyses. Although not an "easy read" (at least not in the opinions of our students), the book has established that there is a need at the elementary level for a rigorous treatment of this material, one to be studied by those preparing to become spatial scientists.

Recent texts in geostatistics (Isaaks and Srivastava, 1989; Carr, 1995) cover a set of techniques important to earth sciences, but without examining the full range of techniques that are of interest to geographers (for example, spatial autoregressive models). Other recent texts on spatial data analysis cover multivariate analysis in one or two chapters (Haining, 1990) or as sections of a few chapters (Bailey and Gatrell, 1995). While very effective in meeting the ambitious goals set by their authors (both books cited here are very good and highly recommended), these texts do not devote sufficient space to the necessary detailed treatment we currently feel is lacking in the literature. Unfortunately, other texts that purport to cover spatial data analysis (see, for example, Walford, 1995) do not go much beyond centrographic measures in their treatment of spatial statistics (Geary and Moran do not even appear in their respective indexes). For others, the vital information of spatial autocorrelation is relegated to the status of a curiosity: "How one can tell statistically if there is [spatial] autocorrelation in a particular situation is a procedure that is peripheral to our discussion" (Earickson and Harlin, 1994, p. 207). Of note is the title of Earickson and Harlin's book: *Geographic Measurement and Quantitative Analysis*! Finally, while the introductory statistics level is becoming well served (a very encouraging trend) by rigorous treatments of spatial statistics (see Burt and Barber, 1996, for the most recent good example), there remains an absence of such synthesis and methodological discussion in the multivariate arena.

Our specific concern, and a primary motivation for writing this book, may be summarized by Bailey and Gatrell (1995, p. 314):

> Specifically spatial forms of multivariate analysis are not well developed and those that do exist are somewhat specialized and rather advanced. What is often done is to use conventional multivariate techniques for purposes of reducing the multivariate attributes to one or two dimensions (significant components) and then to look at these dimensions from a spatial perspective, seeking spatial patterns, relationships and classifications.

We feel strongly today that sufficient conceptual development exists in spatial statistical theory and methodology and that geographers routinely have access to sufficient computing power for us to write this introductory multivariate text in which spatial analysis is the central theme, not merely a special case of classical aspatial techniques and thinking. Our goal in this book is to fill this gap in the literature, a gap that has continued to grow since the publication of even Johnston's now-outdated 1986 edition of his text!

This book follows directly from our first book, *Statistical Analysis for Geographers* (Griffith and Amrhein, 1991). We use as our primary data set the same data set from Puerto Rico featured in our first book. While some students have commented about learning far more about Puerto Rico than they ever imagined, we feel that there is important value added by repeatedly analyzing the same data set from many perspectives. In addition, as an island, with its absence of certain confounding influences (e.g., edge effects), Puerto Rico allows us to simplify some of our analyses. While our simplifications may not be obvious in a first reading, the reader should realize that the complex ideas reviewed in this text could have been presented in a far more complicated manner! In adapting Ackoff's (*Management in Small Doses*, 1986) famous passage to our circumstances:

> I once [knew] ... a professor, who wrote a highly technical [book]. I asked him to assume that I was an ordinary [researcher]. Would he explain his [text clearly, simply and concisely]?
>
> He went to the blackboard and began to cover it with mathematical symbols. I stopped him to remind him that I was an ordinary [researcher], not a mathematician. After a long pause he said, "I don't understand what I've done well enough to explain it in nontechnical language."

Unlike Ackoff's professor, we believe that we express ourselves well in ordinary English, with a minimum of technical jargon, to convey what we are talking about. We paraphrase considerable mathematical and statistical knowledge as rules of thumb and substitute computer demonstrations for theorems and proofs.

To broaden the range of data to which the analyst has access, we introduce in the problem sections an important source of spatial data in digital form. Andrews and Herzberg have created a site on the Internet that permits the downloading of every data set described in their 1985 book which has a rich array of explicitly spatial data, as well as data sets that deal with explicitly spatial problems (e.g., biological populations, marketing, public health). We encourage the exploration of this resource, for the category of spatial data as well as other categories that might contain additional data sets useful to the spatial analyst. Since Andrews and Herzberg reference the original statistical analyses that use these data sets, our selection of example data sets and applications is effectively increased. At the time of writing, access to the Andrews/Herzberg digital tabulations is provided through the anonymous ftp site address utstat.toronto.edu in the directory `pub/david/data-collect`. If this Internet access fails, contact the Department of Statistics, University of Toronto, phone (416) 978-3452, fax (416) 978-5133. Alternatively, contact Amrhein and he will link you to the current location.

Acknowledgments

The assistance of many people made this book possible. Both of us enjoyed the complete support of our departments. Jonathan Hancock and Peter Fellows, from Syracuse University, reviewed early draft versions of parts of this manuscript and provided a number of important suggestions and corrections that improved the final product. Harold Reynolds, from the University of Toronto, contributed the initial draft of §3.0. Michael Kirchoff, of the Cartography Laboratory at Syracuse University, provided a very high level of professional assistance (with remarkable turn-around time) with the illustrations and figures. Neely Law, at the University of Toronto, offered expert assistance with word processing, particularly in forcing the equation editor to serve our needs (rather than the other way around!). And, Kay Steinmetz, at Syracuse University, provided expert copy-editing advice. Finally we would like to thank the referees of the first draft of the text; their thorough reviews and insightful comments greatly helped in shaping our final draft.

Needless to say, of course, we alone assume full responsibility for any errors in writing, logic, calculations, and content.

Daniel A. Griffith
(griffith@maxwell.syr.edu)

Carl G. Amrhein
(carl.amrhein@utoronto.ca)

References

Ackoff, R. 1986. *Management in Small Doses.* New York: Wiley.

Andrews, D. and H. Herzberg. 1985. *Data: A Collection of Problems from Many Fields for the Student and Research Worker.* Berlin: Springer-Verlag.

Bailey, T. and A. Gatrell. 1995. *Interactive Spatial Data Analysis.* Harlow: Longman Scientific Publishers.

Burt, J. and G. Barber. 1996. *Elementary Statistics for Geographers.* New York: Guilford Press.

Carr, J. 1995. *Numerical Analysis for the Geological Sciences.* Englewood Cliffs, N.J.: Prentice Hall.

Earickson, R. and J. Harlin. 1994. *Geographic Measurement and Quantitative Analysis.* New York: Macmillan.

Griffith, D. and C. Amrhein. 1991. *Statistical Analysis for Geographers.* Englewood Cliffs, N.J.: Prentice Hall.

Haining, R. 1990. *Spatial Data Analysis in the Social and Environmental Sciences.* Cambridge: Cambridge University Press.

Isaaks, E. and R. Srivastava. 1989. *An Introduction to Applied Geostatistics.* Oxford: Oxford University Press.

Johnston, R. 1978. *Multivariate Statistical Analysis in Geography.* Harlow: Longman Scientific Publishing.

Walford, N. 1995. *Geographical Data Analysis.* New York: Wiley.

About the Authors

Daniel A. Griffith has been at Syracuse University since 1988 and was previously a faculty member in the Geography Department at the University at Buffalo for ten years. He received his doctorate (1978) in geography from the University of Toronto and an M.Sc. in statistics from the Pennsylvania State University. He currently is Professor and Chair of his department and earlier was the Director of the Applied Statistics Program at Syracuse. In addition to numerous articles and chapters, he has (co-)authored/edited 12 other books, including *Spatial Autocorrelation: A Primer* (1987), *Advanced Spatial Statistics* (1988), and *Spatial Regression Analysis on the PC: Spatial Statistics Using SAS* (1993).

Carl G. Amrhein has been at the University of Toronto since 1986 and was previously a faculty member in the Geography Department at the University of Kentucky for two years. His doctorate in 1984 is from the University at Buffalo. He currently is the Chair of Toronto's Department of Geography and the Graduate Program in Planning. He has played a central role in the development of academic computing infrastructure, including negotiating several large gifts to the University of Toronto. He has published numerous articles and book chapters.

Previous collaboration between the authors includes, in addition to various articles and reports, one book entitled *Statistical Analysis for Geographers* (1991). Both authors are winners of the Nystrom Dissertation Award from the Association of American Geographers.

LIST OF ABBREVIATIONS

AHM	agglomerative hierarchical method
ANCOVA	analysis of covariance
ANOVA	analysis of variance
AR	autoregressive response (model)
CA	cluster analysis
CCA	canonical correlation analysis
CNT	count
CV	canonical variate
df	degrees of freedom
DFA	discriminant function analysis
DHM	divisive hierarchical method
DUD	doesn't use derivatives
FA	factor analysis
GIS	geographic information systems
GLIM	generalized linear interactive model
GLM	general linear model
GR	Geary Ratio
L-A	location-allocation
MANOVA	multivariate analysis of variance
MC	Moran Coefficient
ML	Maximum likelihood
MSE	mean squared error
MVN	multivariate normality
OLS	ordinary least squares
PCA	principal components analysis
SAR	simultaneous autoregressive (model)
SSE	error sum-of-squares
S-W	Shapiro-Wilk
VIF	variance inflation factor
WLS	weighted least squares

Multivariate Statistical Analysis for Geographers

PART I

INTRODUCTION AND REVIEW

CHAPTER 1

ELEMENTARY STATISTICS BACKGROUND

Data Preview. One of the sets of data that is used throughout *Statistical Analysis for Geographers* (Griffith and Amrhein, 1991), namely, the density of selected agricultural production in Puerto Rico for 1974, is employed for illustrative purposes throughout this textbook. These geo-referenced data represent a range of topics geographers commonly address using multivariate statistical techniques. A tabulation of these data appears in Appendix A. These particular data are utilized here because of the extensive analysis to which Griffith and Amrhein already have subjected them. In keeping with the first book, examples appearing in this book that are based upon these data are designated by labels having the prefix PR. This data set has been compiled from the 1974 *U.S. Census of Agriculture* and the Commonwealth of Puerto Rico *Facts and Figures on Puerto Rico's Agriculture*. Supplemental data also are used in this book but only sparingly, selectively, and mostly in the last few chapters.

There are 73 areal units (observations), called *municipio*s, in this data set (Map 1.1). For each municipio there are two labeling variables followed by three classification variables and eight agricultural variables, which historically have constituted the primary types of agricultural production on the island of Puerto Rico. Note that, unless otherwise indicated, all density measurements have been computed for a square-mile parcel. In addition, to facilitate certain illustrations used in this book, the 73 municipios have been grouped into their respective agricultural administrative regions and then arranged in alphabetical order, within these regions, by name. Following is a description of each variable together with an acronym-like mnemonic that appears in Appendix A, as well as in the examples and SAS code contained in ensuing chapters:

Label variable 1 — ordinal number referring to the alphabetical order position of the municipio name (**IDPR**)

Label variable 2 — name of the municipio (**NAME**)

Classification variable 1 — agricultural administrative region (**ADM**)

ADM = 1 if a municipio is located in the San Juan region
ADM = 2 if a municipio is located in the Arecibo region
ADM = 3 if a municipio is located in the Mayaguez region
ADM = 4 if a municipio is located in the Ponce region
ADM = 5 if a municipio is located in the Caguas Region

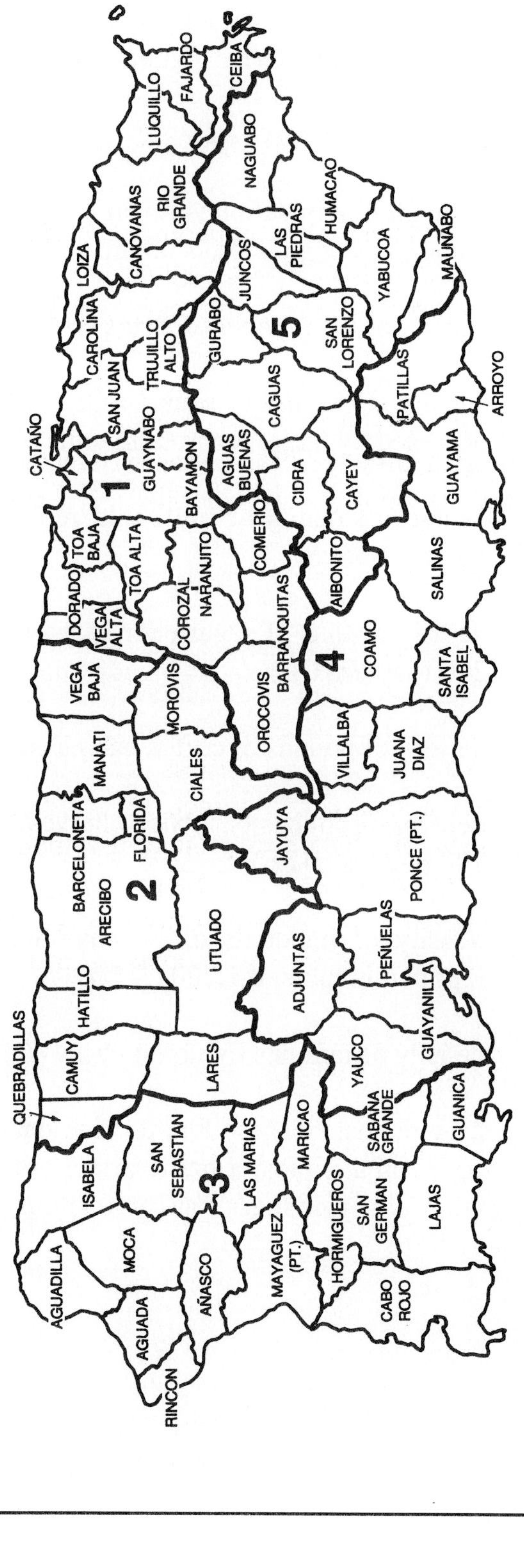

Map 1.1. The partitioning of Puerto Rico into municipios and agricultural administrative regions.

Classification variable 2 generic landform (***LND***)

LND = 0 if a municipio is located in coastal lowlands
LND = 1 if a municipio is located in interior highlands

Classification variable 3 rural/urban dichotomy (***RU***)

RU = 0 if ≤ 50% of a municipio's population is urban
RU = 1 if > 50% of a municipio's population is urban

Agricultural variable 1	density of the number of farms (***DFRM***)
Agricultural variable 2	density of *cuerdas* (1 square mile = 658.94 *cuerdas*) of land occupied by farms reported in units of 10 (***DCD***)
Agricultural variable 3	density of quarts of milk sold reported in units of 4,000 quarts (***DMLK***)
Agricultural variable 4	density of tons of sugar cane harvested reported in units of 200 tons (***DSGR***)
Agricultural variable 5	density of pounds of coffee harvested (***DCF***)
Agricultural variable 6	density of pounds of tobacco harvested reported in units of 200 pounds (***DTB***)
Agricultural variable 7	density of hundreds of bananas and plantains harvested reported in units of 1,000 (***DBN***)
Agricultural variable 8	density of families living on farms ***(DFAM)***

While a univariate problem illustrated by the use of these Puerto Rico data might determine the average yield of milk (DMLK) for the 73 municipios, a multivariate problem might relate the density of milk production of a farm to the density of sugar cane production (DSGR), farm density (DFRM), and landform classification (LND).

1.0. Introduction

The techniques of multivariate statistics provide a rich and powerful array of tools for analyzing large and complex geo-referenced data sets. When they are properly employed, important insights into the nature and degree of any relationship(s) between (among) geographic distributions can be examined. Over the past two decades, the increasing power and availability of computers, coupled with the increasing sophistication of statistical software packages (that is, statistical power combined with user friendliness and ease), has put complex statistical analysis capabilities within the reach of spatial analysts in many fields but who come from a wide range of backgrounds. Access to this analytical power, while able to provide great rewards in the form of insights, also imposes important responsibilities on the analyst, who must not be ignorant of what is being computed and who must resist the temptation to blindly execute whatever arithmetic operations a software package offers. These responsibilities require that the analyst

(1) not misuse geo-referenced data to explore meaningless (or spurious) relationships or relationships the data cannot support,

(2) not misuse statistical techniques, either intentionally or unintentionally, by failing to check the assumptions and requirements of each chosen technique (i.e., know what to compute and when to compute it), and

(3) base results and conclusions on the data, being careful to present additional speculation as such (see Searle, 1989).

These are momentous responsibilities, and our best advice is to follow faithfully a systematic approach to data analysis. In other words, do not jump immediately into a complex multivariate analysis without first exploring the basic, descriptive characteristics of a geo-referenced data set.

The first three chapters of this book present the tools required for this approach, with Chapter 1 reviewing basic descriptive and univariate statistics (presumably learned in an earlier statistics course). Chapter 2 examines the basic components of spatial data, and Chapter 3 reviews the elementary *language of multivariate statistics*—linear algebra.

We turn attention first to an overview of basic conventional statistical concepts.

1.1. A Review of the Basics

In *Statistical Analysis for Geographers* (Griffith and Amrhein, 1991), there is repeated emphasis that the best approach to any statistical problem is to start with the basics. By this we mean that the analyst must first "crawl" with the data before "walking" or "flying" with it. We are not alone in our views, as can be seen in two recent (and excellent) books that we use throughout this text as supporting references (Bailey and Gatrell, 1995; Hair et al., 1995). We continue an approach adopted in our first book: presenting statistical wisdom and (we think) proper procedure in the form of rules of thumb. Our treatment here is brief, and any good introductory book should be consulted for more extensive material (see Griffith and Amrhein, 1991; Burt and Barber, 1995).

> *RULE OF THUMB 1.1.* Select a measurement scale for each areal unit attribute and decide whether or not it is valid and reliable (see Velleman and Wilkinson, 1993, for a counterargument against beginning from this vantage point).

Griffith and Amrhein (1991, pp. 36–42) extensively discuss the various measurement scales (i.e., nominal, ordinal, interval, ratio) that can be used to quantify attribute information. Measurement scales can be ordered, in terms of increasing information content, as nominal, ordinal, interval, and ratio. The interval and ratio scales can be, and frequently are, treated as a single metric scale, while nominal measures constitute a categorical scale, and ordinal measures furnish the foundation for nonparametric analyses (in the absence of ties, these measures always conform to a uniform distribution). A special (and important) categorical scale case is the binomial variable, which, although nominal in nature, can be converted to the interval/ratio scale by calculating the logarithms of probability ratios. Whenever possible it is best to use metric data. However, if there is a lack of confidence in the quality of metric scale data, it is possible to down-scale these data to rankings or categories (note: it is far more difficult to rescale ranked or categorical data to metric scale data); selection of an appropriate measurement scale type helps ferret out nonsense that might be suggested by a naive statistical analysis.

An analyst may lack confidence in data for many reasons. Poor measurement instruments (e.g., malfunctioning, obsolete or improperly calibrated equipment; poorly constructed and/or administered surveys) provide one source of uncertainty. Who collected the data provides another source of concern: primary data (collected by the analyst her/himself) is likely thought of as being most reliable, while secondary and tertiary data may foster less confidence in the accuracy and precision of the collected data. Sample variability is yet a third source of uncertainty, the one statistics is best suited to handle.

> *RULE OF THUMB 1.2.* Define the sample (determine sample size and ensure a random component) and identify its parent statistical population.

The statistical population is composed of the set of all possible areal units that might appear in a sample. For example, a sample of households in New York State could be selected from a statistical population that includes every geo-referenced household located within the boundaries of the state of New York. A sample of trees in a forest stand could be selected from a statistical population of every geo-referenced tree in the forest stand. Ultimately, the goal of inferential statistics is to say something useful about the parent population based only on a sample. An important assumption of virtually every technique discussed in this book is that the sample has a random component. As a result, ensuring that the sample, in fact, has a clearly identifiable random component is important. In its purest sense, a random sample is one in which every element in the population has an equal chance of being selected. However, for many sound and convincing reasons, complex sampling strategies that constrain the probabilities of a particular areal unit being selected have been developed (e.g., see Levy and Lemeshow, 1991; Thompson, 1992). At this point, suffice it to say that the analyst must be fully informed of the sampling design employed in the collection of a sample and fully aware of the random component it should embrace. To control the random component arising from sampling variability, the sample should be as large as is practical (affordable in terms of collection time and costs). On the one hand, the sample should be larger than 30 (some say 100) so as to be firmly in the realm of large-sample theory. Small samples introduce a number of additional and usually unnecessary complications into an analysis. On the other hand, there is no point in selecting a sample that equals the population in size, because then there is little need for either sampling or inferential statistics! Unfortunately, many geographical problems involve a fixed number of areal units, thus predetermining the sample size, and a conceptual or superpopulation that actually cannot be observed in its entirety.

> *RULE OF THUMB 1.3.* Define procedures for dealing with missing data.

Missing data interject further complications into an analysis. At the very least, if observations for which some measurements are missing are excluded from an analysis, the sample size shrinks. In addition, if there is some pattern to the areal units with missing values, then eliminating these observations may seriously undermine the sampling design (hence corrupting the random component) and may well insert a serious bias into the sample data. If a pattern in the missing data is detected, the analyst may have to resample the population or modify the structure of his/her analysis to compensate for this trend.

Recently a great deal of effort has been expended on developing techniques to estimate missing data values (see Little and Rubin, 1987). This estimation, if done rigorously, is greatly preferred to dropping an areal unit from an analysis; in fact, removing areal units from maps creates holes in them, and hinders the calculation of spatial autocorrelation statistics. However, missing data estimation techniques are still being developed, and indeed add certain complexities to an analysis (over and above additional computational burdens). Sophisticated procedures for coping with missing spatial data are outlined in Griffith, Bennett, and Haining (1989), among others.

> *RULE OF THUMB 1.4.* Visualize the geo-referenced data, one variable at a time, using histograms or bar charts, boxplots, maps, pie charts, and so on.

It is said that a picture is worth a thousand words. In statistics, a good graphic is likely worth at least a thousand calculations! Nothing is more important in a statistical analysis than for the analyst to "get a feel" for the data. Getting a feel for data begins with data visualization. Many embarrassing conclusions can be avoided if the analyst has a good sense for the nature of the data as revealed in data pictures. The method of visualization chosen is a function of both the data and the objective of a particular analysis. Obviously, data that are geo-referenced should be mapped, with an eye toward identifying spatial patterns (often focusing on spatial autocorrelation). Spatial patterns might be revealed through clustering of similar values, dissimilar values (such as a chessboard), or randomness.

In the case of data that are not spatially referenced, the popular range of visualization options includes pie charts, histograms, bar charts, boxplots, and scatterplots. In this context, the shape of the distribution (for metric scale data) is of primary concern. In most instances, a bell-shaped distribution is desired. Even if an empirical frequency distribution fails to conform closely to a normal frequency distribution, it is useful for this empirical frequency distribution to have symmetry about some point. Fortunately, progress has been made recently in dealing intelligently and directly with nonnormal distributions, such as the binomial and the Poisson (see Chapter 12).

In every instance, one of the key reasons for visualizing data is to identify unusual, atypical, aberrant, abnormal, anomalous, deviant, rogue, or extreme observations (referred to as outliers and in some contexts as influential or leverage points).

> *RULE OF THUMB 1.5.* Examine identified outliers, double-checking for errors in the recording or the subsequent encoding of measurements, on the one hand, or the possibility of having sampled from multiple populations, on the other hand.

The identification of spatial or statistical distribution outliers (unusually relatively large or small values or mismatched signs) initiates a secondary procedure to verify that the observation is, in fact, extreme and not an instance of incorrectly coded data. Common coding problems include switching the positions of digits (190 is encoded as 910), misplacing decimals (20.90 is encoded as 2090 or 209.0), and using an incorrect metric (recording some measures in meters and other measures in kilometers). If all the possibilities for common data entry mistakes are eliminated and the numerical magnitude of the anomaly in question is feasible, then the observation probably should be assumed to be a true outlier.

There are two schools of thought on what to do with an outlier. Some analysts argue that under certain conditions the outlier can be deleted from a sample data set, whereas others argue that such an action is never justified. It is possible that the outlier, at the end of the analysis, will prove to contain vitally important information. We subscribe to the latter view; that unless an outlier can be demonstrated to be a gross error, it is a legitimate areal unit attribute value and should not be removed from the analysis.

> *RULE OF THUMB 1.6.* Calculate measures of *central tendency*, *dispersion*, statistical distribution shape (e.g., symmetry—*skewness*—and peakedness—*kurtosis*), and *spatial autocorrelation.*

In undertaking manual calculations, especially historically, most analysts have stopped once means and variances have been calculated. However, with the aid of computers it is easy to generate a complete set of descriptive statistics, including quantiles. These descriptive statistics summarize and index the complexity of data and augment visualizations with quantitative gauges (statistics). From this point on, as the statistics become more sophisticated, it is important that the analyst not lose a feel for the data.

EXAMPLE PR-1.0. Compute univariate summary statistics and construct simple graphical portrayals for the original Puerto Rican data set attributes.

For comparison purposes with results reported throughout *Statistical Analysis for Geographers* (Griffith and Amrhein, 1991), the following results have been obtained with the SAS procedure PROC UNIVARIATE:

variable/region	n	$\bar{y}$	$\tilde{y}$	s	x_{min}	x_{max}	skew-ness	kurto-sis
DFRM/island	73	8.85	8.30	5.12	0.7	21.2	0.440	-0.782
San Juan	19	7.66	5.50	5.69	0.7	18.6		
Arecibo	11	9.75	8.90	5.23	3.0	18.3		
Mayaguez	16	10.34	9.55	5.19	2.4	21.2		
Ponce	14	6.94	6.00	4.62	0.7	15.6		
Caguas	13	10.08	10.00	4.14	4.3	15.4		
DCD/ island	73	36.17	37.30	11.96	2.9	62.9	-0.282	0.226
San Juan	19	26.44	24.50	13.38	2.9	53.6		
Arecibo	11	39.64	36.90	8.82	25.6	53.4		
Mayaguez	16	38.83	39.65	10.61	21.6	62.5		
Ponce	14	44.29	42.50	9.17	26.7	62.9		
Caguas	13	35.48	36.80	6.73	18.5	43.1		
DMLK/island	73	27.09	8.60	49.83	0.0	368.6	4.830	30.725
San Juan	19	28.06	19.20	29.53	0.0	94.0		

Arecibo	11	67.54	40.90	103.58	0.0	368.6		
Mayaguez	16	5.96	2.80	10.57	0.0	39.6		
Ponce	14	3.83	3.45	3.54	0.0	10.9		
Caguas	13	42.48	23.70	38.89	0.0	134.4		
DSGR/island	73	6.05	3.60	7.02	0.0	30.0	1.399	1.968
San Juan	19	2.30	0.80	3.16	0.0	10.0		
Arecibo	11	3.99	3.10	4.28	0.0	10.9		
Mayaguez	16	12.10	13.60	7.84	0.0	27.7		
Ponce	14	7.91	7.90	8.35	0.0	30.0		
Caguas	13	3.80	0.90	5.33	0.0	16.5		
DCF/ island	73	5781.58	683.60	11159.19	0.0	48874.9	2.736	7.120
San Juan	19	1239.07	31.10	3070.20	0.0	13265.8		
Arecibo	11	8464.12	610.40	15713.61	45.1	48874.9		
Mayaguez	16	9745.53	5690.50	13316.28	7.0	48140.1		
Ponce	14	9746.53	4633.85	13587.67	0.0	44081.6		
Caguas	13	1002.15	346.90	1513.97	13.0	4505.6		
DTB/ island	73	6.34	0.30	20.47	0.0	153.7	5.789	38.681
San Juan	19	11.28	0.00	35.50	0.0	153.7		
Arecibo	11	5.32	1.50	7.60	0.0	23.7		
Mayaguez	16	1.28	0.05	2.46	0.0	8.3		
Ponce	14	0.31	0.05	0.46	0.0	1.5		
Caguas	13	12.71	3.60	20.16	0.0	63.0		
DBN/ island	73	3.18	1.20	5.67	0.0	32.4	3.335	12.653
San Juan	19	2.40	0.40	4.55	0.0	19.3		
Arecibo	11	4.58	0.60	7.79	0.1	25.6		
Mayaguez	16	2.40	1.25	3.88	0.0	14.3		
Ponce	14	5.04	2.65	8.39	0.0	32.4		
Caguas	13	2.10	1.60	2.64	0.1	10.0		
DFAM/island	73	2.22	1.90	1.47	0.1	7.2	1.091	0.946
San Juan	19	1.60	1.10	1.23	0.1	4.4		
Arecibo	11	2.70	1.90	2.05	0.5	7.2		
Mayaguez	16	2.34	2.05	1.10	0.9	4.8		
Ponce	14	2.06	1.65	1.39	0.6	5.2		
Caguas	13	2.78	2.50	1.53	0.9	5.7		

These summary statistics are enlightening because they reveal that the original attribute values have means ($\bar{y}$), medians ($\tilde{y}$) and variances (s^2) that appear to be nonconstant across the island. The skewness and kurtosis statistics for a normal distribution equal 0. In all eight attribute cases the skewness statistics indicate a presence of moderate to marked asymmetry. For all but attribute DFRM, the kurtosis statistics indicate a presence of peakedness that is moderately to markedly less than that for a normal distribution (i.e., the shapes of the frequency distributions for these attributes are flatter than is the shape of the ideal standard bell-shaped curve); the frequency distribution for DFRM is more peaked than is its normal distribution counterpart. Histograms for these variables are presented in Figure 1.0; their boxplots appear later in this chapter in Figure 1.1.

In many applications, the ideal situation is encountered when the mean, median, and mode coincide in value (indicating a normal-looking distribution), with a small variance value (relative to the mean value). When the mean, median, and mode are markedly different, an analyst often must select one of these measures of central tendency as the guiding statistic for her/his analysis. In the absence of any other guidelines, the mean possesses the best set of statistical properties (i.e., unbiasedness, sufficiency, efficiency, consistency; see Griffith and Amrhein, 1991, pp. 48-50) for this purpose and usually is selected. If more sophisticated techniques (such as those reviewed in Chapters 4–12) are to be employed, the mean will have to be used because it is a key statistic (with the variance) for estimating the parameters of the statistical population under study. Unfortunately, in the presence of highly skewed data, the mean may be particularly inappropriate and the analyst may wish to report more than one central tendency at the descriptive stage of analysis (for example, because it is influenced far less by outliers, the median is usually used, with or without the mean, in studies that include economic data such as wages, housing prices, and income).

These first six rules of thumb describe a systematic procedure for undertaking a spatial statistical analysis that is applied to each attribute measured for each areal unit (if there are, for example, six characteristics for each geo-referenced household selected in a sample, then there will be six tasks to complete). Materials appearing in Chapters 4-12 extend this procedure, placing considerably more emphasis on the role played by spatial autocorrelation in geo-referenced data analyses, as well as special geographic problems.

> *RULE OF THUMB 1.7.* Construct bivariate scatterplots (that is, visualize the joint dispersion of pairs of attributes), where appropriate, and describe the nature of the relationship displayed.

At this point in a study, the analyst (sufficiently familiar with each variable for each areal unit) should begin to explore pairs of variables for bivariate relationships. While such an exploration may be undertaken regardless of the measurement scale on which data have been measured, we cast our discussion in terms of metric data for simplicity. Here the goal is to examine the nature (positive, negative, random) and the strength or degree (tight clustering about an invisible line of best fit, loose clustering, no clustering) of any relationship revealed by the data. Whether the direction of the relationship is positive or negative, ideally the values for the two variables should cluster tightly around an imaginary line that is well behaved (linear, sinusoidal, exponential, and so on).

> *RULE OF THUMB 1.8.* Calculate correlation coefficients.

In the presence of a straight line relationship between pairs of variables, the last step in a preliminary analysis is to index the nature and degree of this linear relationship with correlation statistics. If the bivariate scatterplots suggest a nonlinear relationship, then a simple linear correlation statistic may well reveal little. If the suggested relationship is linear, then a statistic such as Pearson's product moment correlation coefficient (r) will provide important insights for the next likely step in the analysis, namely addressing multiple attributes in the presence of redundant information.

Regardless of the level of sophistication involved in an analysis, stating generalizations almost always is a final intention, and hence the analyst wants to draw inferences about the parent population from which his/her sample was selected. Before exploring the techniques that form the core of this book, we next review the elementary aspects of inferential statistics, which also can be attached to these more sophisticated multivariate techniques.

1.2. Statistical Inference versus Statistical Description

One of the objectives of statistics is to infer statements about a parent population from one of its subsets, the sample data; more specifically, the goal is to state something numerically about population parameters once sample statistics have been calculated. This objective has two closely

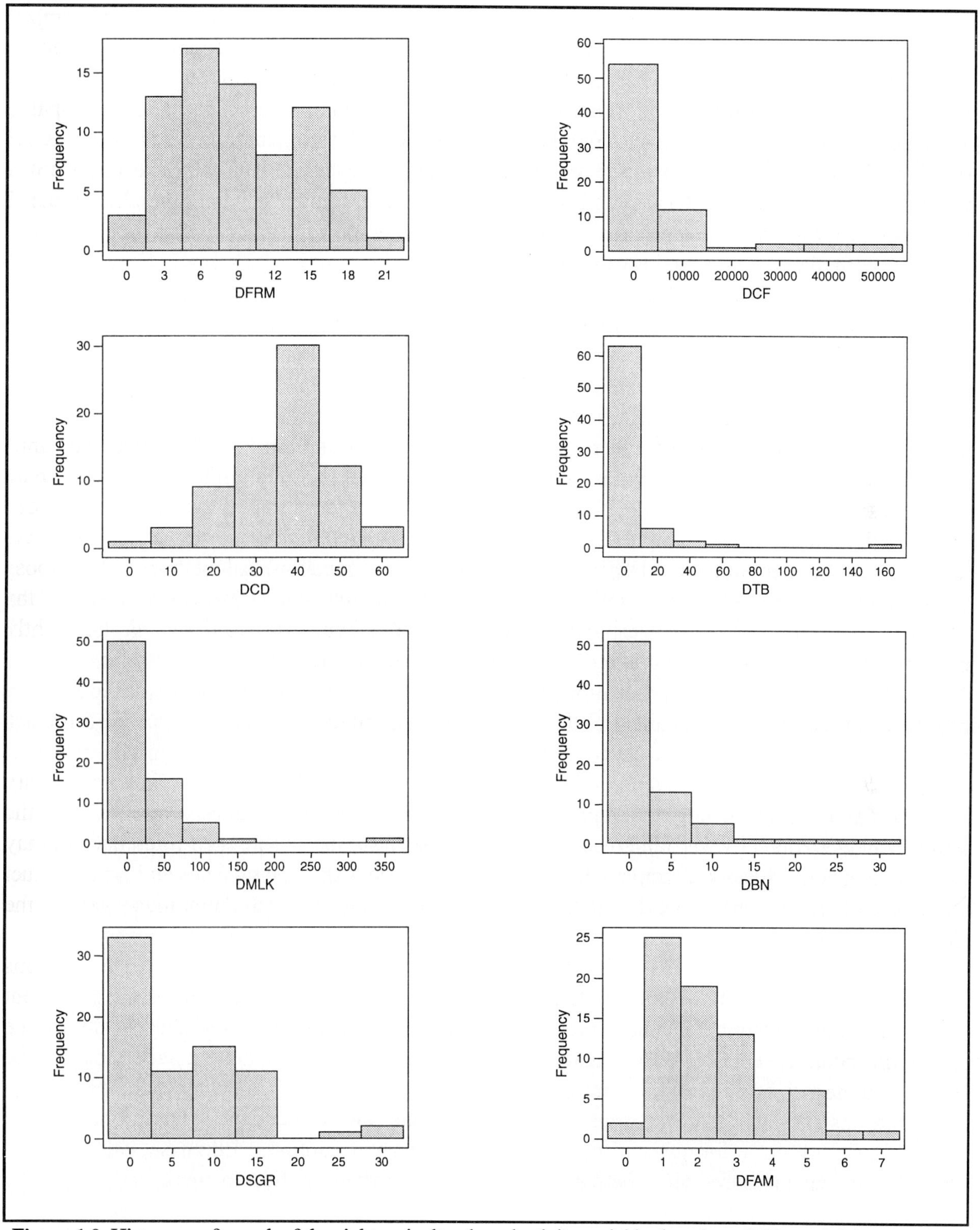

Figure 1.0. Histograms for each of the eight agricultural productivity variables based on all 73 municipios constituting the island of Puerto Rico.

related dimensions. First, two features of such statements are of interest, namely, the numerical value itself, and some declaration of the degree of confidence one has in this computed numerical value. Second, an analyst may be interested in evaluating a claim about the true value of a parameter; based upon sample information, one wants to ascertain whether the claim is correct, given the presence of sampling error. Estimation and hypothesis testing are two types of *statistical inference*. Such inference refers to the technique of interpreting numerical values resulting from sampling and then making decisions and stating generalizations about the parent population based upon statistical results that express the uncertainty due to sample variability in data.

The universal underpinnings for drawing statistical inferences is summarized in Hahn and Meeker (1993). The two components of this foundation may be labeled *designed based* and *model based*. In a design-based analysis, whose name is derived from use of a sampling design, interest centers on a finite, identifiable, and unchanging collection of areal units that can be enumerated. This forms the statistical target population from which samples are selected and to which desired inferences are drawn. This is the justification for much of geographic analyses of census data. The correctness of statistical inference here requires some sort of random sample; often with geo-referenced census data, the design may be viewed as a stratified (by areal unit) random sample of households, farms, businesses, or the like. This assumption of random sampling is critical, because only this type of sample-to-sample variability is quantified by *standard error* (i.e., the standard deviation of the sampling distribution of the parameter in question) calculations. Statistical inference for design-based studies relies heavily on the *Central Limit Theorem* (see Griffith and Amrhein, 1991, pp. 226–30). A critical requirement here is being able to define explicitly and precisely the target statistical population and enumerate its elements (i.e., the sampling frame). Differences between the target population and the enumerated one from which samples are drawn introduce uncertainties into an analysis that are beyond sampling error and that may well impact upon conclusions of a geographic study. Dramatic differences between the target and the sampled populations can result in misleading or incorrect conclusions.

Adaptations of design-based approaches to geo-referenced data sampling are reviewed in Stehman and Overton (1995). With regard to data analysis issues, sampling geo-referenced data is more like selecting clusters of grapes (the stems channeling spatial dependencies and representing geographic configuration), rather than individual numbers in a lottery (or the infamous statistician's urn):

> ... Data of geographic units are tied together, like bunches of grapes, not separate, like balls in an urn. Of course, mere contiguity in ... space does not of itself indicate lack of independence between units in a relevant variable or attribute, but in dealing with social data, we know that by virtue of their very *social* character, persons, groups and their characteristics are interrelated and not independent. [Traditional s]ampling error formulas ... must be used with great caution. Likewise, other statistical measures must be carefully scrutinized when applied to these data ... (Stephan, 1934, 165–66).

Carrying this analogy further, grapes in a given cluster tend to be more similar than those from different clusters; but, of course, none of these grapes are identical. Another physical example is provided by the magnetic sculptures (e.g., the magic box) that can be found in novelty stores or gift shops. Magnetism contained in the base of these sculptures permeates the metal flakes or ball bearings that constitute the sculpture medium. Hence, moving a ball bearing or metal flake to a new position drags others touching it along. In either of these two examples, either individual grapes, ball bearings, or metal flakes can be selected independently and at random. Nevertheless, information about other grapes in a cluster or those ball bearings or metal flakes in contact with the sampled ones still accompany the sampled individuals. These physical analogues exemplify what spatial dependency (i.e., spatial autocorrelation) is all about. Random sampling designs can mitigate and even neutralize these spatial dependency effects; but they do not remove spatial dependency from geo-referenced data (Cressie, 1991, p. 7).

The model-based foundation for drawing statistical inferences no longer deals with a finite, well-defined, unchanging statistical population and presumes that an observed realization of some geographic process results from random sampling of a super- or conceptual population. Two critical assumptions here are (1) an underlying normal distribution for the posited error, and (2) that the map realization is representative of the conceptual geo-referenced population, the process itself is stable, and the attribute values come from independent and identically distributed variables. Statistical inferences drawn in this context apply only to the specific process for which a sample has been observed. Differences between the model assumptions and the empirical behavior of attributes introduce uncertainties, often as specification error, into an analysis that are beyond sampling error and that may well impact upon conclusions of a study. These assumptions should not be glossed over, for their violations can result in seriously flawed inferences.

Ignoring these assumptions of design-based and/or model-based analyses can furnish a false sense of confidence in statistical findings. In spatial analysis, model-based studies are far more prevalent than are designed-based studies.

1.2.1. The Concept of Estimation: Points and Intervals

Estimation attempts to solicit a good guess for the unknown, but true, value of population parameters. If one reports a sample statistic, such as the sample mean (i.e., $\bar{y}$), the outcome will be a *point estimate*, which is the best single value that can be used to estimate a parameter. In other words, the population parameter is estimated by one point on the real number line. The point estimate for a population parameter is the value of its corresponding sample statistic. Multivariate point estimation involves simultaneously estimating parameters for a number of geo-referenced attributes. As in the univariate case, for multivariate point estimators the properties of sufficiency, consistency, efficiency, and unbiasedness provide criteria that help distinguish between good and poor point estimators.

Although a point estimate might be the best single guess for a parameter value, the presence of sampling error means that an entire interval of possible values exists. Even when a point estimator is the preferred minimum-variance unbiased estimator, nothing is known about the estimate that is rendered by a particular sample. The presence of a continuum of possibilities suggests that point estimates may be close to parameter values but more than likely never will actually equal them. In addition to reporting a single value, or point estimate of a parameter, some knowledge of its precision is useful to have on hand. For example, one might like to know the degree of confidence that can be attached to a guess of this sort. Believing that $\bar{y}$ (the sample mean) is close to μ (the population mean) is relative, is more than likely arbitrarily defined, and suggests that any numerical value within a certain distance of μ would be considered an equally good guess; accordingly, estimation can be thought of in terms of an interval rather than a point. In univariate analyses, as well as some multivariate analyses, the usual measuring unit of closeness is based upon the standard error of the estimate; this quantity provides a measure of deviation from some population parameter attributable to sampling error. A *confidence interval* estimate employs this concept, establishing two values for the interval boundaries and assigning a probability measure to the interval's reliability in estimating the parameter in question. Use of a confidence interval amounts to replacing a point estimate — a single number—by an entire interval of plausible values, with the degree of plausibility being specified by a probability measure, or *confidence level*. Conventional probabilities used to construct confidence intervals are 90%, 95%, and 99%. In other words, interval estimation means that two points on the real number line are employed to establish a range within which, with a certain probability, the population parameter is thought to lie.

The multivariate extension of these ideas must handle any correlation amongst attributes as well as adjust confidence levels for the simultaneous treatment of multiple attributes. This first broadened dimension of statistical analysis relates to attribute information contained in a set of variables. Its limits are "nothing new," if all of the attributes are perfectly correlated ($r_{hk} = \pm 1$ for all X_h and X_k), on the one hand, and "considerable supplemental information," if all of the attributes are uncorrelated ($r_{hk} = 0$ for all X_h and X_k, $h \neq k$), on the other hand. This second broadened dimension relates to the notion of binomial distributions (see Griffith and Amrhein, 1991, pp. 170–75)—where one could obtain p failures to reject null hypotheses, and (P - p) rejections of null hypotheses—with the intent of establishing the correct individual attribute confidence level that corresponds with the joint level of 90%, 95%, or 99%.

Good interval estimators, like good point estimators, need to be distinguished from poor ones. The two principal requirements of a good interval estimator are that its accompanying width be as narrow as possible for a given sample size and that it have a very high probability of containing the unknown parameter value. These goals are achieved in part by use of standard errors to calculate interval boundaries, because as is mentioned previously, the distance separating a sample statistic from its parent population parameter value should be measured with this yardstick. Unfortunately, these two requirements usually conflict, with the probability increasing only when the interval width increases (the sample size n being held constant).

1.2.2. The Meaning of a Confidence Region

The construction of a univariate confidence interval begins with the selection of a probability level. This task requires knowledge of an appropriate probability model for describing the associated sampling distribution; this issue will be addressed in the next section. In the presence of pure ignorance (the sample size is n = 0) or when one wants to establish a probability of 1, then the confidence interval is $(-\infty, \infty)$. This interval means that the population parameter value lies somewhere along the real number line, and hence all guesses for its value are equally reasonable (this range can be restricted by eliminating values that are physically impossible). Clearly such a confidence interval is meaningless! The classic position of statistical inference outlined in the preceding section stresses that when a sample is drawn, information contained in the sample suggests that a narrower interval than (-∞, ∞) is reasonable to posit. One does not know for certain, however, whether or not the resulting narrower confidence interval constructed from this sample contains the value of the parameter (sampling error is present), unless the bounds are maintained at plus/minus infinity. Using a probability of, say, 0.90 when constructing a confidence interval translates into an expectation that in the long run (the selection of many, many samples) 90% of the confidence intervals constructed from sample data will contain the population parameter value. Moreover, the interval (-∞, ∞) must shrink to have 10% of the possible intervals failing to capture the population parameter value. Extending this notion, as the interval width shrinks to a single value [C, C], the probability that intervals will embrace the population parameter value will tend to decrease to 0. The multivariate extension of this confidence interval concept is the ***confidence region***, which tends to be elliptical in shape for the simultaneous plotting of individual attribute confidence intervals. If the attributes are uncorrelated ($r_{hk} = 0$ for all X_h and X_k, $h \neq k$), then these regions are circular for two attributes, or hyperspherical for three or more attributes. If the attributes are completely and perfectly correlated ($r_{hk} = \pm 1$ for all X_h and X_k), then this region collapses to a single line and forms a univariate confidence interval.

The confidence region level, therefore, is not a probability statement involving some fixed region whose bounds have been computed from a given sample. Rather, it is a statement about what would happen if a very large number of similar regions were to be constructed. Attributes' population parameter values are fixed, whereas the regions vary from sample to sample. As a result, the probability that the population parameter values will fall into a given region is either 1 or 0; either they are or they are not there, with the presence of sampling error causing a researcher to be uncertain about which is the case. Confidence thus refers to the procedure for constructing regions, not to any particular sample result.

1.2.3. The Role of the Sampling Distribution in Statistical Estimation

Suppose a sample is representative and good estimators of population parameters have been selected. How does one determine those confidence interval/region bounds corresponding to a given probability level? The analysis here is cast in terms of the underlying sampling distribution. As such, a characterization of the affiliated sampling distribution for the statistic in question needs to be obtained, without resorting to complete enumeration of all possible samples (which can never be done with the model-based approach to statistical inference!). If one had to draw all possible samples of size n (which is impossible if the population is infinite in size) and then determine the standard error of the affiliated sampling distribution, the amount of work involved would more than defeat the purpose of drawing samples to conserve resources. Mathematical statistical theory has shown that certain probability models (frequently the normal and t distributions supply appropriate probability models for univariate inferential situations, whereas F and chi square distributions often supply appropriate probability models for multivariate inferential situations) accurately describe sampling distributions. These theoretical models supply information about the distributions, such as standard error, and thus enhance the usefulness of substituting samples for populations. The *Law of Large Numbers* (see Griffith and Amrhein, 1991, pp. 147–49) maintains that as n increases, these probability models provide increasingly better descriptions. Not only do these models supply necessary shorthand information for constructing interval estimates, but they also provide information necessary for evaluating the quality of point estimators.

The relationship between a sampling distribution and a probability model is articulated by theorems from mathematical statistics. One of the most famous is the Central Limit Theorem for $\bar{y}$, which applies to the sampling distribution of the sample mean of Y. This theorem has been adapted for many parameter estimators, both univariate and multivariate, identifying the appropriate probability model for a given sampling distribution as well as pertinent information about that distribution to be used to construct confidence intervals/regions. Theorems like the Central Limit Theorem yield theoretical generalizations for the sampling distributions of various statistics, allowing inferences to be drawn without having to totally enumerate all possible samples.

In general, confidence intervals/regions can be constructed from a sample statistic by extracting *critical values* (those values beyond which sample statistics are considered to be unlikely; the tail probabilities affiliated with these values are conventionally 10%, 5%, or 1%) for the appropriate probability distribution and then combining these values with the associated standard error of the sampling distribution.

1.2.4. Large- versus Small-Sample Theory

Often an analyst is faced with situations where a sample has been drawn from a population for which the variance is not known or where n is less than 100 or even 30. In such circumstances results tend to be far more volatile from sample to sample, and the normal approximation posited by the Central Limit Theorem for $\bar{y}$ usually is poor (specification error). The question of concern here is how one goes about making inferences when only the sample information is available and n is relatively small (asymptotic convergence fails to be helpful). The answer routinely is to employ other probability models.

To compensate for the increased uncertainty that accompanies either the use of one estimator in calculating the value of another estimator or analyses employing small sample sizes (often $n < 30$), the concept of *degrees of freedom* (df) comes into play. In univariate analyses, the number of degrees of freedom, adjusted for constraints placed on a statistical analysis, almost always is defined as being the sample size minus the number of other estimates already made. This notion mutates in multivariate analysis, where redundant information may exist and simultaneous treatment of attributes alters measures of the total amount of information contained in a data set; these additional complications allow for the presence of fractional degrees of freedom.

1.2.5. Hypotheses and Hypothesis Testing

The four preceding subsections deal with point and interval estimation. Also of interest are evaluations of claims about a true value of a parameter; any such claim is stated as an hypothesis. A single-value expectation for each attribute is posited by the *null hypothesis* (denoted by H_0), which is the hypothesis on which a researcher wishes to focus attention. The term null may be defined as "of no value, effect, or consequence," which insinuates that H_0 should be identified with the hypothesis of no difference (a precise value), no improvement (in quality control), no change (from current opinion), no departure (from prevailing beliefs), no relationship, and the like. A second hypothesis about the parameter value, the *alternate hypothesis* (denoted here by H_a), must accompany this null hypothesis and is constructed in such a way that the two hypotheses are mutually exclusive and collectively exhaustive of all parameter-value possibilities. Most often, then, the alternate hypothesis is a statement about the same population parameter(s) differing in value from that/those specified in the null hypothesis. This formulation of pairs of hypotheses allows rejection of the null logically to imply acceptance of the alternative.

The computed sample statistic is evaluated to see whether it is a likely value from the sampling distribution that is associated with the hypothesized (under H_0) parameter; the null hypothesis allows a sampling distribution to be constructed. The sample statistic is evaluated under the assumption that the null hypothesis is a true statement. The sampling distribution that can be constructed under H_0 is partitioned into two sections. The first, which corresponds to a confidence interval/region, identifies statistics that are consistent with the null hypothesis and has probability $(1 - \alpha)$. The second, called the *critical region*, refers to those values that are unlikely under the null hypothesis, references the tail regions (extremes) of the sampling distribution and has probability α. The extreme regions, therefore, contain the set of statistics consistent with H_a. The probability associated with the critical region is called the *level of significance* and is denoted by α (10%, 5%, or 1%). In univariate or conditional multivariate situations, the critical region might be divided into two smaller regions, one placed in each of the two tails of a sampling distribution. For a left-hand one-tailed situation the probability is $\alpha_1 = \alpha$ ($\alpha_2 = 0$). For a right-hand one-tailed situation the probability is $\alpha_2 = \alpha$ ($\alpha_1 = 0$). For a two-tail situation the probability is $\alpha_2 = \alpha_1 = \alpha/2$. If the null hypothesis is true, then the probability that the sample statistic falls into the critical region is called the level of significance. Critical values identify the boundaries of the relevant critical regions. Usually in multivariate analyses, the test statistics are either F or chi square ones, resulting in one-sided critical regions, and hence the level of significance and the critical region are equivalent.

1.2.6. Errors in Hypothesis Testing

A sample is drawn, and the pair of hypotheses partitions all possible sample outcomes into two distinct groups. Thus, the sample comes from either a population described by the null hypothesis or a population described by the alternate hypothesis. Because the sampling distribution under H_0 is divided into a critical region and a confidence region, sampling error that leads to an unlikely statistic under H_0 can lead to an error when deciding from which population the sample was drawn. In this context the null hypothesis is the favored claim, with the burden of proof resting with H_a. There is no reason to disbelieve H_0 unless the sample evidence strongly contradicts it.

Errors in the inferential process are of two types. If a sample is drawn from the population described by H_0 and the conclusion reached is that it in fact was drawn from this population, then a correct decision has been made. But if an unlikely statistic is obtained under H_0, then an incorrect decision has been made. This category of error is called a *Type I error*; the researcher does not believe that the null hypothesis is true when in fact it is true. The probability of a Type I error is given by the size of the critical region (level of significance), α. In contrast, if a sample is drawn from a population described by H_a and the conclusion reached is that it in fact was not drawn from the population described by H_0, then again a correct decision has been made. But, if an unlikely statistic is obtained under H_a that is likely under H_0, then an incorrect decision has been made. This category of error is called a *Type II error*; the researcher believes that the null hypothesis is true when in fact it is false.

The interpretation that accompanies these statistical decisions has a subtly important phraseology. Formal logic behind this required phrasing involves the existence of a null hypothesis and an alternate hypothesis population. This situation leads to the following four logical cases:

> First, it is possible that both of these populations exist, meaning both H_0 and $\sim H_0$ (read not H_0) are true. In this instance, the two sampling distributions could overlap, and one would be unable to know for certain which population yielded a selected sample.
>
> Second, it is possible that only the population under $\sim H_0$ exists and that the population under H_0 is a figment of one's imagination. In other words, when H_a is true, H_0 can be either true or false.
>
> Third, it is possible that only the population under H_0 exists and that a population under H_a is a figment of one's imagination.
>
> But because a sample has been drawn, then both populations cannot be a figment of one's imagination, since they partition the universe into two mutually exclusive and collectively exhaustive possibilities.

Moreover, when H_a is false, then H_0 must be true. Consequently, when an unlikely statistic is obtained under the null hypothesis, the researcher concludes that H_0 is to be rejected. However, if a likely statistic is obtained under the null hypothesis, then because both H_0 and $\sim H_0$ are possible, a researcher's conclusion must be to fail to reject H_0; the null hypothesis cannot be accepted. Moreover, the question being addressed is whether a statistic is significantly different from the expected value posited by H_0 or whether the difference from this expected value is more than likely due to sampling error. Spatial scientists rarely seem to be concerned about Type II errors but perhaps not justifiably so. Often it is the costs associated with committing a Type I error that are deemed important.

1.2.7. Critical Values and Test Statistics

Once a statistic has been computed for some sample, a statistical decision may be made by comparing it to the reference sampling distribution probability model associated with the null hypothesis. This is done by converting the observed statistic to a *test statistic*, which in the case of elementary univariate statistics often means subtracting the proposed population parameter value under H_0 from the observed statistic, and then dividing this difference by the standard error of the sampling distribution. The critical value has been defined as the first value in the critical region. If a test statistic lies beyond the critical value (it falls into the critical region), then one decides in favor of H_a; if a test statistic lies closer to the mean of the sampling distribution than does the critical value, then one fails to reject the null hypothesis.

As mentioned in §1.1.2 and §1.1.5, while univariate analyses almost always involve the z and t distributions, multivariate analyses most often involve the F and chi square distributions. All of these distributions are related through the normal distribution, the "mother of all distributions." Z-scores are for the standard normal distribution. The t distribution converges on the normal distribution as n increases (see, for an example, Griffith and Amrhein, 1991, p. 232). A chi square distribution is the sum of squares of independent standard normal random variables. And the F distribution is the ratio of two independent chi square variables, each divided by its corresponding number of degrees of freedom.

If the entire population is sampled, statistical inference is both unnecessary and inappropriate. Graphical methods and summary statistics, including probability statements about the population, continue to be of value, though. However, when a design-based or model-based study is being conducted, a complete interpretation of statistical findings, based on test statistics coupled with critical values, involves three statements. The first simply states whether the null hypothesis is rejected. The second statement refers to whether evidence has been found in the sample that pertains to H_a; accordingly, there either is or is no evidence to indicate the statement posited in H_a. The third statement refers to the probability of a Type I error: $100\times(1 - \alpha)\%$ of the time (when the decision is to fail to reject the null hypothesis) one would expect to get a test statistic of the observed order of magnitude when in fact H_0 is true; or only $100\times\alpha\%$ of the time (when the decision is to reject the null hypothesis) one would expect to get a test statistic of the observed order of magnitude when in fact H_0 is true.

1.3. Variable Transformations

The selection of a scale of measurement is only the first step in quantification, though (*RULE OF THUMB 1.1*); appropriateness of the units of measurement also is of concern to an analyst. This idea is alluded to when a nominal binary variable is measured with 0 and 1; it could be measured with any two distinct numbers, but these two integers render the most convenient results for interpretation purposes. The goal of establishing an informative scale of measure is to simplify the structure latent in a data set and by doing so enhance interpretation, expedite detection of anomalous areal units, and facilitate prediction, interpolation, and extrapolation. But transforming data to such a scale, even when it proves to be beneficial, also can be accompanied by disadvantages. The new measurement scale may well be less familiar and, as such, may compromise intuitive understanding and an ability to interpret the new numerical values rendered for areal units. Hence transformations should be employed only when benefits outweigh the inconveniences associated with implementing them. Choices here often at least partly are a matter of judgment, and are justified with empirical trends.

The issue of appropriate measurement units interfaces with the validity of many of the assumptions posited for statistical analysis, especially those supporting the drawing of statistical inferences. Three of these assumptions addressed in this section are symmetry (i.e., normality), straightness (i.e., linearity), and constant spread (i.e., variance homogeneity),[1] which relate quantification most directly to the interval/ratio measurement scale. If a sampling design-based statistical analysis is being undertaken, arguably the Central Limit Theorem can be relied upon to ensure especially that the normality and constant variance assumptions do not corrupt an inferential basis. Otherwise, one approach is to change the measurement scale by converting data values to their ordinal scale counterparts (i.e., rank the values) and then undertake nonparametric statistical analyses (for an introduction to this topic see, for instance, Griffith and Amrhein, 1991, pp. 310–14; §2.1.1). This specific approach is favored by data analysts who argue that data values should not be altered in an attempt to force them to conform to the best inputs for statistical techniques, and is further discussed in §1.4. Another change of measurement scale approach is to compute a transformation of the original scale to reexpress the quantification so that it better mimics the posited assumptions than does the original scale used for data collection. This approach is favored by data analysts who argue that data values are always measured on arbitrarily chosen scales, and changing these scales merely changes the spacings between ordered attribute values. Such improvements may enhance data interpretability as well as promote desirable statistical characteristics of the data. The Box-Cox (1964) family of power transformations has been promoted for bringing data that represent the variable of interest (which is referred to as the mean response or Y variable in regression) into better compliance with normality and variance homogeneity assumptions. The Box-Tidwell (1962) family of power transformations has been promoted for bringing mean regressor variables (which are referred to as X variables in regression) into better compliance with the linearity assumption. In either case, the general transformation is of the form

$$f_1(Y) = \alpha + \beta(Y + \delta)^{\gamma} \quad \text{for } \gamma \neq 0 \text{ , or} \qquad (1.1^2)$$

$$f_2(Y) = \alpha + \beta LN(Y + \delta) \quad \text{for } \gamma = 0 \text{ ,} \qquad (1.2)$$

where X would be substituted for Y if a mean response variable is being evaluated. Clearly equations (1.1) and (1.2) are bivariate regression–type specifications. The term δ has been added here, although frequently not explicitly stated in the transformation literature, because a number of geo-referenced variables take on the value of zero. In either case, the parameters of interest that need to be estimated are δ and γ, with the selected estimates being those that minimize the error sums-of-squares (see Hoaglin, Mosteller, and Tukey, 1983). Estimation of these parameters can follow a trial-and-error approach, which is extremely time consuming and risky, or an automated approach guided by the methodology outlined in Chapter 10 of this book. Of particular additional concern is that the use of the resulting transformed values can impact upon the assumed data error structure (see §10.1).

---

[1]Variable transformations also are used in attempts to suppress cross-classification interaction terms in two-way ANOVA (see Chapter 5).

[2]The term $\beta(Y + \delta)^{\gamma}$ should be divided by the Jacobian of the response variable transformation being sought, namely $\gamma\tilde{y}^{\gamma-1}$, where $\tilde{y}$ denotes the geometric mean of the set of y_i data values. Interestingly, the impact on estimates of δ and γ of setting this Jacobian term to 1 seems negligible, allowing the simpler equation (1.1) to be estimated.

When various criteria, such as normality, linearity, and variance homogeneity, are being optimized, trade-offs can occur. In other words, there is no guarantee that employing a transformation to better satisfy one assumption will improve compliance with any of the other assumptions; in fact, there is no guarantee that a transformation exists leading to results that adequately imitate any single desired feature (i.e., symmetry, straightness of a scatterplot, or constant spread). Fortunately, and serendipitously, experience suggests that a re-expressed measurement scale that improves on one of these criteria tends to improve on the others, too. This same outcome characterizes results for univariate relative to multivariate normality (see Johnson and Wichern, 1992, pp. 164–71) as well, although again there is no guarantee that improving the symmetry of a (marginal) univariate frequency distribution will lead to improvement of multivariate normality for its parent joint frequency distribution. Extensions to the multivariate case involve sophisticated computations and hence are beyond the scope of this book; the interested reader should begin such an undertaking by consulting Looney (1995). Even for purely descriptive statistical studies, the presence of symmetry, linearity, and homoscedasticity in attribute values enhances interpretability, whereas attributes that are nonnormal in very different ways obscure interpretability of a data analysis.

1.3.1. The Shapiro-Wilk Statistic and Symmetry

The ***Shapiro-Wilk*** (S-W) statistic generally is recognized as the best diagnostic test statistic for evaluating normality. In the case of a perfect normal distribution, this statistic takes on a value of 1. As its modified implementation in MINITAB reveals, it essentially is the correlation between the ranked attribute data values and their corresponding theoretical normal curve values for each rank. These approximate theoretical values are given by the normal curve z-score values for the percentiles

$$(r_i - 3/8)/(n + 1/4) \quad , i = 1, 2, ..., n,$$

where r_i denotes the rank of the attribute value for areal unit i; these values are obtained with SAS using the function command

```
Z = PROBIT(RANKI - 3/8)/(N + 1/4))
```

in a DATA step, where the Y values have been ranked using PROC RANK.

EXAMPLE PR-1.1. Estimate the power transformation parameter values that maximize the symmetry of each Puerto Rican attribute's frequency distribution.

Through trial-and-error, various ranges of values of δ and γ were evaluated in the DATA STEP, and then PROC REG was used to compute the error sums-of-squares for each trial-end-error pair of guesses. SAS results for this exercise yield the second entry in the following tabulation, while SAS is optimal parameter estimates appear in the first entry in this tabulation:

Variable:	S-W	equation (1.1) δ	γ	S-W	equation (1.2), γ=0 δ	S-W
DFRM:	0.94132	-0.69024	0.63975	0.96802	6.78388	0.95877
trial-and-error		-0.7	0.8	0.95950		
DCD:	0.98310	-2.32193	1.22150	0.98458	***	***
trial-and-error		28.1	1.6	0.98528		
DMLK:	0.55023	0.11392	0.21524	0.94888	1.30621	0.93510
trial-and-error		0.0	0.45	0.90994		
DSGR:	0.81474	0.00001	0.49155	0.89608	3.65448	0.86782
trial-and-error					3.0	0.86778
DCF:	0.56654	0.05250	0.11021	0.96192	1.68783	0.94624
trial-and-error		0.75	0.3	0.89437		
DTB:	0.35961	0.00039	0.16128	0.84636	0.02577	0.83562
trial-and-error		1.0	-1.0	0.77288		
DBN:	0.57648	0.04845	-0.00227	0.97153	0.04841	0.97153
trial-and-error		0.75	-0.5	0.93392		
DFAM:	0.90750	1.06134	-0.13216	0.97735	0.77255	0.97809
trial-and-error		0.25	0.2	0.97959		

Significant deviations from a normal distribution for the selected trial-and-error transformations are indicated by double-underlining of their respective S-W statistics. Symmetry impacts of these transformations can be portrayed with boxplots (Figure 1.1). For the most part, the trial-and-error approach has produced results that are consistent with the sophisticated parameter estimation procedure; most outliers have been removed, interquartile spreads have been expanded, and means and medians have been moved closer together. The DCD attribute example suggests that when the estimated power exponent exceeds 1 there seems to be no reasonable equation (1.2) counterpart to equation (1.1). Moreover, as γ approaches zero, equations (1.1) and (1.2) should give increasingly similar results; the case of DBN illustrates this contention, showing that results for an exponent γ very close to zero in equation (1.1) are equivalent to those for the logarithmic transformation defined by equation (1.2). DCD and DFAM also suggest that one must be careful with a trial-and-error approach, because the optimal results (appearing in the first line for each variable) appear to yield slightly lower S-W statistics values. The S-W statistic for DCD is modestly better than its optimal counterpart, and both are consistent with what is expected in the presence of negative skewness, which suggests that the exponent is $\gamma > 1$. But, the skewness statistic[3] for the optimal transformation (= 0.008) is much lower than it is for the trial-and-error selection (= 0.100). This discrepancy in the case of DFAM, which exhibits positive skewness, is more puzzling. While its histogram suggests that a transformation is needed to reduce values in the upper tail relative to those in the lower tail, the difference in sign of the exponent defies simple explanation. The skewness statistic for the optimal transformation (= -0.014) again is bet-ter than that computed with the trial-and-error transformation (= -0.037).

The trial-and-error transformations are utilized in subsequent chapters of this book, in part because they generally tend to deviate little from their optimal counterparts and in part because they are ones that a novice would tend to stumble upon first. It is unlikely that markedly better results would be obtained if these univariate transformations were replaced with ones optimizing multivariate normality.

[3]The skewness target value for a normal distribution is 0.

1.3.2. Bivariate Linearity

When relationships between attributes are of interest, pairwise scatterplots can be constructed and correlation coefficients can be calculated. Scatterplots portray the nature and degree of a relationship between two variables. When the relationship is nearly linear, the correlation coefficient provides a useful and enlightening quantitative index of it. Other advantages arising with a linear relation include a better ability to evaluate the fit of a regression equation and the availability of a sophisticated body of statistical theory (some of which is surveyed in this book) for data analysis purposes. When linear relations are not apparent in scatterplots, transforming either the Y variable, the X variable, or both (in fact, it usually is best to simultaneously explore transformations of X and Y) may result in the scatterplot trend being straighter than some curved trend displayed by the original, raw data values. The goal here is to promote straight line relationships between two attributes. In an ideal situation, the trend line should be a straight line forming a 45° angle with the horizontal axis. Accordingly, equations (1.1) and (1.2) may be rewritten as

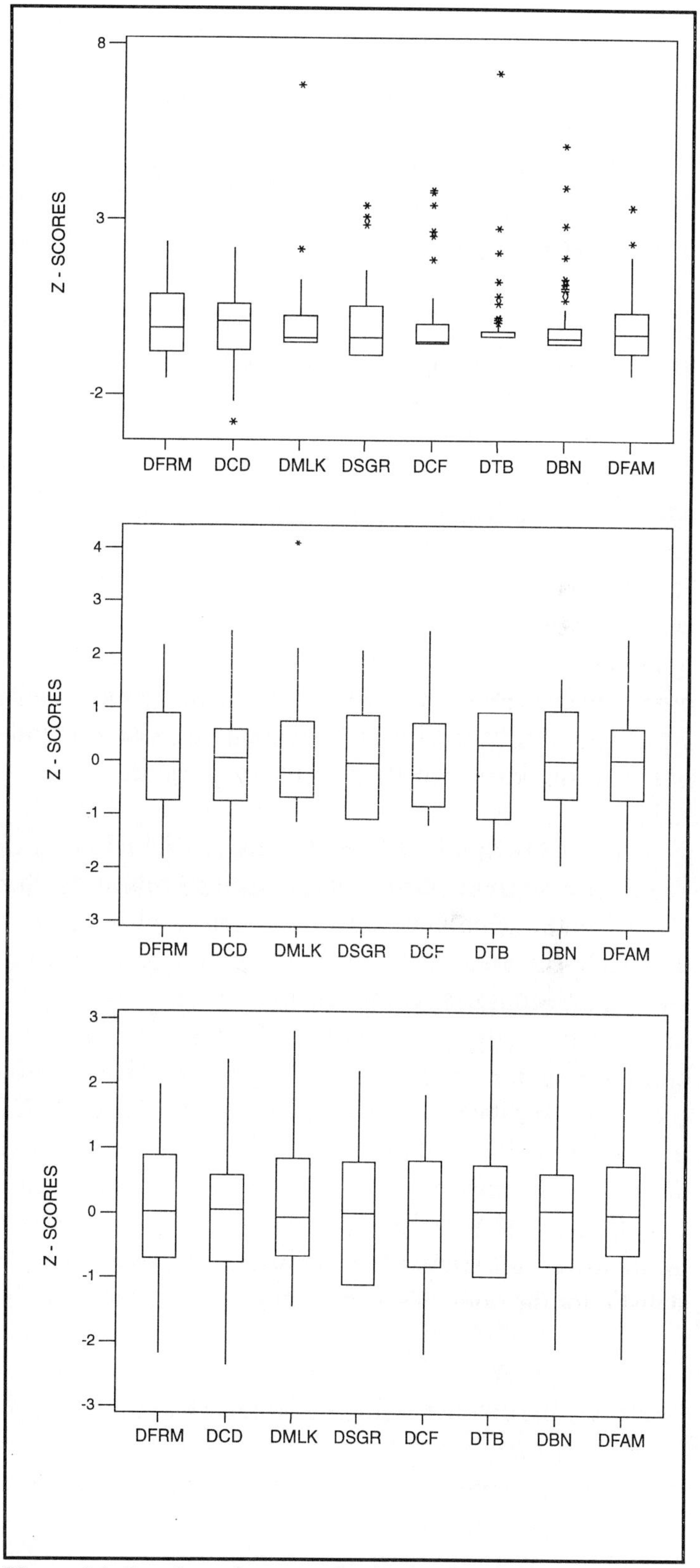

Figure 1.1. Boxplots of z-scores for the raw data (top), the trial-and-error transformed data (middle), and the optimally transformed data (bottom).

$$(Y + \delta_Y)^{\gamma} = \alpha + \beta(X + \delta_X)^{\eta} \quad \text{for } \gamma \neq 0 \text{ and } \eta \neq 0 , \tag{1.3}$$

$$(Y + \delta_Y)^{\gamma} = \alpha + \beta LN(X + \delta_X) \quad \text{for } \gamma \neq 0 \text{ and } \eta = 0 , \tag{1.4}$$

$$LN(Y + \delta_Y) = \alpha + \beta(X + \delta_X)^{\eta} \quad \text{for } \gamma = 0 \text{ and } \eta \neq 0 , \text{ and} \tag{1.5}$$

$$LN(Y + \delta_Y) = \alpha + \beta LN(X + \delta_X) \quad \text{for } \gamma = 0 \text{ and } \eta = 0 , \tag{1.6}$$

As before, estimates of the parameters (i.e., δ_Y, δ_X, γ, and/or η) of any one of these four equations may be done by implementing the methodology outlined in Chapter 10.

Again the serendipitous effects of transformations may result in selection of a mathematical expression, such as equation (1.1), to improve the symmetry (normality) of a frequency distribution, which fortuitously improves the straightness of the trend (linearity) between the pair of variables under study. One strategy for implementing equations (1.3)–(1.6) is to select a benchmark attribute (perhaps one without need of transformation), and then a transformation for each of the remaining attributes that optimizes the respective linear relations between this battery of variables and the benchmark variable.

EXAMPLE PR-1.2. Estimate the power transformation parameter values that maximize the linearity of each pairwise relationship in the Puerto Rican data set, using DFRM as the benchmark variable.

Locally computed trends between DFRM and each of the remaining agricultural productivity attributes (Figure 1.2) reveal that for a number of these attributes, the relationship between DFRM and each variable can be better straightened with a logarithmic transformation of the other attribute. The equational relations in question may be written as

DFRM $\approx$	$(\text{DFRM} - 0.69024)^{0.63975} \approx$
$\alpha + \beta\text{DCD}$	$\alpha + \beta\text{DCD}$
$\alpha + \beta LN(\text{DMLK} + 0.42500)$	$\alpha + \beta LN(\text{DMLK} + 0.26601)$
$\alpha + \beta\text{DSGR}$	$\alpha + \beta\text{DSGR}$
$\alpha + \beta LN(\text{DCF} + 0.77658)$	$\alpha + \beta LN(\text{DCF} + 0.01923)$
$\alpha + \beta LN(\text{DTB} + 0.32036)$	$\alpha + \beta LN(\text{DTB} + 0.25742)$
$\alpha + \beta LN(\text{DBN} + 0.06354)$	$\alpha + \beta LN(\text{DBN} + 0.01985)$
$\alpha + \beta\text{DFAM}$	$\alpha + \beta\text{DFAM}$

These results, obtained with methodology outlined in Chapter 10, reveal that

(1) transforming DFRM to make it more symmetric tends not to alter noticeably the relation between DFRM and each of the other agricultural productivity variables (i.e., DCD, DMLK, DSGR, DCF, DTB, DBN, and DFAM);

(2) reasonably linear relationships exist in the raw measures of DFRM versus DCD and DFAM; and,

(3) the transformed version of DFRM consistently displays lower estimates of δ_X.

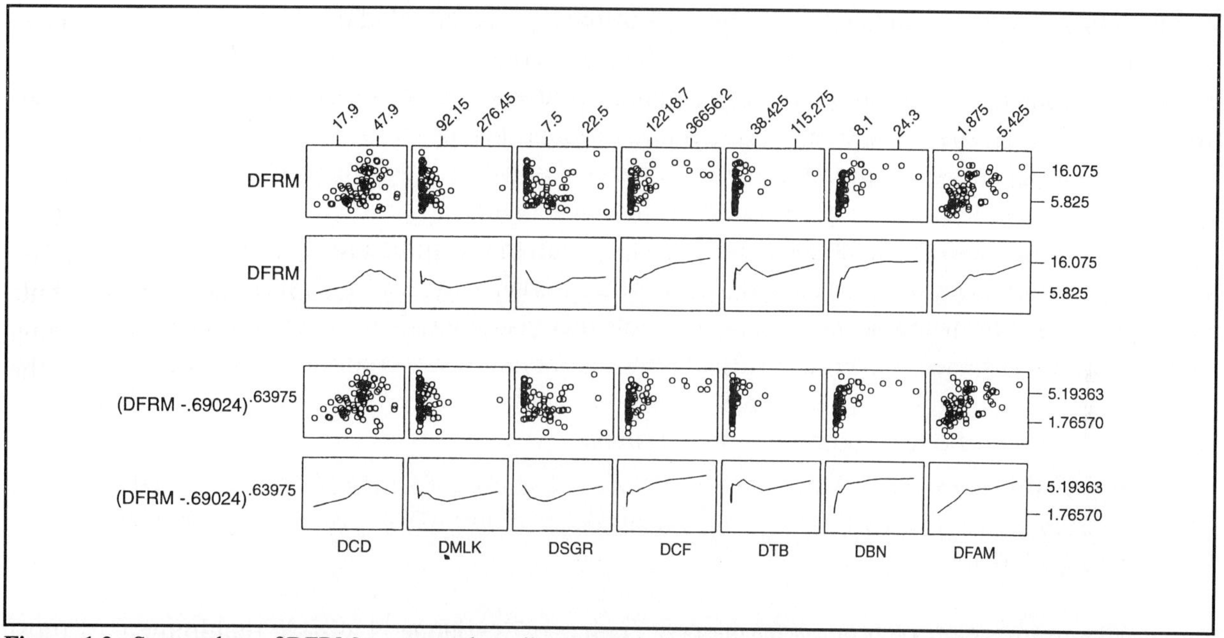

Figure 1.2. Scatterplots of DFRM versus each attribute, and the scatterplot trend lines for DFRM (top 2 rows). Scatterplots of the transformed variable (DFRM - 0.69024) $^{0.63975}$ versus each attribute, and the scatterplot trend lines for (DFRM - 0.69024) $^{0.63975}$ (bottom 2 rows).

The associated correlation coefficients are as follows, where the lower left triangular section of the tabulation refers to correlations between (DFRM - 0.69024)$^{0.63975}$ and the selectively transformed attributes, and the upper right triangular section of the tabulation refers to correlations between (DFRM - 0.69024)$^{0.63975}$ and the original, untransformed attributes:

	DFRM	DCD	DMLK	DSGR	DCF	DTB	DBN	DFAM
DFRM		0.342	-0.101	-0.257	0.490	0.269	0.504	0.509
DCD	0.342		0.057	0.239	0.337	0.195	0.382	0.456
DMLK	-0.233	-0.118		-0.206	-0.246	-0.073	-0.198	0.220
DSGR	-0.257	0.239	-0.269		-0.178	-0.228	-0.303	-0.233
DCF	0.664	0.339	-0.451	-0.193		-0.075	0.837	0.518
DTB	0.496	0.236	0.140	-0.326	0.257		0.038	0.090
DBN	0.742	0.361	-0.373	-0.316	0.754	0.339		0.560
DFAM	0.509	0.456	0.003	-0.233	0.355	0.239	0.479	

In this example, those coefficients for pairs of variables in which one or both have been transformed almost always are greater in absolute value; this tendency is even more pronounced when both attributes are transformed. One notable exception is for *LN*(DBN + 0.01985) and *LN*(DCF + 0.01923), where r decreases from 0.837 to 0.754, highlighting a trade-off that might be encountered when transforming a set of variables, namely, that transforming one pair of attributes to straighten their relationship may lead to a more curved relationship between either of these variables and some other attribute(s) of interest.

Before turning attention to variance homogeneity, analytical guidance in selecting an appropriate transformation to promote straightness merits some attention. This guidance has been developed from algebraic series expansions and differential calculus. The classic paper on this topic was penned by Box and Tidwell (1962), who outline a procedure that depends on the relationship between X and Y being quite conspicuous. Their procedure encounters convergence problems in cases where the error variance is large (a common occurrence in geo-referenced data) or where the range of Y is very small relative to its mean. A second technique, summarized in Emerson and Stoto (1982), exploits the regression slope between Y and X through their median coordinate pair and generates a scatterplot that needs to be linear while involving X and X^2. Experience with geo-reference data suggests that such data rarely comply with the requirements of these two procedures. Rather, geo-referenced data are amenable to the approach outlined in Chapter 10, with heuristic guidance gleaned from simple pairwise scatterplots that visually indicate the presence of a nonlinear relationship (for instance, see Figure 1.2). This situation is further complicated by a frequent need to have linear relationships between a pair of attributes only after the effects of a third variable have been regressed out of one or both of these variables.

1.3.3. Constant Variance

The assumption of homogeneous (i.e., constant) variance is a basic requirement of most traditional statistical techniques. Although it does not compromise the unbiasedness of estimators, it does inflate the variance of their sampling distributions, causing confidence intervals to be less precise. One common reason leading to the violation of this assumption is that the mean response and spread of an attribute are correlated. Accordingly, a scatterplot of variance versus mean response is a useful instrument for guiding the selection of an appropriate power transformation, if one exists for the data under study. An accompanying goal is to reduce or eliminate the dependency between mean and variance, thus acquiring numerical values that are better suited for comparison purposes (such as those of ANOVA). An alternative strategy that avoids the use of power transformations is to specify the nonconstant variance (which often is a function of areal unit size for geo-referenced data) and then, much like calculating a z-score, to divide each areal unit value by its corresponding individual (and perhaps unique) standard deviation. This implementation is referred to as ***weighting*** and will convert the variance to a constant across areal units, by construction.

Generally speaking, remedial action such as the use of a transformation is warranted when the range of the data for a given attribute is quite sizable. Experience in this context implies the following rule of thumb:

> *RULE OF THUMB 1.9.* If the ratio of the extreme data values is such that $y_{max}/y_{min} > 3$ and if a dependency is displayed between the mean and the variance of attribute Y across a geographic landscape, then this dependency may well be reduced or eliminated by employing a power transformation of Y.

Of note is that a transformation applied over a relatively narrow range of values tends to have little effect on the results of an analysis, whereas a transformation applied over a wide range of values can have a dramatic effect. In either case, the magnitude of this impact is dictated by how far the exponent of the power transformation deviates from 1.

A second effective tool that supplements *RULE OF THUMB 1.9* is furnished by a plot of the logarithm of the spread versus the logarithm of the mean for subregional pairs of means and variances computed over a geographic landscape. The slope of the regression line (b_1) indicates the value of the power transformation exponent. Montgomery and Peck (1982, p. 89) summarize selected values of γ or η that are suggested by this instrument; the proportionality relationship may be summarized as follows:

$$\text{power transformation exponent} = 1 - [b_1 \text{ from } LN(s^2) \text{ regressed on } LN(\bar{y})]/2 \ . \qquad (1.7)$$

These proportionalities suggest

RULE OF THUMB 1.10. Given a set of regions that partitions a geographic landscape into mutually exclusive and collectively exhaustive regions, preferably compact ones, the bivariate regression slope from regressing $LN(s^2)$ on $LN(\bar{y})$ determines, by substitution into equation (1.7), the value of the power transformation exponent (γ or η) needed to help stabilize geographic variance across these regions.

The regression referred to in *RULE OF THUMB 1.10* is for the model

$$\sigma^2 = \alpha\mu^{\beta}\times\epsilon \ ,$$

where α, β, and ϵ respectively are the two bivariate regression parameters and the error term (see Griffith and Amrhein, 1991, pp. 321–28). This model of dependency between the mean and the variance of an attribute is linearized by taking the logarithm of each side of the equation, hence rendering the log-log plot mentioned in *RULE OF THUMB 1.10.*

EXAMPLE PR-1.3. Determine whether a power transformation should help to stabilize the variance of each Puerto Rican agricultural productivity attribute.

The *LN*(variance)-*LN*(mean) plots for each agricultural productivity attribute, using both the ADM-by-LND (Map 1.2) and ADM-by-RU (Map 5.1) cross-classification groupings, appear in Figures 1.3(a)–(b). The ratio of attribute extremes, coupled with slope estimates for the scatterplots appearing in Figure 1.3, are as follows:

variable	y_{max}/y_{min}	$y_{max}/(y_{min}+1)$	$LN(s^2)-LN(\bar{x})$ slope ADM×LND	ADM×RU
DFRM	30.3	12.5	0.91 (R^2=0.421)	3.28 (R^2=0.567)
DCD	21.7	16.1	-0.21 (R^2=0.005)	-0.06 (R^2=0.000)
DMLK	****	368.6	1.96 (R^2=0.965)	2.37 (R^2=0.803)
DSGR	****	30.0	1.52 (R^2=0.928)	1.04 (R^2=0.355)
DCF	****	48874.9	1.90 (R^2=0.981)	1.89 (R^2=0.985)

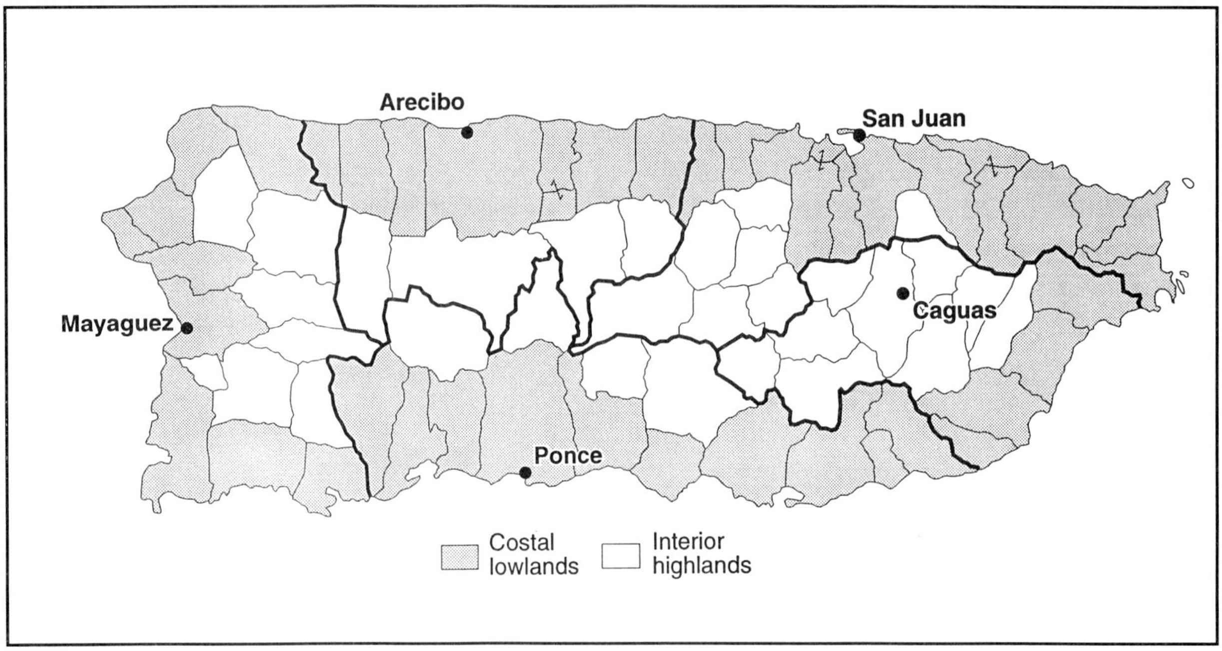

Map 1.2. Geographic disribution of the coastal lowlands–interior highlands dichotomy binary variable, with the ADMs superimposed on the island.

DTB	****	153.7	1.87 (R^2=0.980)	1.89 (R^2=0.984)
DBN	****	32.4	2.10 (R^2=0.982)	2.29 (R^2=0.918)
DFAM	72.0	6.5	1.29 (R^2=0.705)	2.21 (R^2=0.735)

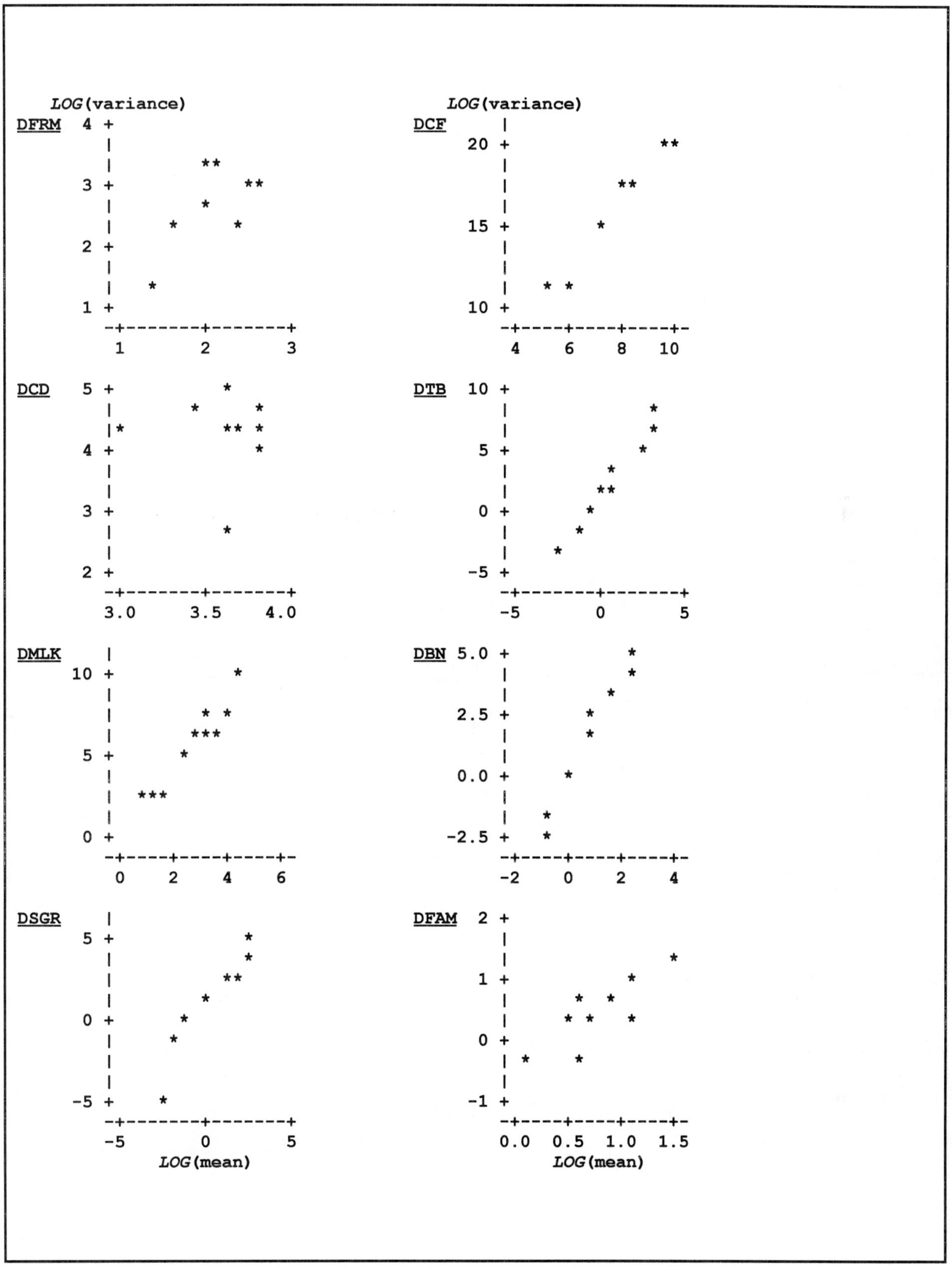

Figure 1.3. (a) *LN* (variance)-*LN* (mean) plots for each agricultural productivity attribute, using the ADM-by-LND cross classification groups.

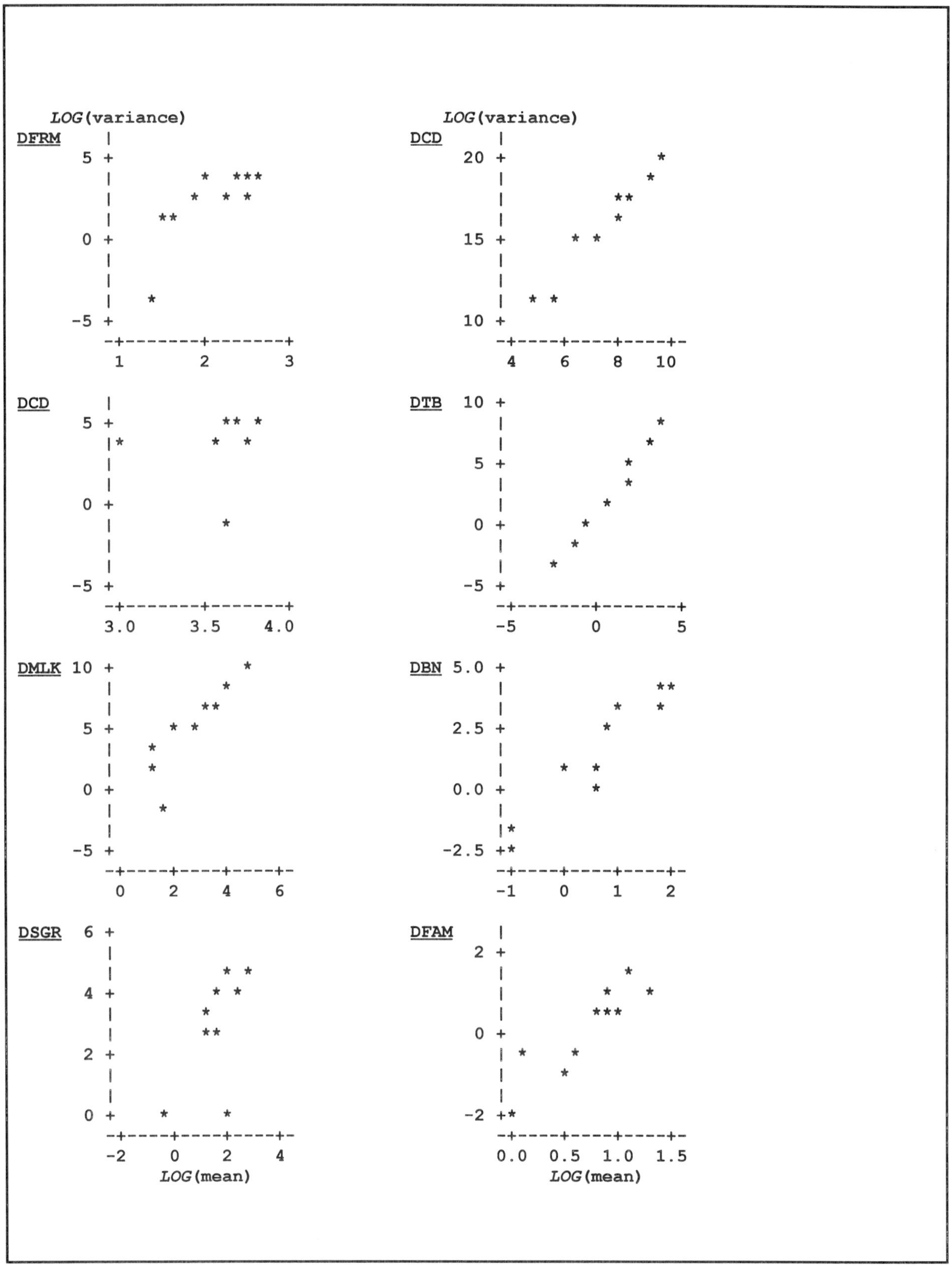

Figure 1.3. (b) *LN* (variance)-*LN* (mean) plots for each agricultural productivity attribute, using the ADM-by-RU cross classification groups.

These results suggest that the variable DCD is not in need of a variance stabilizing transformation. A logarithmic transformation would help to stabilize the variance of attributes DMLK, DCF, DTB, and DBN, and a square root transformation would help to stabilize the variance of variable DSGR. Empirical evidence for DFRM and DFAM is less definitive. The variance of variable DFAM should stabilize with a power transformation whose value falls in the interval (0, 0.5); the suggestion for variable DFRM is a transformation whose exponent lies in the interval (-0.5, 0.5).

These findings may be corroborated by conducting formal tests of equality of variance across geographic regions. Two popular ones are Bartlett's test (see Griffith and Amrhein, 1991, pp. 292–93), which is not robust to the frequency distribution of geo-referenced data departing from a normal distribution, and Levene's test, which requires an interval/ratio measurement scale but is robust for underlying frequency distributions that are not markedly skewed.

EXAMPLE PR-1.3 (continued). Document the presence of nonconstant variance across the island of Puerto Rico for each agricultural productivity attribute, and determine whether a power transformation helps to stabilize these variances.
The power transformations reported in Example PR-1.1 are used to illustrate whether such transformations help to stabilize these variances (note: those test statistics that are significant at the 5% level, and hence signify the presence of nonconstant variance, are underlined, with those for the selected trial-and-error transformations being double-underlined).

variable	ADM×LND		ADM×RU	
	Bartlett	Levene	Bartlett	Levene
DFRM	10.902	0.532	12.722	1.294
trial+error	6.645	0.376	9.691	1.050
optimal	4.600	0.358	8.416	0.926
DCD	11.933	1.067	9.594	0.662
trial+error	13.911	1.102	11.915	0.798
optimal	13.467	1.118	11.405	0.775
DMLK	113.709	1.340	110.947	3.016
trial+error	15.590	0.656	19.021	1.436
optimal	5.314	0.308	8.131	0.689
DSGR	71.164	3.225	27.777	1.518
trial+error	33.396	2.296	10.713	1.135
optimal	20.834	1.882	8.674	1.074
DCF	163.172	7.145	131.872	2.075
trial+error	23.046	2.667	25.497	2.552
optimal	13.038	1.129	9.267	0.773
DTB	255.763	2.077	267.464	2.715
trial+error	20.407	1.283	17.508	0.893
optimal	14.170	1.689	13.888	1.807
DBN	121.549	3.034	92.330	1.247
trial+error	12.346	0.829	12.597	1.555
optimal	8.840	0.573	11.847	1.308
log-version	8.842	0.574	11.852	1.308
DFAM	10.111	0.865	17.750	1.423
trial+error	2.647	0.243	6.707	0.783
optimal	2.408	0.246	6.307	0.792

Except for DCD, each of the spread stabilizations is quite dramatic when the symmetry-enhancing power transformations are employed. As emphasized before and illustrated here by DCD, there is no guarantee that improvement in one statistical feature (e.g., symmetry) will be accompanied by improvement in others (e.g., constant spread). Meanwhile, as DSGR illustrates, there is no guarantee that a transformation will remove completely any detected differences in variances across regions. Further, DBN illustrates that equations (1.1) and (1.2) achieve the same variance stabilizing results when γ is very close to zero. Undisputable candidates for power transforming here are DCF and DTB. DCF, being one of the agricultural productivity attributes displaying a clear need for transformation, again can have transforming impacts, visualized with boxplots (Figure 1.4). For both cross-classification regional groups, the transformation employed here removes the outlier nature of a number of areal units, shrinks large regional spreads, swells small spreads, and even induces better symmetry in the within regions frequency distributions.

1.3.4. Implications

As mentioned in §1.2.1, the transformations uncovered by trial-and-error have been employed to generate results reported in ensuing chapters of this book. For the most part, these selections make sense in a quest to balance the objectives of optimizing symmetry, pairwise straightness, and constant regional spread. A power transformation exponent value of 0.8 applied to DFRM lies close to the optimal symmetry value and is in the same direction away from 1 as are those exponents that should stabilize variance across the two sets of island subregions. This exponent of 0.8 yields a modest improvement in constancy of variance for both cross-classification schemes, too. A power transformation exponent value of 1.6 applied to DCD is roughly of the same order of magnitude as that needed to improve symmetry and, while perhaps too radical for promoting straightness, does not severely erode the observed roughly constant spread. A power transformation exponent value of 0.45 applied to DMLK is much closer to the optima than is 1 (the exponent of an untransformed attribute), and instills a dramatic improvement in variance constancy, as well as a slight improvement in straightness. This is essentially the exponent Griffith and Amrhein (1991, pp. 311–12) use to obtain comparable statistical decisions from both ANOVA and its Kruskal-Wallis nonparametric counterpart across the five ADMs. A power transformation exponent value of 0 (hence, a logarithmic transformation) applied to DSGR overall is closer to the optima and renders improvements to the three statistical properties, although not necessarily sufficient conformity to eliminate assumption violations. A power transformation exponent value of 0.3 applied to DCF is very close to each of the optima and induces a dramatic but not necessarily complete improvement in symmetry and constancy of spread and a noticeable improvement in straightness. A power transformation exponent value of -1 applied to DTB is excessive, but in the appropriate direction, and markedly improves symmetry and constancy of spread while noticeably improving straightness. Zero is much closer to the global optimum underscoring one of the risks

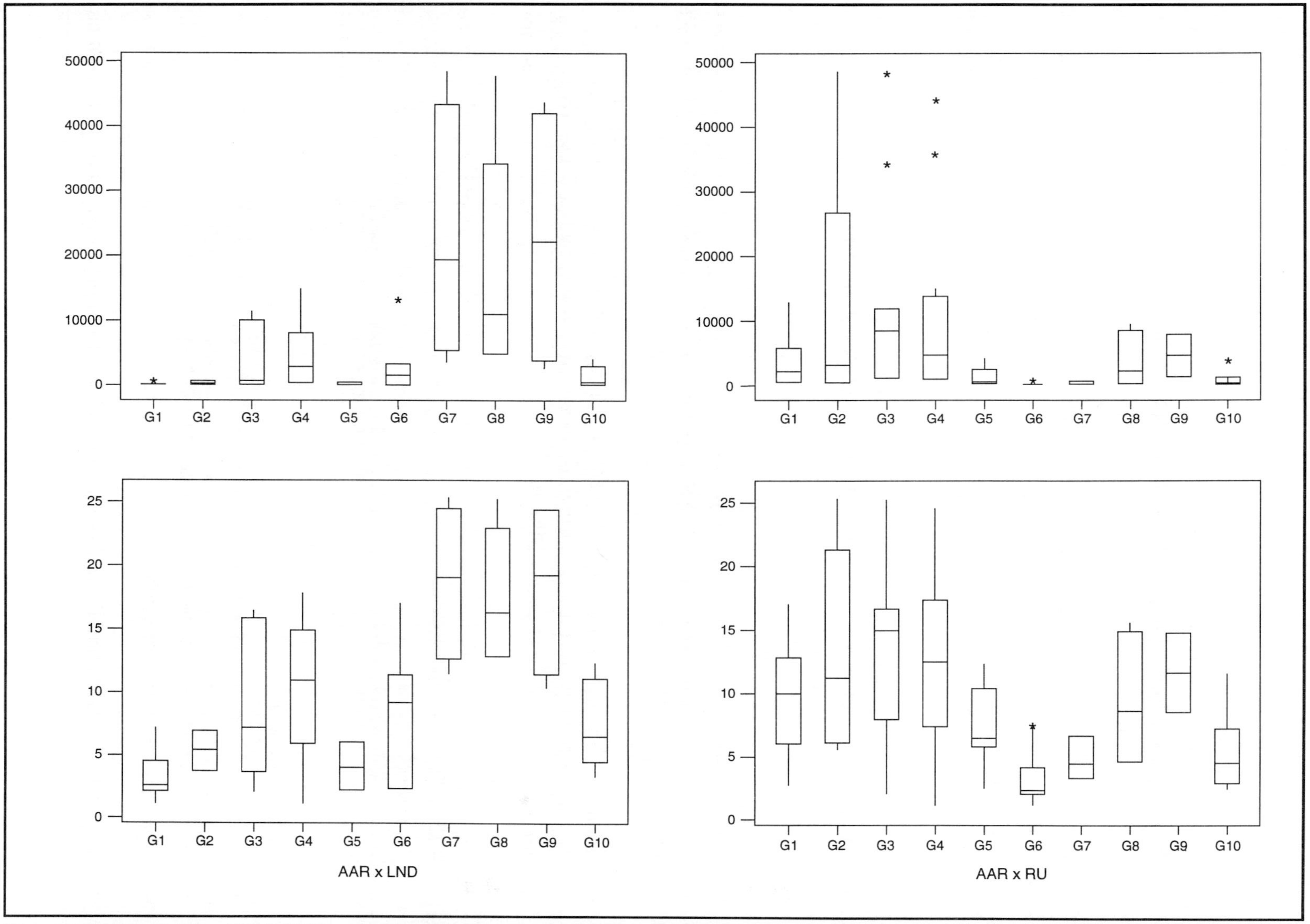

Figure 1.4. Boxplots by ADM×LND cross-classification groups for the raw data (top left) and data transformed with the trial-and-error transformations (bottom left); boxplots by ADM×RU cross-classification groups for the raw data (top right) and data transformed with the trial-and-error transformations (bottom right).

of pursuing unsupervised, trial-and-error power transformation identifications. A power transformation exponent value of -0.5 applied to DBN is somewhat excessive, but also in the appropriate direction, and dramatically improves symmetry, homogeneity of variance, and linearity. Finally, a power transformation exponent value of 0.2 applied to DFAM lies very close to each optimum but produces only modest improvement in symmetry and constancy of variance.

1.4. One More Viewpoint: Special Normality and Constant Variance Assessments in Preparation for Multivariate Analyses

Multivariate analyses encounter constraints from parameter estimates for each variable included. The resulting loss of degrees of freedom means that sample size often must exceed the number of variables, sometimes by a multiple of 2 or 3. In §1.2.1 individual variable appraisals of normality are for the entire collection of areal unit values. But both ANOVA and its multivariate extension require frequency distribution conformity within each region, dramatically reducing the sample size from which degrees of freedom are lost. Consequently, because several of the cross-classifications—those used in §1.2.3 to evaluation stability of variance for individual variables across the geographic landscape of Puerto Rico—have subsample sizes that are as small as 2, findings of supplemental assessments reported in this section are based solely on the five agricultural administrative regions. This framework serves as the foundation for a number of the ensuing multivariate assumption appraisals.

EXAMPLE PR-1.4. Determine how well each of the eight transformed agricultural productivity attributes mimics a normal distribution within each of the five agricultural administrative regions of Puerto Rico.

Computing the S-W statistic with PROC UNIVARIATE in SAS, using a BY ADM option, yielded results that include

variable	San Juan	Arecibo	Mayaguez	Ponce	Caguas
TDFRM	0.90055	0.94554	0.97235	0.97627	0.91217
TDCD	0.93245	0.93887	0.95792	0.96092	0.87807
TDMLK	0.96499	0.88816	0.87269	0.87334	0.95462
TDSGR	0.80993	0.86598	0.88795	0.91536	0.82139
TDCF	0.82160	0.82724	0.95031	0.97907	0.89592
TDTB	0.65280	0.84394	0.73511	0.80449	0.84695
TDBN	0.87300	0.86180	0.94124	0.92292	0.95480
TDFAM	0.95868	0.96717	0.96637	0.94463	0.96368

Note: significant values are double-underlined.

All S-W statistics that are significant at the 1% level (which is a Bonferroni adjustment equivalent to an overall 5% level; see §4.4.1) are double-underlined. While each of the transformed agricultural productivity attributes conforms reasonably well to a normal distribution (Example PR-1.1), this property is not perfectly preserved across the subregional evaluations. Frequency distributions for DSGR, DCF, and DTB deviate from a bell-shaped frequency distribution in the San

Juan region, while DTB deviates from a normal curve in the San Juan, Mayaguez, and Ponce regions. Fortunately, nearly 90% of the empirical frequency distributions do imitate a normal distribution. Discussion of a multivariate normal assessment of these variables is presented in §6.8 for the entire island of Puerto Rico and in §7.1.2 for each of the five agricultural administrative regions.

Results reviewed in Example PR-1.0 (the background univariate statistical summaries discussed in *Statistical Analysis for Geographers*) suggest that the variance is not constant across the five island regions. This is not a surprising finding, given the judgments rendered in §1.2.3.

EXAMPLE PR-1.5. Examine the presence/absence of nonconstant variance across the agricultural administrative regions of the island of Puerto Rico for each of the transformed agricultural productivity attributes.

variable	Bartlett	Levene	skewness
TDFRM	1.680	0.894	0.20343
TDCD	5.800	0.996	0.07554
TDMLK	20.011	2.163	1.17369
TDSGR	2.861	0.627	0.29350
TDCF	10.625	1.661	0.84322
TDTB	5.669	0.615	-0.59364
TDBN	1.457	0.402	-0.23531
TDFAM	3.900	1.040	-0.03858

Note: significant values are double-underlined.

Bartlett's test for homogeneity of variance is very sensitive to deviation from normality of the underlying frequency distribution. In this instance, DMLK, DSGR, DCF, DTB, and DBN all exhibit significant deviation from a normal distribution (Example PR-1.1). The presence of nonnormality for these attribute frequency distributions suggests that caution should be exercised in interpreting especially the homogeneity of variance test statistics for DMLK and DCF. Levene's test is very sensitive to the presence of skewness; none of the skewness statistics reported in Example PR-1.5 appears to be significantly different from 0, the value of skewness for a perfectly symmetric distribution. This finding reinforces the idea that the Bartlett test statistics for DMLK and DCF, whose skewness statistics are the two largest, may well be detecting considerable specification error (i.e., violation of one of its underlying assumptions) rather than nonconstant (i.e., heterogeneous) variance. Discussion of a multivariate homogeneity of variance assessment of these variables is presented in §5.1.1.

1.5. A Brief Excursion into Nonparametric Statistics

When assumptions supporting parametric inference are problematic (and perhaps the use of power transformations is not desirable) or when data measured on the interval/ratio scale are deemed too inaccurate or imprecise, an alternative statistical methodology is supplied by nonparametric statistics. Such statistics involve ranking the data values and performing analyses based on the accompanying uniform distribution (each rank appears once, and hence all ranks are equally likely).

Griffith and Amrhein (1991, pp. 310–13) furnish an example in which one-way ANOVA results for DMLK are compared with their Kruskal-Wallis nonparametric counterpart. Three aspects of a nonparametric approach are noteworthy. First, regardless of the power transformation used (these transformations are monotonic—they change the interval distances between consecutive numerical values, but they preserve the initial ordering of the original values), the rankings of the resulting values remain the same. Second, the sum of n consecutive integers (i.e., the rankings) is $\frac{n(n+1)}{2}$, while the sum of n consecutive squared integers is $\frac{n(n+1)(2n+1)}{6}$ (these two analytical results are why rank ties are problematic in nonparametric statistical analyses and merit special treatment when they occur). In other words, regardless of the set of interval/ratio numbers, the mean and the variance of a set of distinct rankings are always known. And, third, all possible samples of size n, regardless of whether a population is finite or infinite, yield a sampling distribution that can be constructed simply by enumerating all possible permutations (see Griffith and Amrhein, 1991, pp. 150–52) of the n rankings, because there is only a finite number of possible rankings of areal units.

Nonparametric descriptive statistics primarily are parametric statistics computed with ranked data. The same statistical formulae can be applied, as highlighted in the relationship between Pearson's product moment and Spearman's rank correlation coefficients outlined in Griffith and Amrhein (1991, pp. 109–11). An important difference, though, is that merely substituting ranked data into a parametric formula fails to identify the correct sampling distribution for a nonparametric statistic. Again, as emphasized by Griffith and Amrhein (1991, pp. 268–73) with regard to correlation coefficients, the standard error for Spearman's rank correlation coefficient is $\frac{1}{\sqrt{n-1}}$, rather than $\frac{1}{\sqrt{n-3}}$ for its transformed Pearson's product moment counterpart. This calculation equivalency notion can be illustrated with the Mann-Whitney U statistic, a nonparametric alternative for the difference between two independent means t statistic. Let variable Y be the ordinal scale ranking of some attribute, and let X be a nominal scale binary indicator variable whose value is 1 if an areal unit is located in the first of two geographic regions, and whose value is 0 if this areal unit is located in the second of the two regions. Executing a bivariate regression of Y (the rankings) on X (the 0-1 indicator variable) yields a regression slope coefficient, b, that can be converted to the Mann-Whitney U statistic. Accordingly, $\sum_{i=1}^{n} x_i = n_1$, $\sum_{i=1}^{n} x_i^2 = n_1$, $\sum_{i=1}^{n} y_i = \frac{n(n-1)}{2}$, and $\sum_{i=1}^{n} x_i y_i = R_1$, where R_1 is the sum of the rank numbers (i.e., $\sum_{i=1}^{n_1} r_{i1}$) for those areal units located in the first region. Moreover, substituting these four terms into the equation for b (Griffith and Amrhein, 1991, p. 325) renders

$$b_{R_1} = \frac{n\sum_{i=1}^{n_1} r_{i1} - \frac{n(n+1)}{2} n_1}{n n_1 - n_1^2} .$$

Given this equation the Mann-Whitney U statistic may be expressed as

$$\frac{n_1 n_2}{2}\ [1\ -\ \frac{2}{n} b_{R_1}]\ .$$

Next suppose that the indicator variable X is defined in such a way that the regression slope coefficient, b, gives a direct difference of means test–i.e., X takes on a value of 1 if an areal unit is located in the first region, and a value of -1 if the areal unit is located in the second region. Then

$$b_{R_1 - R_2} = \frac{n(\sum_{i=1}^{n_1} r_{i1} - \sum_{i=1}^{n_2} r_{i2})\ -\ \frac{n(n+1)}{2}(n_1 - n_2)}{n \times n\ -\ (n_1 - n_2)^2}\ .$$

But since $R_2 = \frac{n(n+1)}{2} - R_1$ and $n_2 = n - n_1$, this coefficient may be rewritten as

$$b_{R_1 - R_2} = \frac{n_1 n_2\ +\ n_1(n_1 + 1)\ -\ R_1}{4 n_1 n_2} n\ ,$$

and hence the Mann-Whitney U statistic may be expressed as

$$\frac{n_1 n_2}{2}\ [1\ +\ \frac{4}{n} b_{R_1 - R_2}]\ .$$

Since labelling of the two regions is arbitrary, the indicator variable X can be defined so that R_1 is the smaller of the two sums of rank numbers. This result directly links the calculation of the Mann-Whitney U statistic to regression and to the calculation of a difference of means t statistic.

The important relationship in the case of one-way ANOVA is

$$(n - 1)\times R^2 = (\text{Kruskal-Wallis H statistic})\ .$$

This particular connection can be illustrated with the two-regions ANOVA problem. Suppose there are n_1 areal units in the first region and n_2 areal units in the second region ($n = n_1 + n_2$). A bivariate regression of Y on X (see Griffith and Amrhein, 1991, Chapter 12), where X is a categorical variable whose value is 1 when an areal unit is located in the first region, and 0 when an areal unit is located in the second region, and r_{i1} and r_{i2} respectively are the individual rankings assigned to the areal units in regions 1 and 2, yields

$$R^2 = \frac{\sum_{i=1}^{n} y_i \times (a + b x_i)\ -\ n \times \bar{y}^2}{\sum_{i=1}^{n} y_i^2\ -\ n \times \bar{y}^2}\ ,\ \text{where}$$

$$a = \frac{n(n\ +\ 1)\ -\ 2\sum_{i=1}^{n_1} r_{i1}}{2 n_2}\ ,\ \text{and}$$

The right-hand side of this result, indeed, is the Kruskal-Wallis H statistic divided by (n - 1). Again,

$$b = \frac{n[2\sum_{i=1}^{n_1} r_{i1} - n_1(n + 1)}{2n_1n_2} .$$

$$\therefore R^2 = \frac{\frac{(\sum_{i=1}^{n_1} r_{i1})^2}{n_1} + \frac{(\sum_{i=1}^{n_2} r_{i2})^2}{n_2} - \frac{n(n + 1)^2}{4}}{\frac{(n - 1)n(n + 1)}{12}} .$$

then, a nonparametric test statistic can be calculated with the bivariate regression technique; but its associated measures of sampling variation are not forthcoming from the parametric results.

EXAMPLE PR-1.6. Using DMLK, examine relationships between parametric ANOVA calculations and Mann-Whitney (two regions) and Kruskal-Wallis (two or more regions) calculations, respectively, with the LND and ADM regionalization schemes for Puerto Rico.

Two-groups one-way ANOVA (using PROC ANOVA in SAS; see Chapter 4) for DMLK, using LND as the classification variable, yields F = 0.37 (p = 0.5434), with $R^2 = 0.00523$. Repeating these computations for the ranked DMLK values yields F = 0.26 (p = 0.6144), with $R^2 = 0.00359$. The nonparametric counterpart (using PROC NPAR1WAY) renders the Kruskal-Wallis chi square value of 0.25874 (p = 0.6110), or its z-score Mann-Whitney equivalent of 0.50866 (again with p = 0.6110). Now,

$$(73 - 1)\times 0.00359 = 0.25848 \approx 0.26 .$$

For the transformed version of DMLK, namely TDMLK, conducting a one-way ANOVA here yields F = 0.28 (p = 0.5965), with $R^2 = 0.00397$.

Meanwhile, five-groups one-way ANOVA, using ADM as the classification variable, yields F = 4.26 (p = 0.0039), with $R^2 = 0.20033$. Repeating these computations for the ranked DMLK values yields F = 9.42 (p = 0.0001), with $R^2 = 0.35656$. The nonparametric counterpart (again using PROC NPAR1WAY) renders the Kruskal-Wallis chi-square value of 25.672 (p = 0.0001). Now,

$$(73 - 1)\times 0.35656 = 25.67232 \approx 25.672 .$$

For the transformed version of DMLK, namely TDMLK, conducting a one-way ANOVA here yields F = 8.61 (p = 0.0001), with $R^2 = 0.33625$.

For both cases in Example PR-1.6, the relationship between R^2 obtained from an ANOVA of ranked Y values and the Kruskal-Wallis (and by default, the Mann-Whitney) test statistics holds. The power transformation ANOVAs yielded R^2 values almost identical to those of the ANOVAs performed on ranked Y values, with the accompanying pairs of F-ratio values also being very similar.

In concluding this section, as well as this chapter, the analyst should keep in mind that any of the parametric techniques reviewed in this book can be applied to ordinal data as well as interval/ratio data. There is a loss of statistical efficiency when these techniques are substituted for their parametric counterparts, though (e.g., the efficiency for Spearman's rank correlation coefficient is 0.91 and for the Mann-Whitney statistic, 0.95). It needs to be stated emphatically that the statistical inference calculations included in the affiliated computer output, in their reported form, are of no value to the analyst; failure to heed this caution can result in misleading or incorrect inferences. However, resampling statistics methodology (see §2.3) exhibits considerable potential for circumventing this restriction.

1.6. References

Bailey, T. and A. Gatrell. 1995. *Interactive Spatial Data Analysis.* Harlow: Longman Scientific Publishers.

Box, G. and D. Cox. 1964. "An Analysis of Transformations." *Journal of the Royal Statistical* Society. 26: 212-43.

Box, G. and P. Tidwell. 1962. "Transformations of the Independent Variables." *Technometrics* 4: 531-50.

Burt, J. and G. Barber. 1995. *Elementary Statistics for Geographers.* New York: Guilford Press.

Clark, W.A.V. and P. Hosking. 1986. *Statistical Methods for Geographers.* New York: Wiley.

Commonwealth of Puerto Rico. 1975. *1974/74 Facts and Figures on Peurto Rico's Agriculture.* Santurce, P.R.: Department of Agriculture.

Cressie, N. 1991. *Statistics for Spatial Data.* NewYork: Wiley.

Earickson, R. and J. Harlin. 1994. *Geographic Measurement and Quantitative Analysis.* New York: Macmillan.

Emerson, J. and M. Stoto. 1982. "Exploratory Methods for Choosing Power Transformations." *Journal of the American Statistical Association* 77: 103–8.

Griffith, D. and C. Amrhein. 1991. *Statistical Analysis for Geographers.* Englewood Cliffs, N.J.: Prentice Hall.

Griffith, D., R. Bennett, and R. Haining. 1989. "Statistical Analysis of Spatial Data in the Presence of Missing Observations: An Application to Urban Census Data." *Environment & Planning A* 21: 1511–23.

Hahn, G. and W. Meeker. 1993. "Assumptions for Statistical Inference." *The American Statistician* 47: 1–11.

Hair Jr., J., R. Anderson, R. Tatham, and W. Black. 1995. *Multivariate Data Analysis*, 4th ed. Englewood Cliffs, N.J.: Prentice Hall.

Hoaglin, D., F. Mosteller, and J. Tukey, eds. 1983. *Understanding Robust and Exploratory Data Analysis*. New York: Wiley.

Johnson, R. and D. Wichern. 1992. *Applied Multivariate Statistical Analysis*. Englewood Cliffs, N.J.: Prentice Hall.

Levy, P. and S. Lemeshow. 1991. *Sampling of Populations: Methods and Applications*. NewYork: Wiley.

Little, R. and D. Rubin. 1987. *Statistical Analysis with Missing Data*. New York: Wiley.

Looney, S. 1995. "How to Use Tests for Univariate Normality to Assess Multivariate Normality." *The American Statistician* 49: 64–9.

Montgomery, D. and E. Peck. 1982. *Introduction to Linear Regression Analysis*. New York: Wiley.

Searle, S. 1989. "Statistical Computing Packages: Some Words of Caution." *The American Statistician* 43: 189–90.

Stehman, S. and S. Overton. 1995. "Spatial Sampling." In *Practical Handbook of Spatial Statistics*, edited by D. Griffith, 31–63. Boca Raton, Fla.: CRC Press.

Stephan, F. (1934). "Sampling Errors and Interpretations of Social Data Ordered in Time and Space." In *Proceedings of the American Statistical Association, New Series* No. 185A, edited by F. Ross, 165-6.

Thompson, S. 1992. *Sampling*. New York: Wiley.

U.S. Bureau of the Census. 1977. *1974 U.S. Census of Agriculture*. Washington, D.C.: Government Printing Office.

Velleman, P. and L. Wilkinson. 1993. "Nominal, Ordinal, Interval, and Ratio Typologies Are Misleading." *The American Statistician* 47: 65–72.

Walford, N. 1995. *Geographical Data Analysis*. New York: Wiley.

CHAPTER 2

INFORMATION CONTENT IN GEO-REFERENCED DATA

2.0. Introduction

In Chapter 1 issues are discussed in great detail that surround three important assumptions underlying the validity of many standard parametric statistical techniques. Experience and practice suggest that a general rank ordering of these assumptions, in terms of complications introduced into a data analysis by their violation, place nonconstant variance as having the most serious consequences, followed by nonlinearities, with nonnormality in a distant third place. A fourth assumption that is of particular concern in the analysis of geo-referenced data posits that areal unit attribute values are independent. In practice, impacts of the violation of this assumption inserts it into the preceding general rank ordering somewhere between nonconstant variance and nonlinearities.

Numerical values can be arranged into rectangular tabulations for the purpose of reporting them or subjecting them to computer analyses. These rectangular organizations of data consist of rows and columns, with each row-column intersection being called a cell, and may be called ***matrices*** (Griffith and Amrhein, 1991, pp. 362–64); work with matrices is the subject of Chapter 3. Spatial dependency latent in geo-referenced data may be captured, at least in part, by specifying an n-by-n binary connectivity matrix (Griffith and Amrhein, 1991, p. 137), where a cell value of 1 indicates that the corresponding row and column areal units are juxtaposed (i.e., are side-by-side; have a common border). For convenience, the same sequence of areal units should be used to label both the rows and the columns of this table. This particular matrix is denoted throughout this book by **C**; the version used with the Puerto Rican data appears in Appendix B.

Dependency latent in the values of a single geo-referenced attribute is referred to as ***spatial autocorrelation*** (Griffith and Amrhein, 1991, pp. 133–38). This concept may be viewed parallel to the more traditional notion of correlation between two attributes. Consider variables X and Y. If these two variables are uncorrelated ($r=0$), then a bivariate regression of Y on X yields $a=\bar{x}$ and $b = 0$. In other words, there is no duplicate information in X and Y, resulting in the attribute information contained in Y not being predictable from that contained in X. As the correlation between X and Y increases, b moves away from 0, and Y becomes increasingly predictable from X, until for $r=\pm1$ the attribute information in X and Y is identical, and Y can be predicted perfectly from X (see Figure 6.1 for a set theoretic depiction of this idea). Similarly, for a geo-referenced attribute, if spatial autocorrelation is 0, then locations having high values tend to be surrounded by ones having high, medium, and low values. In contrast, as spatial autocorrelation approaches 1, similar values increasingly tend to cluster on a map, meaning locations having high values are surrounded by ones having high values, locations having medium values are surrounded by ones having medium values, and locations having low values are surrounded by ones having low values.

In other words, information in nearby locations is redundant, and hence an attribute value observed for any given location can be predicted from the values of surrounding locations. This duplicate information may be referred to as ***locational information***. And, this parallel between bivariate regression and spatial autocorrelation can be exploited to establish a better understanding of the latter.

2.1. Elementary and Spatial Statistics

The mean, variance, and skewness measures—coupled with histograms, boxplots, and scatterplots—are used in Chapter 1 to help gain a better understanding of geo-referenced data, as well as of how it might have to be preprocessed before multivariate analysis. Bivariate regression (Griffith and Amrhein, 1991, Chapter 12) and correlation serve here as procedures with which this understanding can be deepened. Before initiating this treatment, recall that the bivariate regression coefficients are given by

$$a = \frac{(\sum_{i=1}^{n} y_i)(\sum_{i=1}^{n} x_i^2) - (\sum_{i=1}^{n} x_i)(\sum_{i=1}^{n} x_i y_i)}{n(\sum_{i=1}^{n} x_i^2) - (\sum_{i=1}^{n} x_i)^2}, \text{ and}$$

$$b = \frac{n(\sum_{i=1}^{n} x_i y_i) - (\sum_{i=1}^{n} x_i)(\sum_{i=1}^{n} y_i)}{n(\sum_{i=1}^{n} x_i^2) - (\sum_{i=1}^{n} x_i)^2} .$$

To acquire an intuitive appreciation of what is known as ***no-intercept regression***, suppose an analyst decides that the intercept is zero, and thus sets a = 0. Then after some algebraic manipulation,

$$0 = \frac{(\sum_{i=1}^{n} y_i)(\sum_{i=1}^{n} x_i^2) - (\sum_{i=1}^{n} x_i)(\sum_{i=1}^{n} x_i y_i)}{n(\sum_{i=1}^{n} x_i^2) - (\sum_{i=1}^{n} x_i)^2} \text{ reduces b to}$$

$$\frac{\sum_{i=1}^{n} y_i}{\sum_{i=1}^{n} x_i} = \frac{n(\sum_{i=1}^{n} x_i y_i) - (\sum_{i=1}^{n} x_i)(\sum_{i=1}^{n} y_i)}{(\sum_{i=1}^{n} x_i^2) - (\sum_{i=1}^{n} x_i)^2} = b .$$

Therefore, if variable X is a vector of ones, denoted by **1** (Griffith and Amrhein, 1991, p. 362), then regressing variable Y on this vector **1** while equating the intercept term, a, to zero yields $b = \bar{y}$. This estimate can be obtained with SAS by specifying the PROC REG model option of NOINT (for an illustration of the necessary SAS code to execute this task, see Example PR-2.1). This problem is revisited in Chapter 4.

This no-intercept regression solution is presented here because it is needed to use bivariate regression to compute the Moran Coefficient (MC) measure of spatial autocorrelation. This particular index is very similar to the conventional product moment (Pearson's) correlation coefficient, r: as MC approaches 1, increasingly similar values cluster on a map; as MC approaches -1, increasingly dissimilar values cluster on a map; and, as MC approaches -1/(n-1), map pattern tends to disappear. The differences between MC and r that an analyst needs to keep in mind include (1) the *random map pattern value*, -1/(n-1), only converges on 0 as n increases, and (2) the extreme values may lie beyond ±1. Two bivariate regressions are required to compute MC but only after some preprocessing of the geo-referenced data. First, to simplify calculations, the attribute under study needs to be converted to z-scores (*standard normal curve values*; see Griffith and Amrhein, 1991, pp. 183–86). Second, for each areal unit in turn, the number of neighboring areal units needs to be calculated, as do the sums of the neighboring units' attribute values. Let X0 denote the vector of ones, C1 denote the vector of number of neighboring areal units, and CY denote the vector of sums of the neighboring units' attribute values. Then the two bivariate regressions to be computed regress C1 on X0 (with no intercept), and CY on Y, with MC = $b_{C1|X0}/b_{CY|Y}$ (see §4.2.4). The bivariate regression of CY on Y gives some indication of the extent of redundant information (i.e., $100\times R^2$ = the percentage of variance accounted for in Y by CY) and can be related to a scatterplot.

EXAMPLE PR-2.1. Compute the Moran Coefficient for each of the eight Puerto Rican agricultural productivity attributes and their power transformed counterparts.

Using the following SAS code, where Y in the fourteenth line (marked by ☞) is set to the attribute of interest for a given execution:

```
DATA STEP0;
   INFILE PRDATA;
   INPUT IDPR NAME$ ADM LND RU DFRM DCD DMLK DSGR
          DCF DTB DBN DFAM U V AREA C W;
TDFRM = (DFRM-.7)**0.8;
TDCD = (DCD + 28.1)**1.6;
TDMLK = (DMLK)**0.45;
TDSGR = LOG(DSGR+3);
TDCF = (DCF + 0.75)**(.3);
TDTB = 1/(DTB + 1);
TDBN = 1/SQRT(DBN+0.75);
TDFAM = (DFAM + 0.25)**.2;
X0 = 1;
Y =  TDFRM;                    ☞
RUN;
PROC STANDARD MEAN=0 STD=1 OUT=STEP0 (REPLACE=YES);
   VAR Y; RUN;
DATA STEP1;
   SET STEP0;
   INFILE CONN;
   INPUT NUMCON C1-C73;
```

```
ARRAY CMTX{73} C1-C73;
ARRAY YMTX{73} CY1-CY73;
CSUM=0;
 DO I=1 TO 73;
   CSUM = CSUM + CMTX{I};
   YMTX{I} = Y*CMTX{I};
 END;
RUN;
PROC MEANS DATA=STEP1 NOPRINT;
   VAR CY1-CY73;
   OUTPUT OUT=CYOUT1 SUM=CY1-CY73;
RUN;
PROC TRANSPOSE DATA=CYOUT1 PREFIX=WY OUT=CYOUT2;
   VAR CY1-CY73;
RUN;
DATA STEP1 (REPLACE=YES);
   SET STEP1;
   SET CYOUT2;
CY=WY1;
RUN;
PROC REG DATA=STEP1 OUTEST=BCSUM NOPRINT;
   MODEL CSUM = X0/NOINT; RUN;
PROC REG DATA=STEP1 OUTEST=BCY; MODEL CY = Y; RUN;
PROC MEANS DATA=STEP1 NOPRINT; VAR CSUM X0;
   OUTPUT OUT=CSUMOUT SUM=CSUM N; RUN;
DATA STEP2;
   SET BCSUM (KEEP=X0);
   SET BCY (KEEP=Y);
   SET CSUMOUT;
MC = Y/X0;
ASEMC = SQRT(2/CSUM);
AZ = ABS((MC + 1/(N-1))/ASEMC);
AP = 1 - PROBNORM(Z);
RUN;
PROC PRINT DATA=STEP2; VAR MC ASEMC AZ AP; RUN;
```

produces, after repeated executions,

attribute	raw data		transformed data		
	MC	GR	MC	GR	R^2
DFRM	0.31795	0.74097	0.33049	0.71027	0.391
DCD	0.36097	0.58544	0.34586	0.59445	0.303
DMLK	0.14559	0.73850	0.34777	0.58108	0.353
DSGR	0.38787	0.52304	0.43827	0.55641	0.455
DCD	0.57730	0.67484	0.63596	0.46176	0.676

DTB	0.29903	0.77983	0.33894	0.76050	0.326
DBN	0.35555	0.91377	0.43739	0.62418	0.515
DFAM	0.26176	0.86584	0.30233	0.47309	0.336

Because of the independent observations assumption violation, these R^2 values (as well as the t-statistics—which appear to be about 50% higher than the correct ones for the MC values themselves) that SAS produces are not completely reliable. Geary Ratio (GR) results are also reported here, but the SAS code for computing GR is not included in the previous calculation (see Griffith, 1993, pp. 24–25). It is useful to inspect MC and GR pairs for consistency; in this example, results are consistent. The calculated MC values suggest that in each case a modest to moderate level of positive spatial autocorrelation is contained in each geo-referenced attribute (the extreme MC values in this case are 1.10362 and -0.58724). In all but one case (DCD), the power transformations applied to the variables result in an increase in the detected level of spatial autocorrelation. The accompanying scatterplots for these regressions appear in Figure 2.1 (after Anselin, 1993). Most of these scatterplots appear well behaved; exceptions are for TDTB, which exhibits an apparent curvilinear relationship, TDCF, which displays somewhat of a nonconstant variance, and TDMLK, which has the Hatillo municipio as a spatial outlier.

2.2. Redundant Attribute and Locational Information

Three portions of rectangular tabulations of numerical values, or matrices, are of reference value, namely, the diagonal (those cells for which the ordinal identification number of the row and column are the same), the lower triangle (all cells to the left of the diagonal), and the upper triangle (all cells to the right of the diagonal); see §1.3.2 for an implementation illustration of these table regions. The correlation matrices for both the original, raw data, and the transformed attributes can be obtained with SAS executing PROC CORR. These matrices furnish a summary of linear relationships amongst variables and facilitate an understanding of redundant attribute information latent in the data set under study.

EXAMPLE PR-2.2. Explore the presence of redundant attribute information in the Puerto Rican data by constructing the lower diagonal of the correlation matrix for the raw and the transformed data (note: because $r_{XY} = r_{YX}$, there is no need to report both the upper and the lower triangles of a correlation matrix, and because $r_{YY} = 1$, there is no need to report the diagonal of a correlation matrix).

<u>original data</u>

DCD	0.36489						
DMLK	-0.11941	0.05702					
DSGR	-0.21569	0.23940	-0.20564				
DCD	0.51647	0.33656	-0.24571	-0.17831			
DTB	0.26959	0.19458	-0.07292	-0.22774	-0.07456		
DBN	0.52807	0.38241	-0.19823	-0.30341	0.83734	0.03769	
DFAM	0.50074	0.45637	0.22031	-0.23300	0.51758	0.09019	0.55962
	DFRM	DCD	DMLK	DSGR	DCD	DTB	DBN

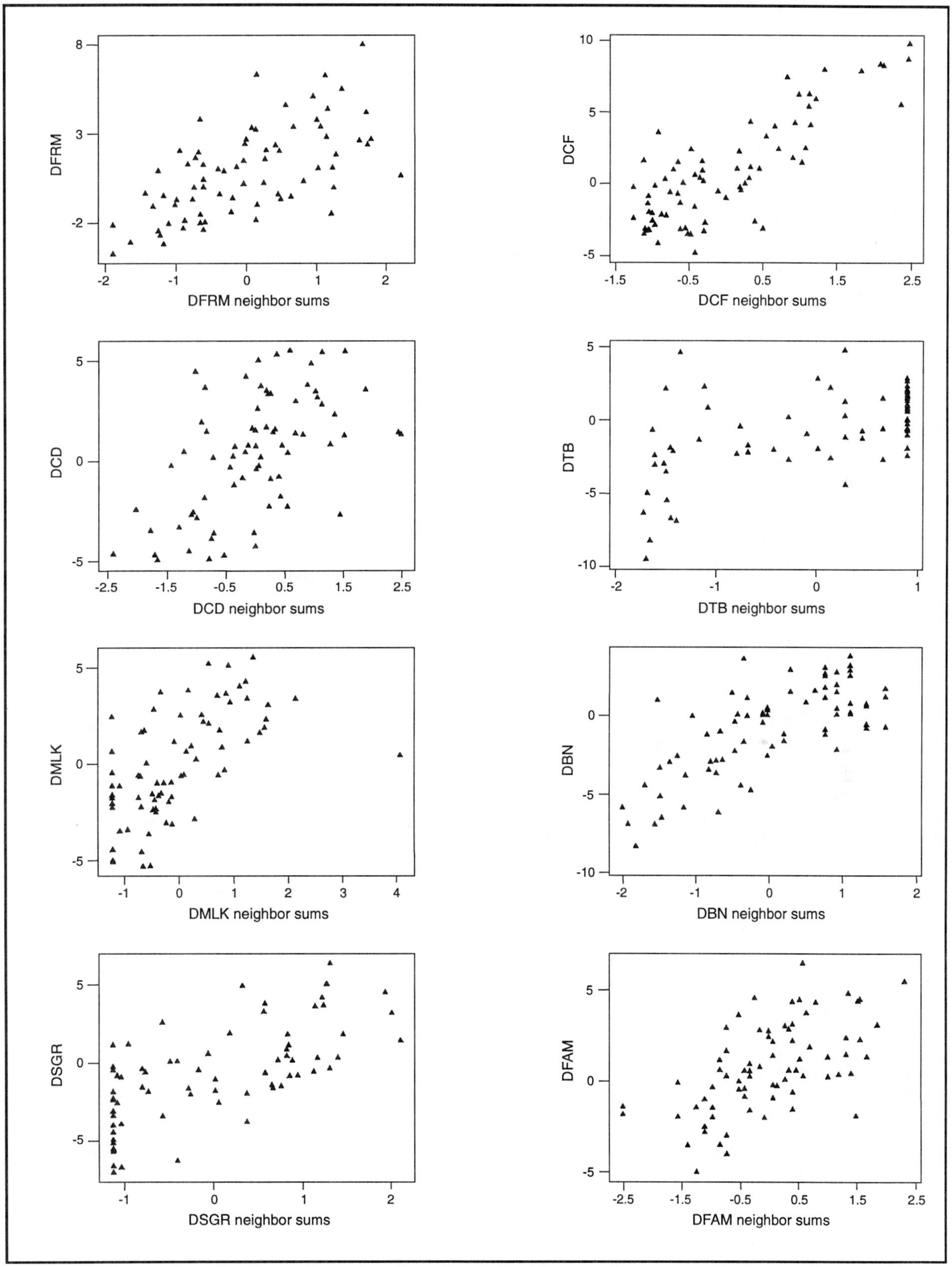

Figure 2.1. Scatterplots of each transformed attribute (vertical axis) versus its corresponding summed geographic neighbors' attribute values (horizontal axis).

transformed data

TDCD	0.33840						
TDMLK	-0.20083	-0.08058					
TDSGR	-0.34582	0.16982	-0.18250				
TDCD	0.62240	0.40669	-0.52994	-0.17477			
TDTB	-0.50121	-0.20804	-0.08181	0.32877	-0.12497		
TDBN	-0.72739	-0.41619	0.40101	0.40411	-0.81417	0.34281	
TDFAM	0.53889	0.49006	0.10192	-0.21178	0.46183	-0.30442	-0.51196
	TDFRM	TDCD	TDMLK	TDSGR	TDCD	TDTB	TDBN

A comparison of these two tables of correlation coefficients corroborates the contention put forth in §1.3.2 that variable transformations to improve symmetry and stabilize spread serendipitously can help improve the straightness of relationships. Nine coefficients are decreased by the employed transformations, with the largest decrement being roughly 0.12, while nineteen coefficients are increased, with the largest increment being roughly 0.31. The greatest degree of conspicuous redundant attribute information exists between TDBN and the remaining seven agricultural productivity variables. Further treatment of this topic is pursued in Chapter 6.

Redundant locational information can be inspected in a similar manner by constructing a matrix of cross-MCs (see Wartenberg, 1985). It can also be examined through bivariate regression analysis, where geo-referenced attributes are regressed on ideal map patterns related to selected levels of spatial autocorrelation. These map patterns, whose construction is discussed in Chapter 3, begin with the maximum level of positive spatial autocorrelation, which relates to a linear trend across a geographic landscape and successively moves toward the maximum level of negative spatial autocorrelation by increasingly fragmenting the map pattern.

EXAMPLE PR-2.3. Explore the presence of redundant locational information in the Puerto Rican data by constructing the lower diagonal of the cross-MC matrix for the transformed data (note: because $MC_{XY} = MC_{YX}$, there is no need to report both the upper and the lower triangles of a cross-MC matrix, and because MC_{YY} is reported in Example PR-2.1, there is no need to repeat the diagonal entries of this cross-MC matrix).

TDCD	0.10537						
TDMLK	-0.11147	-0.18702					
TDSGR	-0.10007	0.06557	-0.13681				
TDCD	0.27957	0.34832	-0.32173	0.02731			
TDTB	-0.17985	-0.06762	-0.09816	0.23156	0.04501		
TDBN	-0.27953	-0.24324	0.22441	0.18570	-0.43993	0.12684	
TDFAM	0.27131	0.23271	-0.05929	-0.06378	0.30768	-0.13838	-0.27074
	TDFRM	TDCD	TDMLK	TDSGR	TDCD	TDTB	TDBN

Griffith (1991) notes that, in the presence of zero spatial autocorrelation, the expected value of a cross-MC is $-\rho_{XY}/(n-1)$; because $\rho_{XX}=1$, this reduces to the familiar $-1/(n-1)$ for individual attributes. This cross-MC expected value indicates that redundant information revealed by a cross-MC matrix is a mixture of redundant attribute information and redundant locational information. Map patterns indexed by the preceding cross-MC matrix suggest, for example, that a modest degree of negative spatial autocorrelation exists between TDBN and TDFRM (-0.27953 versus an expected value of 0.01011), while the geographic distribution of TDBN vis-à-vis TDCD essentially is random (0.04301 versus an expected value of 0.01131).

Cross-MC results can be supplemented by relating, with bivariate regression, each attribute to a selected ideal map pattern that contains a specified level of spatial autocorrelation. These ideal map patterns come from a given geographic connectivity matrix, **C**, are associated with the extreme values of MC, and represent statistically independent map patterns (Griffith, 1996; Tiefelsdorf and Boots, 1995). A bivariate regression in which some attribute is regressed on one of these ideal geographic distributions reveals (1) the percentage of variance accounted for in the attribute by a selected geographic expression of spatial autocorrelation and (2) the decrease in MC for the selected variable that is attributable to removal of the spatial autocorrelation component represented by the selected ideal map pattern (MC is computed for the regression residuals).

EXAMPLE PR-2.3 (continued). Compute bivariate regression results to explore the presence of redundant locational information.

Four ideal map patterns appear in Map 2.1. The first depicts a generally declining trend from west to east across of the island of Puerto Rico. The second depicts a generally declining trend from the southeastern corner of the island. The third depicts a north-south trough across the island. And the fourth depicts a basin located in the east-central part of the island, accompanied by peaks in the northeastern, southeastern, and northwestern parts of the island.

SAS bivariate regression and MC results involving these four ideal map patterns include

attribute	selected ideal map pattern #	selected ideal map pattern MC	R^2	residual MC
DFRM	4	0.50302	0.367	0.23036
DCD	2	0.85621	0.175	0.23758
DMLK	2	0.85621	0.168	0.24537
DSGR	3	0.82180	0.190	0.34843
DCD	1	1.10362	0.261	0.47119
DTB	4	0.50302	0.131	0.31427
DBN	4	0.50302	0.259	0.41448
DFAM	4	0.50302	0.308	0.21306

On average the residual MC values have been reduced by roughly one-third, and the percentages of variance accounted for—while much smaller than those reported in Example PR-2.1—still range from 13% to 37%. The implication here is that a sizable amount of the information contained in each attribute value is made up of both redundant attribute and locational information.

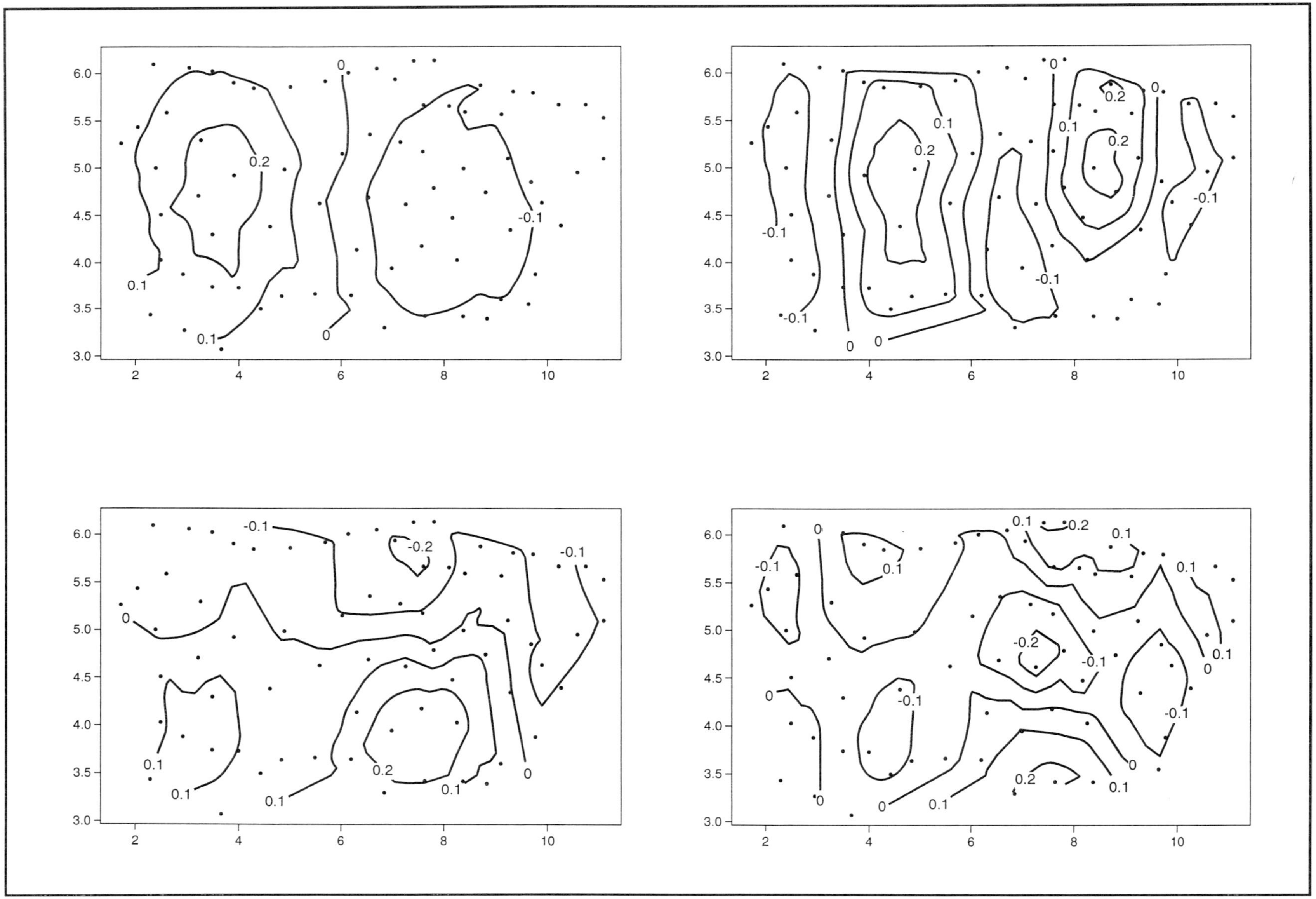

Map 2.1. Ideal map patterns: west-east trend surface across the island (top left); trend surface declining from the southeast corner of the island (bottom left); north-south trough traversing the island (top right); and, funnel-shaped trend surface inclining to the east-central part of the island (bottom right).

2.3. Exploring Residuals in Spatial Analysis

Thomas's (1968) seminal piece outlines how mapping residuals from regression can be used to foster more informed and comprehensive spatial analysis. He argues that spatial outliers can prompt a more in-depth study of their corresponding areal units and that residual map patterns can provide a clue as to what additional attributes might be useful to include in a spatial analysis (Chapter 4 in this volume furnishes the multiple regression methodology for incorporating any additional variables). Residual maps also disclose the possibility of remaining spatial auto-correlation impacts that are not attributable to the mean or nonconstant variance terms.

Another use of residuals is to acquire a better understanding of the sampling distribution of statistics such as MC. This goal may be achieved through ***Monte Carlo simulation***, the ***bootstrap technique*** (see Efron, 1982), or other resampling procedures. The name Monte Carlo simulation, derived from the European gambling resort, appropriately conveys an element of randomness and an element of mock-up. This approach uses a computer to create many samples of artificial data (i.e., pseudorandom numbers[1]) drawn from a selected hypothetical population/frequency distribution invented by the spatial analyst (and hence known to be true), from which a sampling distribution is emulated by calculating a statistic for each of the fabricated samples. Virtually any sampling distribution can be simulated if its properties are well understood. Arithmetic means of empirical sampling distributions constructed by Monte Carlo simulation provide a way to assess statistical unbiasedness, whereas their standard deviations provide a way to assess statistical efficiency. One classical use of Monte Carlo methods is to simulate the sampling distribution for MC given zero spatial autocorrelation. One popular standard error formula reported for MC couples this property of zero spatial autocorrelation with a normal distribution (Cliff and Ord, 1981, p. 44). Drawing $\mathfrak{R}$ (denoting the number of replicates) samples of size n from a normal distribution, allocating each set of these values *at random* to the areal units constituting a map, and then calculating MC for each of these allocations verifies the analytical formula result.[2] The residuals used here equal the difference between each pseudorandom number and its sample mean (recall the regression of Y on a vector of ones discussed in §2.1)—the constant mean model specification. Theoretical properties of MC become less clear, particularly when n is small, if the pseudorandom numbers fail to be randomly allocated to those areal units on a map. Monte Carlo results reported in this section are based upon $\mathfrak{R}$ (= 100) samples of size n drawn from a normal distribution, allocating these values according to their rankings *in the rank order of the observed attribute values* to the areal units on the map of Puerto Rico, and then calculating MC for each of these allocations. Essentially, this Monte Carlo simulation combines the properties of the observed level of spatial autocorrelation and a normal frequency distribution (which is mimicked by applying the appropriate power transformation to the original attribute).

Mathematical reasoning based upon a set of posited assumptions yields theoretical sampling distributions for many classical statistics, ones whose properties are heavily dependent upon these

---

[1] whose generators are equations that can produce large amounts of random data from any theoretical distribution.

[2] Monte Carlo simulation results are not needed to acquire this knowledge. Theoretical arguments render the same conclusions on more general grounds, by exploiting theoretical mathematical statistical properties of MC coupled with the underlying normal distribution of independent and identically distributed errors.

assumptions. Assumptions such as zero spatial autocorrelation and a normal frequency distribution are made *only* for mathematical convenience. Without these assumptions, mathematical reasoning may become infeasible. The resampling technique called bootstrapping, whose name is reminiscent of the phrase "pulling oneself up by one's bootstraps," replaces such knowledge with understanding based upon brute force computing, just like Monte Carlo simulation does. Experience shows that it offers a superior alternative to analytical results when the usual underlying statistical assumptions seem dubious; it is a powerful statistical tool that liberates spatial analysts from the restrictions of unrealistic assumptions. Once again the sampling distribution is simulated but now by using the real data at hand to draw random samples of size n (with replacement) from a set of *independent and identically distributed*[3] values that constitute the sample, without presuming that a true data model is known. No specific form of the parent frequency distribution need be assumed, although the observed sample is expected to be representative of the population. Constant mean and variance and zero spatial autocorrelation are assumed for the sample that is being reused to select the bootstrap samples. These properties can be induced in an attribute by power transformations coupled with a bivariate regression employing a composite variable that yields residuals with zero spatial autocorrelation (see §2.2 for discussion of the nature of this type of variable). The bootstrap actually is applied to these residuals, whose resampled values then are added to the predicted regression values to obtain replicate areal unit attribute values; failing to remove spatial autocorrelation effects first can render misleading bootstrap results. Variability across these bootstrap samples serves as a substitute for sample-to-sample differences that would be obtained if the parent population was repeatedly sampled. Practice indicates that for the bootstrap $\mathfrak{R}$ should be somewhere between 50 and 200; for standard error estimates, $\mathfrak{R}$=100 usually is adequate.

EXAMPLE PR-2.4. Compare analytical standard errors for MC with those computed using Monte Carlo simulation and the bootstrap for each of the Puerto Rican agricultural productivity attributes.

The analytical result assuming an underlying normal frequency distribution is σ_{MC} = 0.07542 [also, see equation (4.13) in §4.2.4]; its asymptotic counterpart, given by $\sqrt{\frac{2}{\sum_{i=1}^{n}\sum_{j=1}^{n} c_{ij}}}$ = 0.07625, is almost exactly the same. Monte Carlo simulation and bootstrapping have yielded the following results (SAS code is not presented here because these procedures are not yet routinely available in commercial statistical software packages):

---

[3]In fact, the mechanism of bootstrap resampling should imitate the way in which a sample was originally chosen from its parent population. Given the model-based inference perspective promoted in this book, we cast all analyses within the context of independent and identically distributed errors. This assumption is not a necessity for employing bootstrapping. But data resampling employing other possibilities often yields results strikingly different from those involving independent and identically distributed values, making such results more difficult to compare with classical multivariate output. When model-based inference assumptions hold, bootstrapping should yield approximately the same standard errors produced by classical mathematical statistical theory; otherwise the results will often be divergent.

attribute	s_{MC} from		$\bar{x}_{MC}$ from	
	Monte Carlo	bootstrap	Monte Carlo	bootstrap
TDFRM	0.01308	0.03729	0.31195	0.41126
TDCD	0.01686	0.05923	0.34747	0.42539
TDMLK	0.02342	0.06045	0.38573	0.41674
TDSGR	0.01856	0.04321	0.43222	0.49308
TDCD	0.03189	0.02744	0.53047	0.65854
TDTB	0.03412	0.04427	0.30597	0.42992
TDBN	0.02386	0.02544	0.39056	0.49273
TDFAM	0.01139	0.03802	0.31750	0.37314

First, the two analytical results are almost identical and are larger in every case than their Monte Carlo and bootstrap counterparts. But is this attributable to their lack of robustness? Are they untrustworthy due to false assumptions? Monte Carlo estimates of σ_{MC} are variable and are less than their analytical companions by a factor of at least 2. Meanwhile, the bootstrap $\bar{x}_{MC}$s are systematically greater than their observed counterparts, whereas the bootstrap s_{MC}s are both variable and tend to lie between their analytical and Monte Carlo companions. On the one hand, these means do tend to center around their respective computed attribute MCs rather than on the corresponding population MCs. On the other hand, their marked deviations from their analytical counterparts suggest that they are probably improved standard error estimates, allowing unnoticed pathologies of geo-referenced data to be more obvious (even if a relatively small number of areal units would have been present). If the analytical standard errors actually are unrealistically high, then a spatial analyst is being led astray by more than mere chance alone in concluding that spatial dependence is absent from geo-referenced data when in fact it is not.

The reliability of these bootstrap results is bolstered by a number of favorable diagnostic results that have been calculated for the regression residuals from which bootstrap samples have been drawn, coupled with each attribute's composite variable being fixed for the given geographic landscape of Puerto Rico (i.e., even if new samples were collected across the island by the U.S. Department of Agriculture, these composite variables would remain the same). In each of the eight attribute cases, roughly two-thirds or more of the variation is accounted for by the artificial map pattern used to remove spatial autocorrelation effects. Each of these eight sets of residuals displays negligible spatial autocorrelation, conforms to a normal frequency distribution, and exhibits constant variance across the geographic landscape. Boxplots reveal the presence of an outlier for TDFRM, TDCD, TDMLK, TDSGR, and two for TDTB; but these outliers are not sufficiently extreme to cause deviation of their respective frequency distributions from normality. Further, these same five transformed attributes display only subtle deviations from constant variance across the magnitude of predicted areal unit values. All in all, then, considering these regression residuals as being *independent and identically distributed* does not seem unreasonable.

2.4. References

Anselin, L. 1993. "Moran Scatterplots as a Means to Visualize Instability in Spatial Autocorrelation." Paper presented to the NCGIA "Exploratory Spatial Data Analysis and GIS" Workshop, U.C.-Santa Barbara, February 25–27.

Cliff, A. and J. Ord. 1981. *Spatial Processes*. London: Pion.

Efron, B. 1982. *The Jackknife, the Bootstrap, and Other Resampling Plans*. Philadelphia: Society for Industrial and Applied Mathematics.

Griffith, D. 1991. "Univariate and Multivariate Modelling of areal data." In *GIS and Spatial Data Analysis: Report on the Sheffield Workshop,* edited by R. Haining and S. Wise. Discussion Paper No. 11. Sheffield: Regional Research Laboratory Initiative.

—. 1993. *Spatial Regression Analysis on the PC: Spatial Statistics Using SAS.* Washington, DC: Association of American Geographers.

—. 1996. "Spatial Autocorrelation and Eigenfunctions of the Geographic Weights Matrix Accompanying Geo-Referenced Data." *The Canadian Geographer*, 40: 351-67.

Griffith, D. and C. Amrhein. 1991. *Statistical Analysis for Geographers*. Englewood Cliffs, N.J.: Prentice Hall.

Thomas, E. 1968. "Maps of Residuals from Regression." In *Spatial Analysis: A Reader in Statistical Geography*, edited by B. Barry and D. Marble, 326–52. Englewood Cliffs, N.J.: Prentice Hall.

Tiefelsdorf, M. and B. Boots. 1995. "The Exact Distribution of Moran's I." *Environment and Planning A* 27: 985–99.

Wartenberg, D. 1985. "Multivariate Spatial Correlation: A Method for Exploratory Geographical Analysis." *Geographical Analysis* 17: 263–83.

CHAPTER 3

INTRODUCTION TO MATRIX ALGEBRA

3.0. Elementary Concepts of Matrices

As discussed in Chapter 2, numerical values can be arranged into rectangular tabulations for the purposes of reporting or furthering analyses. These rectangular organizations of data consist of rows and columns, with each row-column intersection called a cell, and the set of rows and columns called a matrix (Griffith and Amrhein, 1991, pp. 362–64). Matrix notation represents a powerful shorthand notation for representing systems of equations which are created whenever a problem requires that values for more than one variable be computed simultaneously. When all of the equations are linear, they can be expressed in an orderly fashion in a form such as $\begin{bmatrix} 5x + 3y = -6 \\ -2x + 7y = 4 \end{bmatrix}$, and a solution by one of a number of standard methods may be attempted. All of these solutions involve manipulation of the coefficients of the x and y variables, so it is convenient to be able to isolate them and treat them as a single unit. The previous equation is thus rewritten in *matrix form* as

$$\begin{bmatrix} 5 & 3 \\ -2 & 7 \end{bmatrix} \begin{bmatrix} x \\ y \end{bmatrix} = \begin{bmatrix} -6 \\ 4 \end{bmatrix}. \qquad (3.1)$$

Griffith and Amrhein's (1991, pp. 362–64) summary of matrix operations and terminology is expanded in this section to provide a foundation for matrix operations analyzed throughout this book.

In general, a ***matrix*** is a rectangular array that, in this text, contains only real numbers (but in general may contain real or complex numbers). The matrix is treated as a single object for the purpose of manipulation. A general matrix will have *m* rows and *n* columns, and is described as an *m-by-n* matrix. Equation 3.1 shows the product of a 2-by-2 matrix and a 2-by-1 matrix being equal to a 2-by-1 matrix. Matrix multiplication will be discussed in more detail in §3.0.1.

A ***matrix element*** is one of the terms in the rectangular array, and its location within the array is the cell whose location is given by a pair of subscripts (frequently m and n, or i and j). One subscript (i or m) denotes the row containing the element, and a second subscript (j or n) denotes the column containing the element. For example, the matrix

$$\mathbf{A} = \begin{bmatrix} a_{11} & a_{12} & \cdots & a_{1n} \\ a_{21} & a_{22} & \cdots & a_{2n} \\ \vdots & \vdots & \ddots & \vdots \\ a_{m1} & a_{m2} & \cdots & a_{mn} \end{bmatrix} \tag{3.2}$$

has elements a_{ij}, where i is the row and j is the column containing this element. By convention, the symbol for a matrix is written in bold face (e.g., **A**), and the matrix itself is usually enclosed in square brackets. Because a vector is a special case of a matrix, vectors also are written in bold face.

Certain types of matrices occur frequently enough that they are specifically named. A matrix that has an equal number of rows and columns is called a ***square matrix*** or an *order n matrix.* If a matrix has only one row or one column, it is called a ***vector***, and if it has only one row **and** one column, it is called a *scalar.* By definition, the ***main diagonal*** of a square matrix is the set of elements a_{ii}, which runs from the upper left to the lower right corners of the matrix. If a square matrix has nonzero elements only on the main diagonal, it is labeled a ***diagonal matrix***. The diagonal matrix whose elements are all equal to 1 is called the ***identity matrix*** and is denoted by the symbol **I**. A matrix consisting entirely of zeros is called a ***zero matrix*** and is represented by **0**. An ***upper triangular matrix*** is one in which all of the elements to the left of the main diagonal are zero, whereas a ***lower triangular*** matrix has only zero elements to the right of the main diagonal. Other special types of matrices will be defined when they arise.

3.0.1. Matrix Operations

Since matrices may be treated as single mathematical objects, it is important to be able to define mathematical operations that can be performed on them and relate these operations to the standard operations of addition and subtraction, multiplication and division, and equality, which also are performed on scalar quantities. Other operations that are exclusive to matrices also exist.

Two matrices **A** and **B** are equal if and only if they have the same number of rows and columns, and if $a_{ij}=b_{ij}$ for all values of i and j. If two matrices are of different sizes, then equality, as well as addition and subtraction, is impossible.

The matrix $\mathbf{C}=\mathbf{A}\pm\mathbf{B}$ has elements $c_{ij}=a_{ij}\pm b_{ij}$; that is, each element of **C** is simply equal to the sum (or difference) of the corresponding elements of matrices **A** and **B**. Needless to say, if the two matrices are not of the same size, not all elements can be matched, and the addition or subtraction operation will be impossible.

The multiplication of a matrix **A** by a scalar κ is equivalent to adding **A** to itself κ times, which is the same concept as the multiplication of two scalars. Hence matrix $\mathbf{C} = \kappa\mathbf{A}$ will have elements $c_{ij} = \kappa a_{ij}$.

Unfortunately, the multiplication of one matrix by another matrix is not as straightforward as that of multiplying a matrix by a scalar. First, consider the multiplication of a column vector by a row vector,

$$\mathbf{uv} = [u_1, u_2, ..., u_n] \begin{bmatrix} v_1 \\ v_2 \\ \vdots \\ v_n \end{bmatrix} = \left[\sum_{i=1}^{n} u_i v_i \right], \tag{3.3}$$

an operation which is known as the *inner product* of **u** and **v**. It is a scalar quantity and can only be performed if **u** has the same number of columns as **v** has rows. Now consider multiplying two matrices, **A** (m-by-p) and **B** (p-by-n). Matrix **A** has m rows, each of which can be considered as a row vector of length p, and matrix **B** has n columns, each of which can be considered a column vector of length p. Multiplying these two matrices involves the simultaneous multiplication of the m row vectors and n column vectors, and will result in a matrix **C** of size (m-by-n), where each element of **C** is computed by $c_{ij} = \sum_{k=1}^{p} a_{ik} b_{kj}$.

The requirement that the first matrix product have p columns and the second have p rows immediately indicates that in general **AB**≠**BA**, so that one must be careful to maintain the order of the matrices when manipulating expressions involving matrix multiplications. By convention, for the product **AB** we say **A** *premultiplies* **B**, and **B** *postmultiplies* **A**. As an aside, even though matrix multiplication is not commutative, it is distributive [i.e., (**A**(**B**+**C**) = **AB** + **AC**)] and associative [i.e., (**A**(**BC**)=(**AB**)**C**=**ABC**)].

If **A** is an (n-by-m) matrix with elements a_{ij}, the ***transpose*** of **A**, denoted $\mathbf{A}^T$, is defined by interchanging the rows and columns of the matrix. In other words, if $\mathbf{B}=\mathbf{A}^T$, then the elements $b_{ij} = a_{ji}$, and **B** is an (m-by-n) matrix. As with matrix multiplication, the transpose has some special properties: $(\mathbf{A}^T)^T = \mathbf{A}$, $(\mathbf{A} + \mathbf{B})^T = \mathbf{A}^T + \mathbf{B}^T$, and $(\mathbf{AB})^T = \mathbf{B}^T\mathbf{A}^T$. A matrix is ***symmetric*** if $\mathbf{A}^T=\mathbf{A}$ (i.e., $a_{ij}=a_{ji}$) and ***antisymmetric*** if $\mathbf{A}^T=-\mathbf{A}$ (i.e., if $a_{ij}=-a_{ji}$), but of course these terms only apply to square matrices. A matrix is ***orthogonal*** if $\mathbf{A}^T\mathbf{A}=\mathbf{I}$.

The ***trace*** of a matrix, denoted by *TR*(**A**), is the sum of the elements of its diagonal,

$$TR(\mathbf{A}) = \sum_{i=1}^{n} a_{ii} \tag{3.4}$$

The *determinant* of a matrix (which must be square) is a scalar, denoted by *DET*(**A**) or |**A**|, associated with the matrix and computed from its elements. It is defined by the expression

$$DET(\mathbf{A}) = \sum_{j=1}^{n} a_{1j} |A_{1j}| , \tag{3.5}$$

where A_{1j} is the *cofactor* of element a_{1j}. A cofactor of element a_{1j} is the determinant of the matrix formed by the removal of row 1 and column j of **A** multiplied by $(-1)^{1+j}$. This definition can be applied recursively as required until a 1-by-1 matrix is reached, whose determinant is simply the

sole scalar element. The next simplest determinant is that for a 2-by-2 matrix, $\begin{vmatrix} a & b \\ c & d \end{vmatrix} = ad - bc$.
Using equation (3.5), the determinant of a 3-by-3 matrix can be calculated as follows:

$$DET(\mathbf{A}) = |\mathbf{A}| = \begin{vmatrix} a_{11} & a_{12} & a_{13} \\ a_{21} & a_{22} & a_{23} \\ a_{31} & a_{32} & a_{33} \end{vmatrix} = a_{11}\begin{vmatrix} a_{22} & a_{23} \\ a_{32} & a_{33} \end{vmatrix} - a_{12}\begin{vmatrix} a_{21} & a_{23} \\ a_{31} & a_{33} \end{vmatrix} + a_{13}\begin{vmatrix} a_{21} & a_{22} \\ a_{31} & a_{32} \end{vmatrix} .$$

Clearly, manually computing the determinant for any matrix beyond 3-by-3 size using this method is exceedingly cumbersome and should be avoided if at all possible. The determinant has a number of useful properties, most of which are summarized in any introductory linear algebra text (see, for instance, Noble and Daniel, 1977; Strange, 1980). These properties will not be discussed here.

The notion of a scalar inverse is important when dealing with the arithmetic operation of division. For scalar b, this inverse is denoted by $\frac{1}{b}$, so that $b \times \frac{1}{b} = 1$. This definition clarifies why $b \neq 0$, since $\frac{1}{0}$ is undefined (or sometimes defined as ∞).

The *inverse* of a square matrix **A**, denoted $\mathbf{A}^{-1}$, is such that $\mathbf{A}\mathbf{A}^{-1} = \mathbf{A}^{-1}\mathbf{A} = \mathbf{I}$, where the identity matrix is a generalization of the scalar quantity 1. If $DET(\mathbf{A}) \neq 0$, then $\mathbf{A}^{-1}$ exists, and **A** is ***non-singular*** or ***invertible***; if $\mathbf{A}^{-1}$ does not exist, **A** is ***singular***. It is easy to prove the useful properties $(\mathbf{AB})^{-1} = \mathbf{B}^{-1}\mathbf{A}^{-1}$ and $(\mathbf{A}^{-1})^{T} = (\mathbf{A}^{T})^{-1}$. Many methods exist to compute the inverse, all of which are computationally intensive.

3.1. Eigenfunctions and Their Meanings

Ever since publication of the classic *Methoden der Mathematischen Physik* by Courant and Hilbert (1924), ***eigenfunctions*** of matrices have occupied a central position in applied mathematics and engineering. The mathematical principles for solving these functions already had been established by the French mathematicians Sturm and Liouville in the 1830s. Eigenfunctions, which consist of *eigenvalues* paired with their corresponding *eigenvectors*, also are known as *characteristic*, *latent*, *proper*, and *spectral functions*. The prefix "eigen-" is German for "own"; in French these functions are referred to as *fonctions propres.* The notion of a characteristic function relates to the mathematical definition of an eigenfunction being based on the determinant of a square matrix. A matrix determinant is a particular, unique scalar quantity computed from a polynomial function of its individual cell entries, a single ordinary number that quantitatively "characterizes" the matrix. Eigenvalues are hidden in this polynomial function and hence are present but invisible, or "latent." Whereas *propre* is the French word for "own," the English word "proper" translates as a noun to this French word. And, as can be seen by inspecting various spectral functions, eigenvalues appear in them. As will be seen in §3.3, eigenvalues sometimes can be directly written in terms of the trigonometric *SIN* and *COS* functions, with the precise spectral decomposition for some matrix **A** being given by $\mathbf{A} = \sum_{j=1}^{P} \lambda_j \mathbf{E}_j \mathbf{E}_j^T$, where λ_j coupled with $\mathbf{E}_j$ constitute the j-th eigenfunction of matrix **A**. This decomposition is a powerful analytical tool for multivariate analysis; the idea of decomposing data into orthogonal components, the coefficients of which measure the components' importance, is an attractive one. The *SIN* and *COS* functions reference angles between variables and eigenfunctions; large-scale variation is summarized in the first few eigenfunctions, whereas data noise is captured by those eigenfunctions having negligible eigenvalues. A geometric representation of these functions is presented in Chapter 6.

Eigenvalues and eigenvectors take on specific meanings in different substantive contexts. In statistical mechanics, eigenvalues are the principal stresses at a given surface point, with their accompanying eigenvectors defining the principal directions of these stresses. In quantum mechanics and computational chemistry, they refer to energy values. In conventional multivariate statistics, they refer to measuring spread or variation perpendicular to a line, like the regression line. In bivariate regression, the OLS (ordinary least squares) principle deals with error deviations along the horizontal or vertical axes (Griffith and Amrhein, 1991, p. 326). The imaginary OLS line superimposed upon a scatterplot of data points also could have error deviations measured perpendicular to it. And in geography, eigenfunctions identify ideal levels and patterns of spatial autocorrelation, as well as supply an index for the degree of connectivity for a geographic graph or surface partitioning.

Mathematically, the eigenfunction problem for any square matrix **A** may be stated as follows:

$$\text{eigenvalues: } DET(\mathbf{A} - \lambda\mathbf{I}) = 0 \text{ , and}$$
$$\text{eigenvectors: } (\mathbf{A} - \lambda\mathbf{I})\mathbf{E} = \mathbf{0} \text{ ,}$$

where the n eigenvalues are denoted by λ_j and their companion eigenvectors are denoted by $\mathbf{E}_j$, which to avoid trivial solutions cannot equal the vector **0**. The matrix determinant generates an n-th order polynomial from matrix **A** in terms of λ (which would be a laborious and tedious task to write out manually for most any real-world problem[1]), with the eigenvalues being the roots of this polynomial equation (hence their being called characteristic or latent "roots"). Meanwhile, since the eigenvector problem is one in which there are n unknowns to be found for each λ_j, one for each element in vector $\mathbf{E}_j$, but where there are actually only (n-1) equations with which to determine these n unknowns, the additional equation $\mathbf{E}_j^T\mathbf{E}_j = 1$, representing a constraint imposed upon the values that can be taken on by vector $\mathbf{E}_j$, must be attached to this eigenvector problem; the resulting constrained vectors are called ***normalized eigenvectors***. Further, since both sides of this eigenvector equation can be multiplied by -1 without changing it [i.e., $(-1)\,(\mathbf{A}-\lambda\mathbf{I})\mathbf{E} = (-1)\,\mathbf{0} \Rightarrow (\mathbf{A}-\lambda\mathbf{I})(-\mathbf{E}) = \mathbf{0}$], computed normalized eigenvectors are only unique to a factor of -1. In other words, nothing changes mathematically if all elements of a given eigenvector have their respective signs switched; clearly $(-\mathbf{E}_j^T)\times(-\mathbf{E}_j)$ still equals 1.

The largest eigenvalue problem also may be rewritten as the following optimization problem:

---

[1]Computing a determinant involves identifying the n! possible products of the cell entries of a matrix (where each of these products contains a cell entry from each row and every column) and then summing these quantities.

$$\text{MAX: } \mathbf{E}_1^T\mathbf{A}\mathbf{E}_1/\mathbf{E}_1^T\mathbf{E}_1 = \lambda_1 ,$$

where the left-hand side of this equation is called a Rayleigh quotient (Borowski and Borwein, 1991, p. 492). Since the constraint of $\mathbf{E}_j^T\mathbf{E}_j = 1$ is imposed in the calculation of normalized vector $\mathbf{E}_j$, this particular optimization problem reduces to $\mathbf{E}_1^T\mathbf{A}\mathbf{E}_1 = \lambda_1$. In fact, any eigenvalue can be rewritten as a Rayleigh quotient: $\mathbf{E}_j^T\mathbf{A}\mathbf{E}_j/\mathbf{E}_j^T\mathbf{E}_j = \mathbf{E}_j^T\mathbf{A}\mathbf{E}_j = \lambda_j$. And, the smallest eigenvalue problem may be rewritten in terms of a Rayleigh quotient as the following optimization problem:

$$\mathit{MIN}\text{: } \mathbf{E}_n^T\mathbf{A}\mathbf{E}_n = \lambda_n .$$

These forms of the eigenfunction problem are useful in establishing the extreme values of the MC (Moran Coefficient), among other things.

The eigenfunction solution also may be expressed more generally as the following matrix decomposition, a useful form when seeking diagonalization of a matrix:

$$\mathbf{A} = \mathbf{E}\mathbf{\Lambda}\mathbf{E}^{-1} \text{ or } \mathbf{E}^{-1}\mathbf{A}\mathbf{E} = \mathbf{\Lambda} ,$$

where now $\mathbf{E}$ is the n-by-n matrix of eigenvectors, $\mathbf{\Lambda}$ is the n-by-n diagonal matrix of eigenvalues, and the order of eigenvalues and their affiliated eigenvectors, as they appear in these two matrices, is the same. While this ordering is arbitrary, for convenience it traditionally is such that the eigenvalues of matrix $\mathbf{\Lambda}$ are arranged in descending order of magnitude. This specific form of the eigenfunction problem aids in the interpretation of much of the output from multivariate techniques.

Finally, eigenvalues can be computed with an iterative computer algorithm, where the problem becomes one of finding values of λ and $\mathbf{x}_0$ such that when vector $\mathbf{x}_0$ is premultiplied by matrix $\mathbf{A}$, the resulting vector $\mathbf{A}\mathbf{x}_0$ is proportional to $\mathbf{x}_0$ itself, with the constant of proportionality equaling λ. In this instance the problem of computing the largest eigenvalue becomes one of iteratively multiplying each side of the following equation by matrix $\mathbf{A}$:

$$\mathbf{A}\mathbf{x}_0 = \lambda_1\mathbf{x}_0 , \; \mathbf{x}_0 \neq \mathbf{0},$$

for an arbitrarily selected vector $\mathbf{x}_0$, until convergence is attained for λ_1. This equation is a rewriting of the initial eigenvalue equation: after subtracting $\lambda_1\mathbf{x}_0$ from both sides of this latter equation, $\mathbf{A}\mathbf{x}_0 - \lambda\mathbf{x}_0 = (\mathbf{A} - \lambda\mathbf{I})\mathbf{x}_0 = \mathbf{0}$. For convenience, the arbitrary vector $\mathbf{x}_0$ can be set equal to the vector of ones, $\mathbf{1}$. This approach results in, as k increases toward infinity,

$$\mathbf{1}^T\mathbf{A}^{k+1}\mathbf{1}/\mathbf{1}^T\mathbf{A}^k\mathbf{1} \rightarrow \lambda_1 , \text{ and}$$
$$\mathbf{A}^{k/2}\mathbf{1}/\sqrt{\mathbf{1}^T\mathbf{A}^k\mathbf{1}} \rightarrow \mathbf{E}_1 .$$

Tinkler (n.d.) describes the geometric interpretation and graphical description of this iterative approach and notes that all eigenvalues can be computed in this fashion, but with considerably more difficulty unless for each new eigenvalue, λ_j sought the initial vector $\mathbf{x}_0$ is judiciously selected so that it forms right (i.e., 90°) angles with all of the previously computed eigenvectors; otherwise convergence will be to one of the already calculated eigenfunctions.

While eigenfunctions can be calculated for any square matrix, in multivariate geo-referenced data analysis, attention can be confined to matrices that are symmetric; after all $r_{XY} = r_{YX}$, and if areal unit *i* is adjacent to areal unit *j* then areal unit *j* is adjacent to areal unit *i*. Thus, for information pertaining to attributes, **A** = **R**, the correlation matrix. For locational information, **A** = **C**, the binary geographic weights or connectivity matrix. In certain problems treated in subsequent chapters, these symmetric matrices are used to compute asymmetric matrices. Fortunately, this conversion is done in such a way that virtually all of the mathematical properties of their symmetric predecessors continue to hold for the resulting asymmetric matrices. For example, frequently the geographic weights matrix is converted to its row-standardized form, say matrix **W**. As such, **W1** = **1** and **W** = $\mathbf{D}^{-1}\mathbf{C}$ where matrix **D** is diagonal with the diagonal entries being given by $<\mathbf{C1}>_{diag}$; in other words, $d_{ii} = \sum_{j=1}^{n} c_{ij}$ and $w_{ij} = \frac{c_{ij}}{\sum_{j=1}^{n} c_{ij}}$. From the permissible mathematical operations than can be performed on determinants, the eigenvalue problem then becomes:

$$DET(\mathbf{D}^{-1}\mathbf{C} - \lambda\mathbf{I}) = 0$$
$$DET(\mathbf{D}^{1/2}) \times DET(\mathbf{D}^{-1}\mathbf{C} - \lambda\mathbf{I}) \times DET(\mathbf{D}^{-1/2}) = DET(\mathbf{D}^{1/2}) \times 0 \times DET(\mathbf{D}^{-1/2})$$
$$DET(\mathbf{D}^{-1/2}\mathbf{C}\mathbf{D}^{-1/2} - \lambda\mathbf{I}) = 0$$

where at first **A** = $\mathbf{D}^{-1}\mathbf{C}$, an asymmetric matrix, and then the eigenvalue problem sequentially is converted to that for matrix **A** = $\mathbf{D}^{-1/2}\mathbf{C}\mathbf{D}^{-1/2}$, a symmetric matrix. Similarly, the accompanying eigenvector problem becomes:

$$(\mathbf{D}^{-1}\mathbf{C} - \lambda\mathbf{I})\mathbf{E} = \mathbf{0}$$
$$(\mathbf{D}^{1/2}) \times (\mathbf{D}^{-1}\mathbf{C} - \lambda\mathbf{I})\mathbf{E} = (\mathbf{D}^{1/2}) \times \mathbf{0}$$
$$(\mathbf{D}^{-1/2}\mathbf{C} - \lambda\,\mathbf{D}^{1/2}) \times (\mathbf{D}^{-1/2} \times \mathbf{D}^{1/2}) \times \mathbf{E} = \mathbf{0}$$
$$(\mathbf{D}^{-1/2}\mathbf{C}\mathbf{D}^{-1/2} - \lambda\mathbf{I}) \times (\mathbf{D}^{1/2} \times \mathbf{E}) = \mathbf{0}\,,$$

which reveals the relationship between the two sets of eigenvectors for matrices $\mathbf{D}^{-1/2}\mathbf{C}\mathbf{D}^{-1/2}$ and **W**. This special result proves useful to acquiring a better understanding of what happens with spatial autocorrelation in geo-referenced data geographically distributed over a lattice (e.g., remotely sensed data) as the number of areal units (e.g., pixels) increases to a very large number.

Salient properties of eigenfunctions for symmetric matrices alluded to in the preceding paragraph are summarized in Table 3.1. Property P2 should be intuitively apparent here, since a symmetric matrix and its transpose are exactly the same matrix. Property P9 means that $\mathbf{E}^T\mathbf{E} = \mathbf{I}$. Property P3 signifies that the sum of the eigenvalues of a correlation matrix always equals the number of attributes, say P, being studied. Property P10 means that the diagonalization form of the eigenfunction problem may be rewritten as

$$\mathbf{A} = \mathbf{E}\Lambda\mathbf{E}^T \text{ or } \mathbf{E}^T\mathbf{A}\mathbf{E} = \Lambda\ .$$

Property P12 means that $\lambda_1 = 1$ for the row-standardized geographic matrix **W**. Property P13 signifies that if $e_{i1} < 0$ for geographic connectivity matrix **C**, then the principal eigenvector must be premultiplied by -1 and hence converted to $-\mathbf{E}_1$. Property P14 indicates that $|\lambda_n| \leq \lambda_1$, a feature that reflects upon the possible range of spatial autocorrelation for a geographic landscape. Property P15 means that $\sigma_{MC} \approx \sqrt{\frac{2}{\mathbf{1}^T\mathbf{C}\mathbf{1}}}$ (see Example PR-2.4) is equivalent to $\sigma_{MC} \approx \sqrt{\frac{2}{\sum_{i=1}^{n} \lambda_i^2}}$. Throughout this book these fifteen properties will be cited to help clarify situations involving eigenfunctions.

3.2. Illustrative Eigenfunction Uses in Multivariate Analysis

The multivariate analysis literature is replete with eigenfunctions (Tatsuoka, 1977). Eigenvalues and their affiliated eigenvectors play a central role in multivariate analysis dealing with redundant attribute information (§2.2) as well as multicollinearity (Griffith and Amrhein, 1991, p. 365) latent in a data set. The primary form of the eigenfunction problem in this context is that of matrix diagonalization (e.g., $\mathbf{E}^T\mathbf{R}\mathbf{E} = \Lambda$). Much of multivariate analysis is concerned with independent variables obtained with the aid of this diagonalization operation. Most often the matrix to be diagonalized is the correlation matrix; sometimes, however, the covariance matrix is the one needing to be diagonalized. In this context, eigenvalues index the degree of multicollinearity and the extent of redundant attribute information, whereas eigenvectors supply those coefficients necessary to compute independent synthetic variates as linear combinations of the original battery of variables.

Consider the simple, two-attribute (i.e., bivariate) multivariate problem. In this instance the correlation matrix is given by $\begin{bmatrix} 1 & r_{12} \\ r_{12} & 1 \end{bmatrix}$, and hence the characteristic equation for the eigenvalue problem becomes

$$DET\left(\begin{bmatrix} 1 & r_{12} \\ r_{12} & 1 \end{bmatrix} - \begin{bmatrix} \lambda & 0 \\ 0 & \lambda \end{bmatrix}\right) = DET\begin{bmatrix} 1-\lambda & r_{12} \\ r_{12} & 1-\lambda \end{bmatrix} = 0 \text{ , or}$$

$$(1 - \lambda)^2 - r_{12}^2 = \lambda^2 - 2\lambda + (1 - r_{12}^2) = 0 \text{ ,}$$

which yields the following roots as the general eigenvalue solutions: $\lambda_1 = 1 + r_{12}$, and $\lambda_2 = 1 - r_{12}$. Referencing these results to Table 3.1 yields the following property linkages:

P1: because any r_{hk} is a real number, both $\lambda_1 = 1 + r_{12}$ and $\lambda_2 = 1 - r_{12}$ are real

P3: the trace of a 2-by-2 correlation matrix equals 2, and $\lambda_1 + \lambda_2 = 1 + r_{12} + 1 - r_{12} = 2$

P6: the determinant of a 2-by-2 correlation matrix equals $1 - r_{12}^2 = (1 + r_{12})(1 - r_{12}) = \lambda_1 \times \lambda_2$

Table 3.1. Selected Useful Properties of Eigenfunctions (Adapted from Griffith, 1988, p.44).

P1:	all of the eigenvalues and eigenvectors of a real symmetric matrix consist of real (rather than complex or imaginary) numbers
P2:	the eigenvalues of a matrix and its transpose are the same
P3:	the sum of the eigenvalues of a matrix equals the sum of that matrix's principal diagonal elements (i.e., its trace)
P4:	if some constant b is added to each element in the diagonal of a matrix, then b is added to each of the eigenvalues of that matrix, but the matrix's eigenvectors remain unchanged
P5:	if a matrix is multiplied by some scalar constant, then its eigenvalues also are multiplied by this constant, but its eigenvectors remain unchanged
P6:	the product of a matrix's eigenvalues equals the value of that matrix's determinant
P7:	the eigenvalues of a matrix and its inverse are inverses of each other, while the eigenvectors are the same
P8:	if a matrix is powered by some positive integer value, each of its eigenvalues is powered by this same positive integer value, but its eigenvectors remain unchanged
P9:	for a symmetric matrix, two eigenvectors associated with two distinct eigenvalues are mutually orthogonal (i.e., $\mathbf{E}_h^T\mathbf{E}_k = 0$, $h \neq k$)
P10:	for a real symmetric matrix, the transpose of the eigenvector matrix extracted from it equals the inverse of this eigenvector matrix (i.e., $\mathbf{E}^T = \mathbf{E}^{-1}$)
P11:	the eigenvalues of a triangular or diagonal matrix are the elements in its principal diagonal
P12:	the principal (i.e., largest or dominant) eigenvalue of a matrix is contained in that interval defined by the largest and smallest row sums for this matrix, where these sums are of the absolute values of the row cell entries
P13:	the principal eigenvector of a nonnegative, symmetric matrix has all nonnegative values.
P14:	the principal eigenvalue of a matrix is positive, and no other eigenvalue of this matrix is greater in absolute value
P15:	the sum of the squared eigenvalues of a matrix is less than or equal to the sum of all of the elements of this matrix, where these sums are of the absolute values of the cell entries, with exact equality achieved for binary matrices

P12: each of the row sums of the absolute values of cell elements for a 2-by-2 correlation matrix equals $(1 + |r_{12}|)$, which defines the upper and lower bounds for the principal eigenvalue, which is $\lambda_1 = 1 + r_{12}$ if $r_{12} > 0$, and $\lambda_2 = 1 - r_{12}$ if $r_{12} < 0$

P14: if $r_{12} > 0$, then $\lambda_1 > 0$ and $\lambda_1 \geq \lambda_2$; otherwise, if $r_{12} < 0$, then $\lambda_2 > 0$ and $\lambda_2 \geq \lambda_1$

P15: the sum of the absolute values of the cell entries of a 2-by-2 correlation matrix equals $(1 + |r_{12}| + 1 + |r_{12}|) = 2(1 + |r_{12}|)$; the sum of the squared eigenvalues equals $(1 + r_{12})^2 + (1 - r_{12})^2 = 2(1 + r_{12}^2)$; and, $2(1 + r_{12}^2) \le 2(1 + |r_{12}|)$, since $|r_{12}| \le 1$

The corresponding eigenvector problem has the characteristic equation of

$$\left(\begin{bmatrix} 1 & r_{12} \\ r_{12} & 1 \end{bmatrix} - \begin{bmatrix} \lambda_j & 0 \\ 0 & \lambda_j \end{bmatrix}\right)\begin{bmatrix} e_{1j} \\ e_{2j} \end{bmatrix} = \begin{bmatrix} 1-\lambda_j & r_{12} \\ r_{12} & 1-\lambda_j \end{bmatrix}\begin{bmatrix} e_{1j} \\ e_{2j} \end{bmatrix} = \begin{bmatrix} 0 \\ 0 \end{bmatrix},$$

which, by invoking the normalization condition, yields the following system of equations when $r_{12} > 0$:

$$\begin{aligned} e_{11} - e_{21} &= 0 \\ e_{11}^2 + e_{21}^2 &= 1 \end{aligned} \qquad\qquad \begin{aligned} e_{21} + e_{22} &= 0 \\ e_{21}^2 + e_{22}^2 &= 1 \end{aligned}$$

Accordingly, from this first pair of equations, $e_{11} = e_{21} \Rightarrow e_{21}^2 + e_{21}^2 = 1 \Rightarrow e_{21} = \frac{1}{\sqrt{2}}$; from this second pair of equations, $e_{22} = -e_{12} \Rightarrow e_{12}^2 + (-e_{12})^2 = 1 \Rightarrow e_{12} = \frac{1}{\sqrt{2}}$. Therefore, this pair of equations yields as the eigenvectors $\mathbf{E}_1 = \begin{bmatrix} \frac{1}{\sqrt{2}} \\ \frac{1}{\sqrt{2}} \end{bmatrix}$ and $\mathbf{E}_2 = \begin{bmatrix} \frac{1}{\sqrt{2}} \\ -\frac{1}{\sqrt{2}} \end{bmatrix}$. Referencing these results to Table 3.1 yields the following property linkages:

P9: $\mathbf{E}_1^T\mathbf{E}_2 = (\frac{1}{\sqrt{2}})(\frac{1}{\sqrt{2}}) + (\frac{1}{\sqrt{2}})(-\frac{1}{\sqrt{2}}) = \frac{1}{2} - \frac{1}{2} = 0$

P10: both matrices $\mathbf{E}^T$ and $\mathbf{E}^{-1}$ equal $\begin{bmatrix} \frac{1}{\sqrt{2}} & \frac{1}{\sqrt{2}} \\ \frac{1}{\sqrt{2}} & -\frac{1}{\sqrt{2}} \end{bmatrix}$.

These analytical results are useful for any supplementary pedagogic explorations pursued by the reader, because the analytical eigenvalues reported here are for *any* bivariate case, and the associated normalized eigenvectors are always those stated here, regardless of the value of the bivariate correlation coefficient, r_{12} (although their order changes when r_{12} switches from positive to negative). In addition, these results help illuminate the extreme cases of attribute information content of data: if X and Y are perfectly correlated (i.e., $r_{12} = \pm 1$) then the eigenvalues are 2 and 0, whereas if X and Y are uncorrelated (i.e., $r_{12} = 0$) then the eigenvalues are 1 and 1 (these two results can be comprehended directly from property P12). This first instance coincides with the presence of perfect multicollinearity, whereas this second instance coincides with the complete absence of multicollinearity. With regard to indexing multicollinearity, then, relative comparisons needed pertain to how close the principal eigenvalue is to the total number of variables included in an analysis, P, whether or not one or more of the smallest eigenvalues essentially equal zero, and how closely the set of eigenvalues conforms to a uniform distribution.

Next consider the three-attribute (i.e., trivariate) multivariate problem. In this instance, the correlation matrix is given by $\begin{bmatrix} 1 & r_{12} & r_{13} \\ r_{12} & 1 & r_{23} \\ r_{13} & r_{23} & 1 \end{bmatrix}$, and hence the characteristic equation for the eigenvalue problem becomes

$$(1 - \lambda)^3 + 2r_{12}r_{13}r_{23} - (1 - \lambda)(r_{12}^2 + r_{13}^2 + r_{23}^2) = 0 .$$

The determinant of this 3-by-3 matrix is given by

$$1 + 2r_{12}r_{13}r_{23} - (r_{12}^2 + r_{13}^2 + r_{23}^2) ,$$

whose terms can be easily recognized in the preceding characteristic equation. Suppose r_{12} = 0.44805, r_{13} = -0.89393, and r_{23} = -0.39907. The roots of this third-order equation, which are the eigenvalues of the 3-by-3 correlation matrix being studied, are $\lambda_1 = 2.19447$, $\lambda_2 = 0.70119$, and $\lambda_3 = 0.10434$. Implementing the iterative procedure to estimate λ_1 and $\mathbf{E}_1$ produced

k	$\lambda_1(k)$	k	$\lambda_1(k)$	k	$\lambda_1(k)$
1	0.43670	6	2.18746	11	2.19445
2	1.70749	7	2.19222	12	2.19446
3	2.00438	8	2.19375	13	2.19447
4	2.12848	9	2.19424	14	2.19447
5	2.17276	10	2.19440		

When combined with properties P3 and P6, this iterative solution allows the remaining two eigenvalues to be quickly calculated. The determinant of the correlation matrix in this case is 0.16055. Thus, solving the following pair of simultaneous equations in two unknowns,

$$2.19447 + \lambda_2 + \lambda_3 = 3$$
$$2.19447 \times \lambda_2 \times \lambda_3 = 0.16055 ,$$

yields, from the first equation, $\lambda_2 = 0.80553 - \lambda_3$, which when substituted into the second equation gives $\lambda_3 \times (0.80553 - \lambda_3) = 0.16055/2.19447$, or the quadratic equation

$$\lambda_3^2 - 0.80553\lambda_3 + 0.16055/2.19447 = 0 ,$$

which can be solved using the standard quadratic solution formula taught in elementary algebra courses. The two roots of this equation are $\frac{-(-0.80553) \pm \sqrt{(-0.80553)^2 - 4\times 1\times \frac{0.16055}{2.19447}}}{2\times 1}$, namely 0.70119 and 0.10434, or λ_2 and λ_3.

The normalized eigenvector matrix for this 3-by-3 correlation matrix case is $\begin{bmatrix} -0.63726 & 0.28019 & -0.71791 \\ -0.44843 & -0.89243 & 0.04976 \\ 0.62674 & -0.34364 & -0.69436 \end{bmatrix}$. The iterative estimation procedure yields $\begin{bmatrix} 17567.0 \\ 12361.8 \\ -17277.1 \end{bmatrix}$ as the unnormalized principal eigenvector, which normalizes to $\begin{bmatrix} 0.63627 \\ 0.44843 \\ -0.62674 \end{bmatrix}$. This result illustrates that each eigenvector is unique to a multiplicative factor of -1. Referencing these results to Table 3.1 yields the following theoretical corroborations:

P12: the upper and lower bounds for the principal eigenvalue, $\lambda_1 = 2.19447$ are given by 2.3420 and 1.8471, respectively

P10: $\mathbf{E}^T = \mathbf{E}^{-1}$

P9: $\mathbf{E}^T\mathbf{E} = \mathbf{I}$

P14: indeed, $|\lambda_3| \leq \lambda_1$

P15: the sum of the absolute values of the cell entries of this 3-by-3 correlation matrix equals 6.48211, which is greater than the sum of the squared eigenvalues, which equals 5.31825

As an aside, complex analytical eigenvalue solutions for the 3-by-3 correlation matrix presently are well known but are not reproduced here because they are too cumbersome.

Findings for the 2-by-2 and the 3-by-3 correlation matrix cases emphasize that the eigenfunction properties listed in Table 3.1 summarize considerable usable mathematical details that have been extracted from formal theorems and proofs. This table is an indispensable "ready reference" aid when seeking to understand and interpret much of the output from multivariate techniques.

3.3. Illustrative Eigenfunction Uses in Spatial Statistical Analysis

The spatial analysis literature, especially those writings about spatial statistics (e.g., Griffith, 1989), also includes numerous references to eigenfunctions. Eigenvalues and their affiliated eigenvectors play an important role in spatial analysis dealing with locational information (see §2.2). The primary form of the eigenfunction problem in this context is that of Rayleigh quotient optimization and the iterative estimation procedure. Much of spatial statistics is concerned with avoiding the double-counting of geographic information contained in geo-referenced data. In this context, eigenvalues are involved in the mathematical transformation needed to convert a spatially autocorrelated into a spatially unautocorrelated attribute space. They disclose the various distinct levels of spatial autocorrelation that can materialize from a given geographic surface partitioning, revealing the extreme levels of positive and negative spatial autocorrelation that are possible. The eigenvectors supply a means for visualizing map patterns corresponding to these distinct spatial autocorrelation possibilities.

Consider a geographic landscape having a linear surface partitioning superimposed upon it. This situation is topologically equivalent to the following configuration of areal units:

The binary geographic connectivity or weights matrix, **C**, depicting this surface partitioning may be written as $\begin{bmatrix} 0 & 1 & 0 \\ 1 & 0 & 1 \\ 0 & 1 & 0 \end{bmatrix}$, and hence the characteristic equation for the eigenvalue problem becomes

$$DET\left(\begin{bmatrix} 0 & 1 & 0 \\ 1 & 0 & 1 \\ 0 & 1 & 0 \end{bmatrix} - \begin{bmatrix} \lambda & 0 & 0 \\ 0 & \lambda & 0 \\ 0 & 1 & \lambda \end{bmatrix}\right) = DET\begin{bmatrix} -\lambda & 1 & 0 \\ 1 & -\lambda & 1 \\ 0 & 1 & -\lambda \end{bmatrix} = 0 \quad \text{, or}$$

$$(-\lambda)^3 - (-2\lambda) = 0 \ ,$$

which yields the following roots as the general eigenvalue solutions: $\lambda_1 = \sqrt{2} = 2COS(\pi/4)$, $\lambda_2 = 0 = 2COS(2\pi/4)$, and $\lambda_3 = -\sqrt{2} = 2COS(3\pi/4)$. These results allude to analytical ones that, again, are useful for any supplementary pedagogic explorations pursued by the reader, because the analytical eigenvalues for *any* linear geographic landscape and binary connectivity matrix **C** are given by

$$\lambda_j = 2COS[(j\pi/(n+1)] \ , j = 1, 2, ..., n \ ,$$

where n is the number of areal units into which a geographic landscape has been partitioned. Referencing these particular results to Table 3.1 yields the following property linkages:

P1: since both 0 and 1 are real numbers, $\lambda_j = 2COS[(j\pi/(n+1)]$, j = 1, 2, ..., n , are real

P3: the trace of this 3-by-3 geographic connectivity matrix equals 0, and $\lambda_1 + \lambda_2 + \lambda_3 = 0$

P4: since each of the diagonal entries of a correlation matrix is one (i.e., $r_{XX} = 1$), a 1 could be inserted into each of the diagonal cells of matrix **C** to trick SAS into treating this matrix as a correlation matrix; each of the resulting eigenvalues calculated by SAS will need to have 1 subtracted from it then

P12: $1 < \lambda_1 < 2$; in fact, as n goes to infinity, $\lambda_1 = 2COS[(\pi/(n+1)]$ goes to 2, moving toward the upper bound of this interval, whose limits remain the same across all values of n

P14: $|-\sqrt{2}| \le \sqrt{2}$

P15: the sum of the values of the cell entries of a binary geographic connectivity matrix, **C**, for a linear surface partitioning always equals 2(n - 1); $2\sum_{j=1}^{n} cos^2(\frac{j\pi}{n+1})$ also equals 2(n - 1)

The corresponding eigenvector problem has the characteristic equation of

$$\left(\begin{bmatrix} 0 & 1 & 0 \\ 1 & 0 & 1 \\ 0 & 1 & 0 \end{bmatrix} - \begin{bmatrix} \lambda_j & 0 & 0 \\ 0 & \lambda_j & 0 \\ 0 & 1 & \lambda_j \end{bmatrix}\right)\begin{bmatrix} e_{1j} \\ e_{2j} \\ e_{3j} \end{bmatrix} = \begin{bmatrix} -\lambda_j & 1 & 0 \\ 1 & -\lambda_j & 1 \\ 0 & 1 & -\lambda_j \end{bmatrix}\begin{bmatrix} e_{1j} \\ e_{2j} \\ e_{3j} \end{bmatrix} = \begin{bmatrix} 0 \\ 0 \\ 0 \end{bmatrix},$$

which, by invoking the normalization condition, yields the following matrix of eigenvectors:

$$\begin{bmatrix} \frac{1}{2} & \frac{1}{2} & \frac{1}{\sqrt{2}} \\ \frac{1}{\sqrt{2}} & -\frac{1}{\sqrt{2}} & 0 \\ \frac{1}{2} & \frac{1}{2} & -\frac{1}{\sqrt{2}} \end{bmatrix}$$

Referencing these particular results to Table 3.1 yields the following property linkages:

P4: this eigenvector matrix will be the same, even if the diagonal entries of matrix **C** are coded with 1s to trick SAS into doing the eigenfunction computations

P9: $\mathbf{E}_1^T\mathbf{E}_2 = (\frac{1}{2})(\frac{1}{2}) + (\frac{1}{\sqrt{2}})(-\frac{1}{\sqrt{2}}) + (\frac{1}{2})(\frac{1}{2}) = 0$, $\mathbf{E}_1^T\mathbf{E}_3 = (\frac{1}{2})(\frac{1}{\sqrt{2}}) + (\frac{1}{\sqrt{2}})(0) + (\frac{1}{2})(-\frac{1}{\sqrt{2}}) = 0$,
and $\mathbf{E}_2^T\mathbf{E}_3 = (\frac{1}{2})(\frac{1}{\sqrt{2}}) + (-\frac{1}{\sqrt{2}})(0) + (\frac{1}{2})(-\frac{1}{\sqrt{2}}) = 0$

P10: both matrices $\mathbf{E}^T$ and $\mathbf{E}^{-1}$ are the same, and hence $\mathbf{E}^T\mathbf{E} = \mathbf{I}$

P13: $\mathbf{E}_1 \ge \mathbf{0}$

Moreover, with regard to property P13, an analytical solution exists for the eigenvector problem that yields for this principal eigenvector the entries $e_{i1} = (\sqrt{\frac{2}{n+1}})\times SIN(\frac{i\pi}{n+1})$, i = 1, 2, ..., n. Because the *SIN* function only takes on values between 0 and 1 for argument angles between 0 and 180°, we know theoretically that this property holds for any size linear geographic landscape.

Repeating this exercise for the corresponding row-standardized geographic weights matrix, **W**, depiction of this very same surface partitioning may be written as $\begin{bmatrix} 0 & 1 & 0 \\ \frac{1}{2} & 0 & \frac{1}{2} \\ 0 & 1 & 0 \end{bmatrix}$, and hence the characteristic equation for the eigenvalue problem becomes

$$(-\lambda)^3 - (-\lambda) = 0 ,$$

which yields the following roots as the general eigenvalue solutions: $\lambda_1 = 1 = COS(0\pi/2)$, $\lambda_2 = 0 = COS(\pi/2)$, and $\lambda_3 = -1 = COS(2\pi/2)$. These results also allude to analytical ones that, once more, are useful for any supplementary pedagogic explorations pursued by the reader, because the analytical eigenvalues for *any* linear geographic landscape and row-standardized binary connectivity matrix **W** are given by

$$\lambda_j = COS[(j\pi/(n-1)] \ , j = 0, 1, 2, ..., n-1 ,$$

where, as before, n is the number of areal units into which a geographic landscape has been partitioned. Now the principal eigenvector is $\mathbf{1}/\sqrt{n} \geq \mathbf{0}$, and $\lambda_1 \equiv 1$ since $\mathbf{W1} = \mathbf{1}$.

Returning to the principal eigenfunction of matrix **C**, this eigenfunction furnishes an index of overall topological connectedness that can be gleaned from the iterative matrix form of the eigenfunction problem (property P8). Specifically, because the iterative solution works with powers of matrix **C**, the meaning of these powered matrices merits attention. Consider matrix $\mathbf{C}^0 = \mathbf{I}$: the cell entries here indicate the number of ways one could move from row areal unit *i* to column areal unit *j* crossing exactly 0 areal unit borders. Since this matrix is the identity matrix, one must remain in the areal unit one begins in, which can happen in only one way (i.e., to be there to begin with!), to cross zero boundaries. Next, consider matrix $\mathbf{C}^1$: one can go from areal unit *i* to any adjacent areal unit crossing exactly one border. Because the requirement of crossing exactly one boundary prevails, then one cannot either remain in or return to any origin areal unit, which means that each of the diagonal entries of this matrix is zero. Considering matrix $\mathbf{C}^2$, which for the preceding 3-by-3 illustration is $\begin{bmatrix} 1 & 0 & 1 \\ 0 & 2 & 0 \\ 1 & 0 & 1 \end{bmatrix}$: one can go from the western areal unit to the western areal unit crossing exactly two borders in only one way, and that is by going from the western to the central areal unit, and then returning to the western areal unit from the central areal unit; one can go from the central areal unit to the central areal unit crossing exactly two borders in two different ways, either by going from the central to the western areal unit and back, or by going from the central to the eastern areal unit and back. The combinatorics underlying these types of configurational possibilities are summarized in the principal eigenvector, which provides a set of relative indices that are directly proportional to how centrally located each areal unit is within the topological arrangement of the geographic surface partitioning. For this n=3 linear geographic landscape, the central areal unit is located in the topological center of the map and, accordingly, has the largest eigenvector value of 0.70711; both the eastern and the western areal units are equally peripheral and further away from the topological center of this map and accordingly, have the same eigenvector value, 0.5. This particular application serves as one example of converting nominal measures to an interval measurement scale.

The notion of the principal eigenvalue supplying an overall index of connectivity for a geographic surface partitioning is best illustrated by inspecting another, more fully articulated n=3 geographic landscape. Consider a geographic landscape in which each areal unit is in juxtaposition with every other areal unit (i.e., maximum geographic connectedness). This situation is topologically equivalent to the following configuration of areal units:

The geographic connectivity or weights matrix, **C**, depicting this surface partitioning may be written as $\begin{bmatrix} 0 & 1 & 1 \\ 1 & 0 & 1 \\ 1 & 1 & 0 \end{bmatrix}$, and hence the characteristic equation for the eigenvalue problem becomes

$$(-\lambda)^3 + 2 - (-3\lambda) = 0 \ ,$$

which yields the following roots as the general eigenvalue solutions: $\lambda_1 = 2$, $\lambda_2 = -1$, and $\lambda_3 = -1$. By better articulating the geographic surface partitioning, λ_1 increases from $\sqrt{2}$ to 2. A sizable, and somewhat demanding, geographic literature exists that discusses these types of interpretations of eigenfunctions in spatial analysis (Boots, 1982, 1984, 1985; Boots and Royle, 1991; Boots and Tinkler, 1983; Griffith, 1984; Tinkler, 1972).

A second routine use of eigenfunctions in spatial analysis concerns spatial autocorrelation. Besides technical requirements ensuring that probabilities for the normal frequency distribution cumulate to 1 both in spatially autocorrelated attribute space and its spatially unautocorrelated counterpart (see Chapter 12, in which property P6 is exploited), eigenfunctions of a geographic weights matrix also reveal valuable information and insights about spatial autocorrelation itself (Griffith, 1996). Consider the Moran Coefficient, which may be written in matrix form as MC = $\frac{n}{\mathbf{1}^T\mathbf{C}\mathbf{1}} \times \frac{(\mathbf{Y}-\bar{y}\mathbf{1})^T\mathbf{C}(\mathbf{Y}-\bar{y}\mathbf{1})}{(\mathbf{Y}-\bar{y}\mathbf{1})^T(\mathbf{Y}-\bar{y}\mathbf{1})}$. But the second term in this product is a Rayleigh quotient. Rewriting $\bar{y}$ in its no-intercept regression form, namely $\bar{y} = \mathbf{Y}^T\mathbf{1}/n$, this Rayleigh quotient may be rewritten as

$\dfrac{\mathbf{Y}^T(\mathbf{I}-\frac{\mathbf{11}^T}{n})\mathbf{C}(\mathbf{I}-\frac{\mathbf{11}^T}{n})\mathbf{Y}}{\mathbf{Y}^T(\mathbf{I}-\frac{\mathbf{11}^T}{n})\mathbf{Y}}$. The optimization problems outlined in §3.1 reveal that this quotient reaches its maximum when vector **Y** is replaced by the principal eigenvector of matrix $(\mathbf{I}-\frac{\mathbf{11}^T}{n})\mathbf{C}(\mathbf{I}-\frac{\mathbf{11}^T}{n})$ and its minimum when vector **Y** is replaced by eigenvector $\mathbf{E}_n$ of this very matrix. In other words, the maximum value of MC is achieved with $\frac{n}{\mathbf{1}^T\mathbf{C1}}\times\lambda_1$, and the minimum value with $\frac{n}{\mathbf{1}^T\mathbf{C1}}\times\lambda_n$. These are the results reported by de Jong, Sprenger, and van Veen (1984). These findings also reveal why MC is not necessarily restricted to the interval [-1, 1]. These results have been generalized by Tiefelsdorf and Boots (1995) to the range of possible distinct levels of spatial autocorrelation associated with a given geographic landscape: $MC_j = \frac{n}{\mathbf{1}^T\mathbf{C1}}\times\lambda_j$, j = 1, 2, ..., n. An interesting conjecture here is that the eigenvectors of matrix $\mathbf{M} = (\mathbf{I}-\frac{\mathbf{11}^T}{n})\mathbf{C}(\mathbf{I}-\frac{\mathbf{11}^T}{n})$ asymptotically converge upon those of matrix **C** (i.e., as n increases, 1/n goes to zero and (n-1)/n goes to 1). One difference in practice is that the eigenvectors extracted from matrix **C** have the property $\mathbf{E}^T\mathbf{1} \neq 0$, whereas the eigenvectors extracted from matrix **M** have the property $\mathbf{E}^T\mathbf{1} = 0$; another difference is that the principal eigenvector of matrix **C**, which serves as a topological connectedness index, is not amongst the eigenfunctions of matrix **M**.

Utilizing Rayleigh quotients also reveals that the eigenvector associated with a given eigenvalue corresponds to a map pattern that contains the level of spatial autocorrelation gauged by its adjusted eigenvalue value, $\frac{n}{\mathbf{1}^T\mathbf{C1}}\times\lambda_j$. These eigenvectors are the ideal spatial autocorrelation map patterns used in Chapter 2 in support of Monte Carlo simulation and residual bootstrapping. Prominent ones, to which attention can be restricted, can be identified by referencing their corresponding eigenvalues to the sampling distribution of MC, which frequently conforms to a bell-shaped curve. Accordingly, employing a two-tailed statistical test, where $z > 1.96$ is indicative of significant positive spatial autocorrelation, and utilizing the asymptotic approximate standard error (Example PR-2.4) given by $\sigma_{MC} \approx \sqrt{\frac{2}{\mathbf{1}^T\mathbf{C1}}}$,

$$\frac{\frac{n}{\mathbf{1}^T\mathbf{C1}}\lambda_j - (-\frac{1}{n-1})}{\sqrt{\frac{2}{\mathbf{1}^T\mathbf{C1}}}} \geq 1.96$$

$$\therefore\ \lambda_j \geq \frac{1.96\sqrt{2}\sqrt{\mathbf{1}^T\mathbf{C1}}}{n} - \frac{\mathbf{1}^T\mathbf{C1}}{n(n-1)}$$

This upper-tail criterion is suitable only when positive spatial autocorrelation—which prevails in an overwhelming number of geo-referenced data cases—is of concern.

Although contemplating a linear geographic surface partitioning is illuminating for pedagogic reasons, it is not very practical. Another type of geographic landscape that exhibits some of the same properties but is far more practical is a regular lattice, the square tessellation surface partitioning associated with remotely sensed data pixels. Consider the following simple 2-by-2 square lattice:

The binary geographic connectivity matrix associated with this surface partitioning is $\begin{bmatrix} 0 & 1 & 1 & 0 \\ 1 & 0 & 0 & 1 \\ 1 & 0 & 0 & 1 \\ 0 & 1 & 1 & 0 \end{bmatrix}$, using a "rook's move in chess" definition of geographic adjacency. Its affiliated eigenvalue problem is

$$DET\left(\begin{bmatrix} 0 & 1 & 1 & 0 \\ 1 & 0 & 0 & 1 \\ 1 & 0 & 0 & 1 \\ 0 & 1 & 1 & 0 \end{bmatrix} - \begin{bmatrix} \lambda & 0 & 0 & 0 \\ 0 & \lambda & 0 & 0 \\ 0 & 0 & \lambda & 0 \\ 0 & 0 & 0 & \lambda \end{bmatrix}\right) = DET\begin{bmatrix} -\lambda & 1 & 1 & 0 \\ 1 & -\lambda & 0 & 1 \\ 1 & 0 & -\lambda & 0 \\ 0 & 1 & 1 & -\lambda \end{bmatrix} = 0 \quad \text{, or}$$

$$(-\lambda)^4 - 4(-\lambda)^2 = 0 \ ,$$

which yields, for this n_1-by-n_2 = 2-by-2 rectangular-shaped region, the following roots: $\lambda_1 = 2 = 2[COS(\frac{\pi}{n_1+1}) + COS(\frac{\pi}{n_2+1})$, $\lambda_2 = 0 = 2[COS(\frac{\pi}{n_1+1}) + COS(\frac{2\pi}{n_2+1})$, $\lambda_3 = 0 = 2[COS(\frac{2\pi}{n_1+1}) + COS(\frac{\pi}{n_2+1})$, and $\lambda_4 = -2 = 2[COS(\frac{2\pi}{n_1+1}) + COS(\frac{2\pi}{n_2+1})]$. Analytically, these results are useful for both pedagogical and research purposes; the analytical eigenvalues of the binary connectivity matrix **C** for any n_1-by-n_2 rectangular geographic region partitioned by a regular square grid are given by

$$\lambda_{n_2(k-1)+j} = 2\left\{COS\left[\frac{j\pi}{(n_1+1)}\right] + COS\left[\frac{k\pi}{n_2+1}\right]\right\} ,$$

These eigenvalues are easily recognizable in spectral density functions connected with those spatial autoregressive models presented in Chapter 12. This analytical solution circumvents the need to solve the eigenvalue problem for this type of geographic surface partitioning. Because all of the row sums equal 2 for this connectivity matrix example, by property P12 we know theoretically that λ_1

= 2, without needing to construct or solve the characteristic equation. The principal eigenvector for this example may be calculated from

$$\begin{bmatrix} -2 & 1 & 1 & 0 \\ 1 & -2 & 0 & 1 \\ 1 & 0 & -2 & 1 \\ 0 & 1 & 1 & -2 \end{bmatrix} \begin{bmatrix} e_{11} \\ e_{21} \\ e_{31} \\ e_{41} \end{bmatrix} = \begin{bmatrix} 0 \\ 0 \\ 0 \\ 0 \end{bmatrix}$$

$$e_{11}^2 + e_{21}^2 + e_{31}^2 + e_{41}^2 = 1 ,$$

which yields $\begin{bmatrix} \frac{1}{2} \\ \frac{1}{2} \\ \frac{1}{2} \\ \frac{1}{2} \end{bmatrix}$. This result reveals that each of the areal units is equally centrally located in the surface topology, which becomes apparent by visually inspecting the map itself.

Extending this example to the 9-by-9 regular square tessellation surface partitioning, for both benchmarking purposes and pedagogic reasons, discloses that two possible geographic patterns render the maximum spatial autocorrelation possible, where MC = 0.99004; one is a northeast-southwest gradient ($\mathbf{E}_1$ of matrix **M**) and the other is a northwest-southeast gradient ($\mathbf{E}_2$ of matrix **M**). These are the types of map patterns uncovered with a canonical correlation analysis in §9.2.1. The fourth eigenvector of matrix **M**, for which MC = 0.89970, generates a negative exponential declining surface from the physical center of the map. And the eigenvector coupled with the smallest eigenvalue, for which MC = -1.06994, exhibits a checkerboard-type geographic map pattern (Map 3.1).

The binary geographic connectivity matrix associated with this n_1-by-n_2 = 2-by-2 rectangular-shaped region surface partitioning is $\begin{bmatrix} 0 & 1 & 1 & 1 \\ 1 & 0 & 1 & 1 \\ 1 & 1 & 0 & 1 \\ 1 & 1 & 1 & 0 \end{bmatrix}$, using a "queen's move in chess" definition of geographic adjacency. Its affiliated eigenvalue problem is

$$DET\left(\begin{bmatrix} 0 & 1 & 1 & 1 \\ 1 & 0 & 1 & 1 \\ 1 & 1 & 0 & 1 \\ 1 & 1 & 1 & 0 \end{bmatrix} - \begin{bmatrix} \lambda & 0 & 0 & 0 \\ 0 & \lambda & 0 & 0 \\ 0 & 0 & \lambda & 0 \\ 0 & 0 & 0 & \lambda \end{bmatrix}\right) = DET\begin{bmatrix} -\lambda & 1 & 1 & 1 \\ 1 & -\lambda & 1 & 1 \\ 1 & 1 & -\lambda & 1 \\ 1 & 1 & 1 & -\lambda \end{bmatrix} = 0 \quad , \text{ or}$$

$$(-\lambda)^4 - 6(-\lambda)^2 - 8\lambda - 3 = 0 ,$$

which from P12 has as part of its solution $\lambda_1 = 3$. Analytical eigenvalue solutions are available for this adjacency definition, too, and in general are given by

$$\lambda_{n_2(k-1)+j} = 2\left\{COS\left[\frac{j\pi}{(n_1+1)}\right] + COS\left[\frac{k\pi}{n_2+1}\right] + 2COS\left[\frac{j\pi}{(n_1+1)}\right] \times COS\left[\frac{k\pi}{n_2+1}\right]\right\},$$

Accordingly the remaining three eigenvalues for this 2-by-2 rectangular-shaped region surface partitioning example are $\lambda_2 = \lambda_3 = \lambda_4 = -1$.

Finally, in terms of the Puerto Rico case, for which maps of eigenvectors $\mathbf{E}_1$, $\mathbf{E}_4$, $\mathbf{E}_6$, and $\mathbf{E}_{12}$ extracted from matrix **M** appear in Map 2.1, maps portraying the geographic distributions of eigenvectors $\mathbf{E}_1$, $\mathbf{E}_2$, $\mathbf{E}_3$, and $\mathbf{E}_{73}$ appear in Map 3.2. The map pattern displayed by eigenvector $\mathbf{E}_1$ is roughly a linear gradient declining from the west to the east across the island. The map pattern displayed by eigenvector $\mathbf{E}_2$ is roughly a mountain in the center of the island, whose slope is steeper in the east than in the west. The map pattern displayed by eigenvector $\mathbf{E}_3$ is roughly a basin centered on the San Juan area. And, the map pattern displayed by eigenvector $\mathbf{E}_{73}$ is a highly fragmented surface, with considerably more fluctuation in the eastern than in the western region of the island.

Finally, as the approximate standard error of MC indicates when coupled with property P15, variability displayed by the sampling distribution of MC under a null hypothesis of zero spatial autocorrelation is contributed to by this entire range of possible spatial autocorrelation map patterns. The extent of this contribution is quantified by the absolute magnitude of an eigenvector's corresponding eigenvalue. Accordingly, very little variation arises from those distinct map patterns that embrace essentially zero spatial autocorrelation, because their respective eigenvalues are nearly zero. The S-W statistic for the set of 73 Puerto Rican eigenvalues is 0.90492, which is markedly significant. This situation can be remedied by applying a power transformation to these eigenvalues; $(\lambda + 2.768)^{0.44095}$ improves the bell-shape appearance of the eigenvalue frequency distribution and increases S-W to 0.97521 ($p = 0.44$).

3.4. References

Boots, B. 1982. "Comments on the Use of Eigenfunctions to Measure Structural Properties of Geographic Networks." *Environment and Planning A* 14: 1063–72.

—. 1984. "Evaluating Principal Eigenvalues as Measures of Network Structure." *Geographical Analysis* 16: 270–75.

—. 1985. "Size Effects in the Spatial Patterning of Nonprincipal Eigenvectors of Planar Networks." *Geographical Analysis* 17: 74–81.

Boots, B. and Royle. 1991. "A Conjecture on the Maximum Value of the Principal Eigenvalue of a Planar Graph." *Geographical Analysis* 23: 276–82.

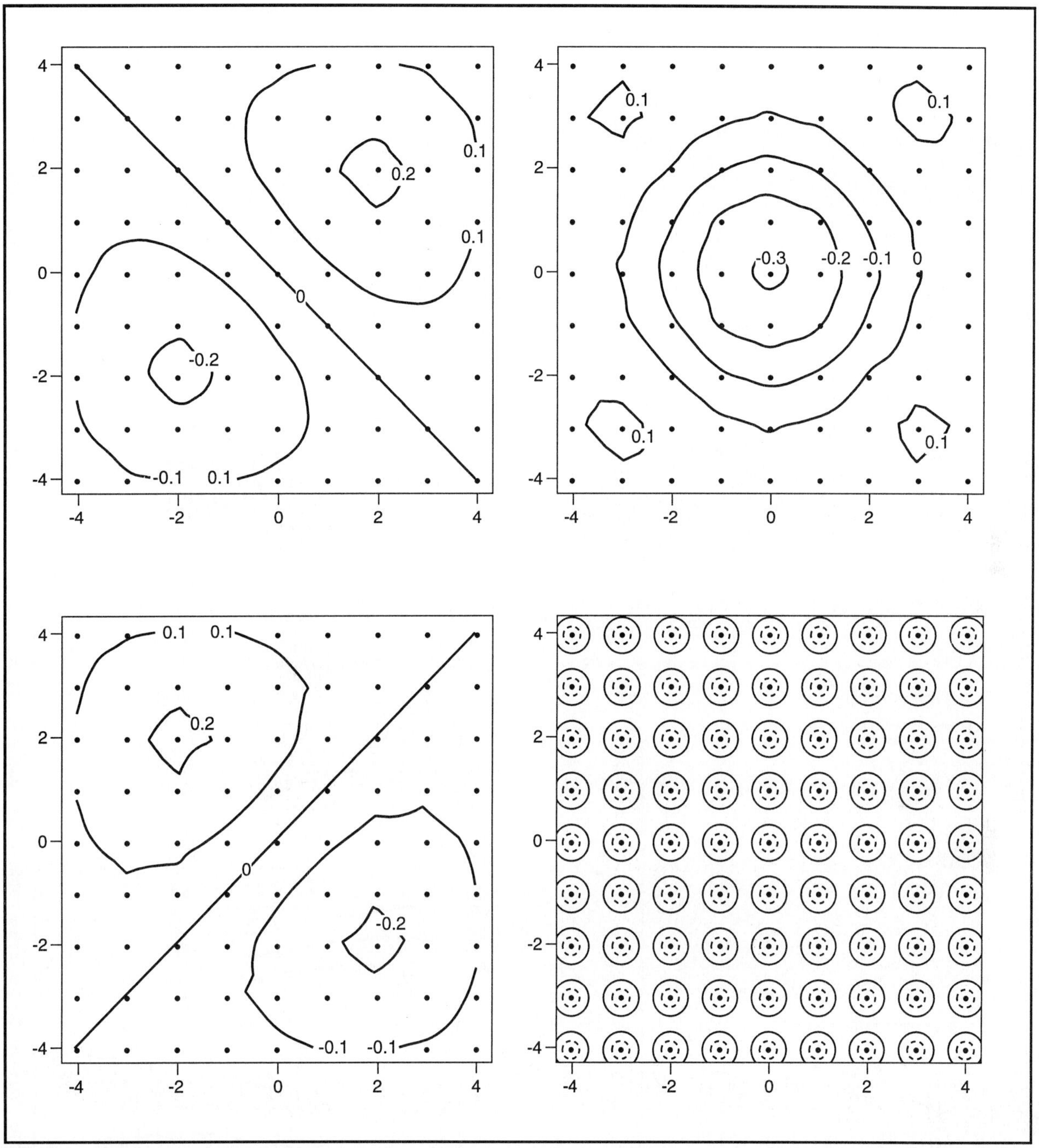

Map 3.1. The geographic distribution of selected eigenvector values: $\mathbf{E}_1$(upper left); $\mathbf{E}_2$ (lower left); $\mathbf{E}_4$ (upper right); $\mathbf{E}_{81}$ (lower right).

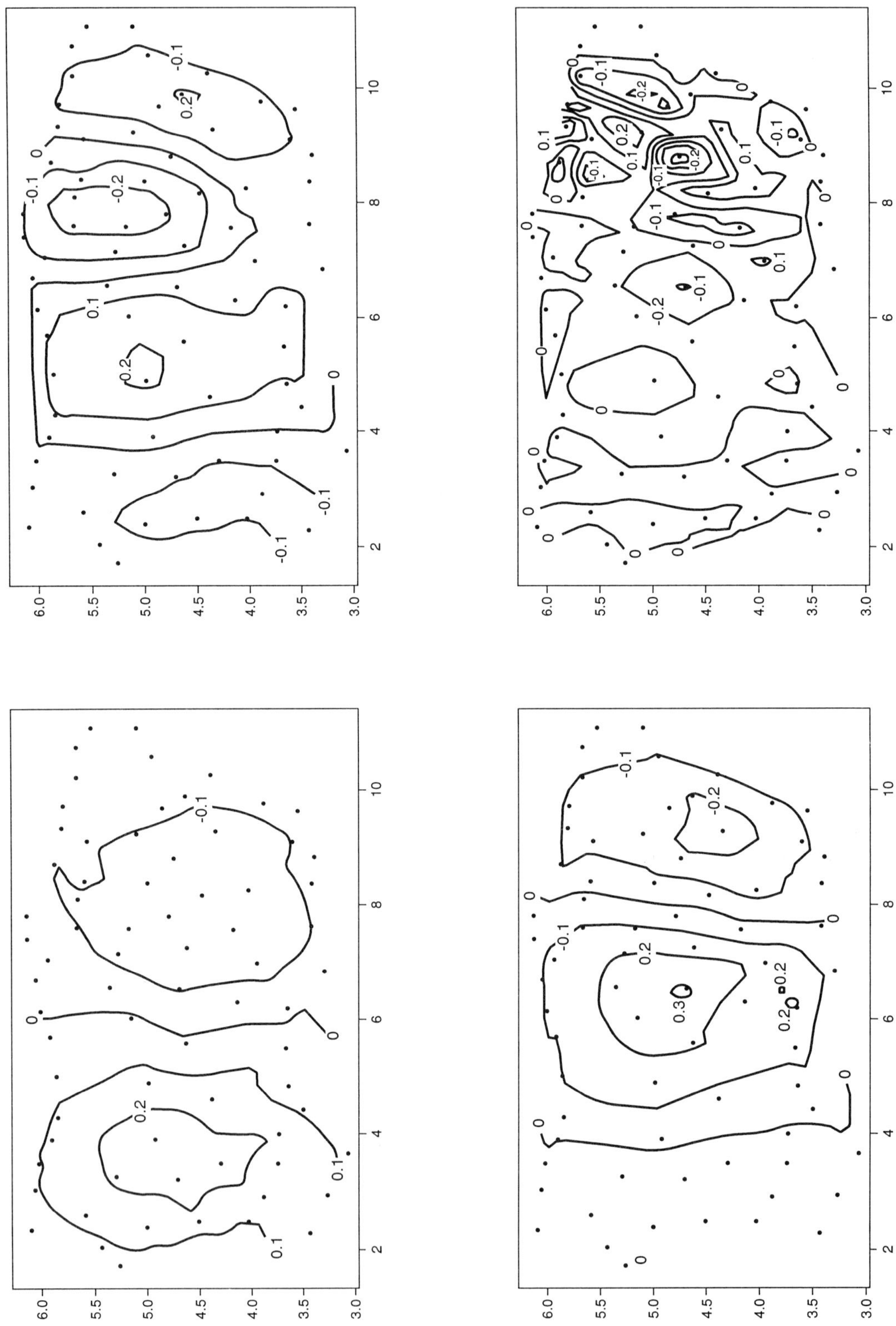

Map 3.2. The geographic distribution of selected eigenvector values: $\mathbf{E}_1$(upper left); $\mathbf{E}_2$(lower left); $\mathbf{E}_3$(upper right); $\mathbf{E}_{73}$(lower right).

Boots, B., and K. Tinkler. 1983. "The Interpretation of Non-principal Eigenfunctions of Urban Structure." In *Proceedings of the 14th Annual Pittsburgh Modelling and Simulation Conference*, edited by W. Vogt and M. Mickle , 715–26. Research Triangle Park, N.C.: Instrument Society of America.

Borowski, E. and J. Borwein, eds. 1991. *Dictionary of Mathematics*. New York: HarperCollins.

Courant, R. 1924. *Methoden der Mathematischen Physiks*. Berlin: Springer.

de Jong, P., C. Sprenger and F. van Veen. 1984. "On Extreme Values of Moran's I and Geary's c." *Geographical Analysis* 16: 17–24.

Gould, P. 1967. "On the Geographical Interpretation of Eigenvalues." *Transactions*, Institute of British Geographers 42: 53-92.

Griffith, D. 1984. "Measuring the Arrangement Property of a System of Areal Units Generated by Partitioning a Planar Surface," in G. Bahrenberg, M. Fischer and P. Nijkamp (eds.), *Recent Developments in Spatial Analysis: Methodology, Measurement, Models*. Aldershot, U.K.: Gower, pp. 191–200.

—. 1988. *Advanced Spatial Statistics*. Dordrecht: Kluwer.

—. 1989. "Spatial Regression Analysis on the PC: Spatial Statistics Using MINITAB." *Discussion Paper* #1. Ann Arbor, Mich: Institute of Mathematical Geography.

—. 1996. "Spatial Autocorrelation and Eigenfunctions of the Geographic Weights Matrix Accompanying Geo-referenced Data." *The Canadian Geographer* 40: 351-67.

Griffith, D. and C. Amrhein. 1991. *Statistical Analysis for Geographers*. Englewood Cliffs, N.J.: Prentice Hall.

Noble, B. and J. W. Daniel. 1977. *Applied Linear Algebra*. 2nd ed. Englewood Cliffs, N.J.: Prentice Hall.

Strang, G. 1980. *Linear Algebra and Its Applications*. Orlando: Academic Press.

Tatsuoka, M. 1971. *Multivariate Analysis: Techniques for Educational and Psychological Research*. New York: Wiley.

Tiefelsdorf, M. and B. Boots. 1995. "The Exact Distribution of Moran's I." *Environment and Planning A* 27: 985–99; "Letters to the Editor" addendum, 28: 1900.

Tinkler, K. n.d. *Introductory Matrix Algebra*. Norwich, Eng: GEO Books, CATMOG #48.

—. 1972. "The Physical Interpretation of Eigenfunctions of Dichotomous Matrices." *Transactions*, Institute of British Geographers 55: 17–46.

—. 1974. "On Summing Power Series Expansions of Accessibility Matrices by the Inverse Method." *Geographical Analysis* 6: 175-8.

—. 1976. "On Functional Regions and Indirect Flows." *Geographical Analysis* 8: 205-13.

CHAPTER 4

MULTIPLE LINEAR REGRESSION AND CORRELATION ANALYSIS ($P \geq 3$)

Chapter Preview. There is little question that "[r]egression analysis is the most widely used and versatile dependence technique..." (Hair et al., 1995, p.79), and as such is the "... workhorse of much non-spatial data analysis" (Bailey and Gatrell, 1995, p. 274). In fact, most every technique examined in this book can be restated as a variation of the regression model. It is not surprising, then, that for the majority of statistical analyses, regression is the first method beyond simple descriptive statistics to which analysts turn.

In the multivariate world, regression analysis is representative of a set of techniques. Our goal in this chapter is twofold. First, we seek to emphasize explicit linkages between linear regression and other elementary concepts, such as correlation analysis; related approaches, such as analysis of variance and covariance; and, the realm of spatial analysis, through trend surface modeling. Second, as the virtual default technique, regression is prone to misapplication, and hence we seek to carefully describe situations and diagnostics that permit the proper application of regression models.

As the first methodological chapter in this text, we take care to define terms and develop notation more fully here than in later chapters. In addition, throughout the text we take time to reference each linear technique back to what might be viewed as the "regression roots" that are established in this chapter.

4.0. Introduction

One problem discussed in introductory statistics is the issue of predicting one set of unknown values from a second set of known values, or one geographic distribution from a second geographic distribution. Prediction as an objective is one of the hallmarks of ***linear regression analysis***, a technique used to specify a linear equation prediction of some variable Y.

REG Question #1: What is the rule of correspondence between variable Y and a battery of X variables that enables Y to be predicted from these Xs?

The principle utilized to solve this problem is called ***ordinary least squares*** (commonly referred to as OLS) and involves drawing a sample of size n, followed by minimizing the sum of the squared differences between the observed values, y_i, and their predicted value counterparts, $\hat{y}_i$, for measures made on variable Y:

$$\text{MIN}: Z = \sum_{i=1}^{n} (y_i - \hat{y}_i)^2 . \tag{4.1}$$

If $\hat{y}_i$ provides a perfect prediction of y_i, then expression (4.1) equals 0. In the worst case, $\hat{y}_i = \bar{y}$, resulting in expression (4.1) equaling $(n-1) \times s^2$. Let

$$Y = \hat{Y} + \epsilon , \tag{4.2}$$

where ϵ is the error (calculation, measurement, specification, sampling, and/or stochastic) affiliated with the problem. Substituting equation (4.2) into expression (4.1) reduces this minimization to

$$\text{MIN}: Z = \sum_{i=1}^{n} [(\hat{y}_i + \epsilon_i) - \hat{y}_i)]^2 = \sum_{i=1}^{n} \epsilon_i^2 . \tag{4.3}$$

Expression (4.3) indicates that the regression problem actually is one of error minimization.

The minimization process employed is a standard calculus procedure. First the partial derivative of Z (i.e., ∂Z) with respect to each parameter used to specify $\hat{Y}$ is derived and set equal to zero. Second, this set of partial differential equations, called the ***normal equations***, is solved in terms of the parameters, yielding parameter estimates. When the deviations $(y_i - \hat{y}_i)$ conform to a normal frequency distribution in the population, this solution to the normal equations also is the one with the greatest probability, or the ***maximum likelihood estimate***. This characteristic links the regression technique to model-based inferential statistics.

4.1. Background

If the second geographic distribution (the distribution of known values) is a vector of ones (i.e., **1**), then the regression problem reduces to that of computing a sample mean. Using matrix notation,

$$\hat{Y} = \mu\mathbf{1} , \text{ or } \mathbf{Y} = \mu\mathbf{1} + \epsilon,$$

and hence

$$\begin{aligned} \text{MIN}: Z &= \sum_{i=1}^{n} (y_i - \hat{y}_i)^2 \\ &= \sum_{i=1}^{n} [(\mu \times 1 + \epsilon_i) - \mu \times 1]^2 \\ &= \sum_{i=1}^{n} \epsilon_i^2 . \end{aligned}$$

From the corresponding calculus problem, the solution sought is for the differential equation $\frac{\partial Z}{\partial \mu} = 0$. This differential equation leads to the normal equation

$$\sum_{i=1}^{n} 2 \times (y_i - \mu \times 1) \times (-1) = 0 \ ,$$

yielding the estimate $\hat{\mu} = \sum_{i=1}^{n} \frac{y_i}{n} = \bar{y}$. Using matrix notation, the normal equation is written as $(-2)\mathbf{1}^T(\mathbf{Y} - \mu\mathbf{1}) = 0$, yielding $\hat{\mu} = (\mathbf{1}^T\mathbf{1})^{-1}\mathbf{1}^T\mathbf{Y} = \bar{y}$. Two points worth noting here are (1) writing a linear regression problem using matrix notation will involve a vector of ones, and (2) because standard commercial statistical software packages include vector **1** by default, explicitly including it in a specification requires the selection of a "no-intercept/no-constant" option.

EXAMPLE PR-4.1. Calculate the mean of the density of farms (DFRM) in the Arecibo agricultural administrative region (Example PR-4.1 of Griffith and Amrhein, 1991, p. 77).

From the previous book, $\bar{x}_{DFRM} = 9.75$. Now, for convenience, we denote this value as $\bar{y}_{DFRM}$, since labeling this variable X or Y often is arbitrary.

Using the SAS procedure

```
PROC REG;
  MODEL DFRM = X0/NOINT;
RUN;
```

produced the SAS output

```
Dependent Variable: DFRM

                         Parameter Estimates

                   Parameter   Standard     T for H0:
Variable    DF   Estimate       Error      Parameter=0   Prob > |T|
X0           1   9.745455   1.57539973        6.186        0.0001
```

The SAS output reported in Example PR-4.1 indicates that computing a sample mean as a regression problem also automatically yields the standard error of $\frac{s}{\sqrt{n}} = \frac{5.23}{\sqrt{11}} = 1.5754$ (using s reported in Example PR-4.6, Griffith & Amrhein, 1991, p. 88). Commercial software packages also offer the advantage of computer calculation of t-statistics and their associated probabilities.

This simple application of the regression technique permits a heuristic examination of the OLS principle. Consider equation (4.1), which can be rewritten without the summation notation, and assuming a constant $\hat{y}_i$, (e.g., $\hat{y}_i = \bar{y}$) as

$$(y_1 - \hat{y})^2 + (y_2 - \hat{y})^2 + ... + (y_n - \hat{y})^2 \ .$$

The aim of the minimization problem is to get this expression as close as possible to zero. Any one of these n terms can be equated to zero by setting $\hat{y} = y_i$. Thus, $y_{min} \leq \hat{y} \leq y_{max}$, indicating that the predicted value in this case must be contained within the range of the sampled values of Y; otherwise, none of the n terms would be zero. Making such substitutions for Example PR-4.1 produces the graph appearing in Figure 4.1.

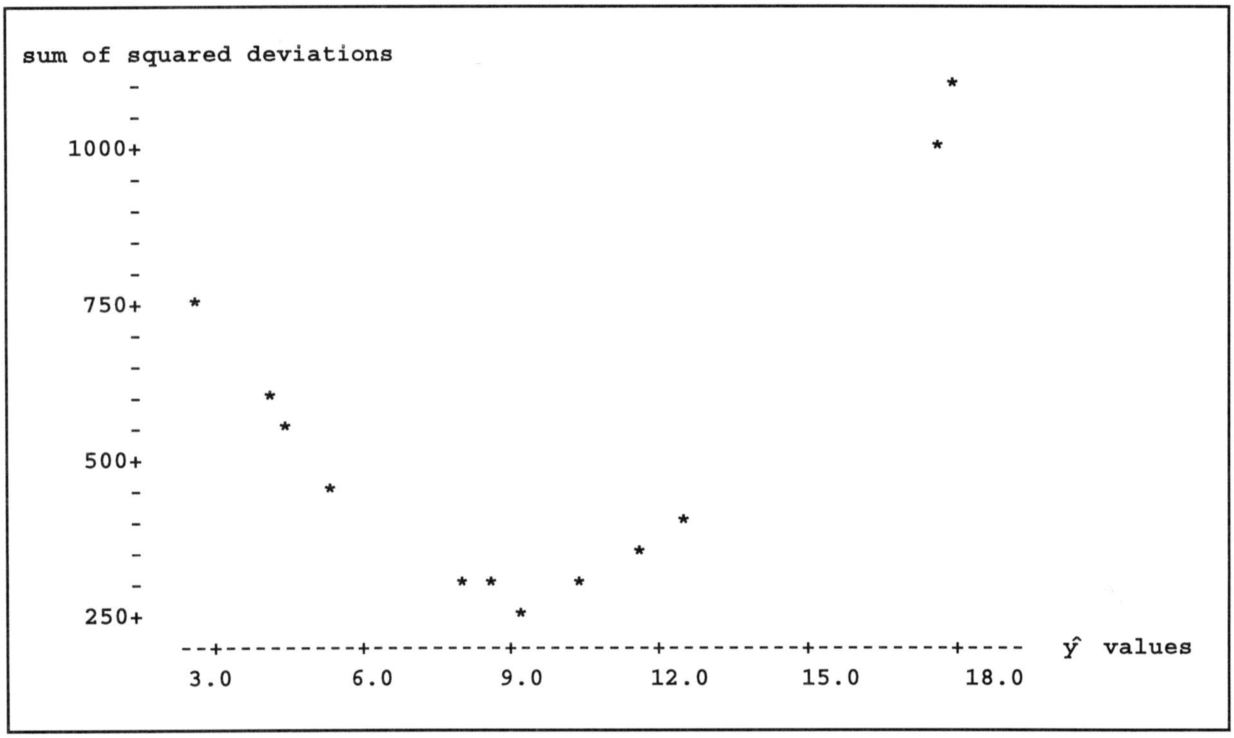

Figure 4.1. Behavior of the sum-of-squared deviations as $\hat{y}$ approaches $\bar{y}$.

The smallest value of DFRM is 3.0, which when substituted for $\hat{y}$ produces a sum-of-squares value of 773.52. As the value substituted for $\hat{y}$ moves toward $\bar{y}_{DFRM}$, the sum-of-squares value decreases, reaching a minimum of 273.01 when $\hat{y} = \bar{y}_{DFRM}$. Beyond $\bar{y}_{DFRM}$ the sum-of-squares value begins to increase again, until for the largest value of DFRM (18.3) it equals 1077.99. The OLS principle traces this curve and identifies the minimum point.

The bivariate linear regression problem augments this constant geographic distribution of ones with some variable geographic distribution. Following convention, this second distribution is denoted by X_1. Often the regression equation representation of this problem is specified as

$$\hat{Y} = \alpha \mathbf{1} + \beta \mathbf{X}_1 \text{, or } \mathbf{Y} = \alpha \mathbf{1} + \beta \mathbf{X}_1 + \epsilon \ ,$$

with the subscript 1 attached to variable X frequently being omitted, but understood. Hence,

$$\begin{aligned} \text{MIN}: Z &= \sum_{i=1}^{n} (y_i - \hat{y}_i)^2 \\ &= \sum_{i=1}^{n} [(\alpha \times 1 + \beta x_i + \epsilon_i) - (\alpha \times 1 + \beta x_i)]^2 \\ &= \sum_{i=1}^{n} \epsilon_i^2 \ . \end{aligned}$$

The minimization process involves two differential equations, because there are two parameters (i.e., α and β). This pair of differential equations ($\frac{\partial Z}{\partial \alpha} = 0$ and $\frac{\partial Z}{\partial \beta} = 0$) leads to the set of normal equations

$$\sum_{i=1}^{n} 2 \times [y_i - (\alpha \times 1 + \beta x_i)] \times (-1) = 0 \ , \text{ and}$$

$$\sum_{i=1}^{n} 2 \times [y_i - (\alpha \times 1 + \beta x_i)] \times (-x_i) = 0 \ ,$$

which render the estimates

$$\hat{\alpha} = \frac{[(\sum_{i=1}^{n} y_i)(\sum_{i=1}^{n} x_i^2) - (\sum_{i=1}^{n} x_i)(\sum_{i=1}^{n} x_i y_i)]}{[n(\sum_{i=1}^{n} x_i^2] - (\sum_{i=1}^{n} x_i)^2]} = a = \bar{y} - b\bar{x} \ , \text{ and}$$

$$\hat{\beta} = \frac{[n(\sum_{i=1}^{n} y_i x_i) - (\sum_{i=1}^{n} y_i)(\sum_{i=1}^{n} x_i)]}{[n\sum_{i=1}^{n} x_i^2] - (\sum_{i=1}^{n} x_i)^2]} = b$$

(see Griffith and Amrhein, 1991, p. 325). This first statistic relates the linear equation intercept to means $\bar{y}$ and $\bar{x}$, just as in the preceding regression analysis. Moreover, it is the mean of Y when matrix **X** contains only a vector of ones, becoming a conditional mean (Y|**X**) as this vector of ones is augmented.

To use matrix notation, the normal equations require a vector representation of the parameters. The standard practice is to use **ß**, which is why the intercept frequently is denoted with β_0 (with the affiliated statistic denoted by b_0, and the vector **1** denoted by $\mathbf{X}_0$) rather than α. And, because variable X has an implicit subscript of 1, its accompanying parameter is denoted by β_1 (with the affiliated statistic denoted by b_1), the second entry in vector **ß**. Following these conventions, the normal equations (obtained from $\frac{\partial Z}{\partial \beta_j} = 0,\ j = 1,...,P$) may be written simultaneously as $(-2)\mathbf{X}^T(\mathbf{Y} - \mathbf{X}\boldsymbol{\beta}) = 0$, yielding

$$\mathbf{b} = (\mathbf{X}^T\mathbf{X})^{-1}\mathbf{X}^T\mathbf{Y}\ , \tag{4.4}$$

where vector **b** commonly is used to denote the estimate of the parameter vector $\boldsymbol{\beta}$. This is one of the most famous results of regression and should be committed to memory! It alludes to a second salient question affiliated with regression analysis:

> *REG Question #2*: What are the coefficients of the linear combination of Xs that constitute a rule of correspondence between Y and the battery of X variables and that yield a linear combination of the Xs for predicting Y?

It also identifies a gateway for computational error to enter regression analysis; as $DET(\mathbf{X}^T\mathbf{X})$ approaches zero, which occurs when Xs become highly correlated, or ***multicollinear***, the inverse $(\mathbf{X}^T\mathbf{X})^{-1}$ becomes increasingly more difficult, numerically, to compute.

EXAMPLE PR-4.2. Estimate the regression equation in which Y = density of banana/plantain production (DBN) and X = density of farms (DFRM) for the Arecibo region of Puerto Rico (Example PR-12.1 of Griffith and Amrhein, 1991, pp. 327–28).

From the previous book, a = b_0 = -5.3815, and b = b_1 = 1.0223. Using the SAS procedure

```
PROC REG;
   MODEL DBN = DFRM;
RUN;
```

produced the SAS output

```
Dependent Variable: DBN

                        Parameter Estimates

                 Parameter     Standard     T for H0:
Variable    DF   Estimate        Error     Parameter=0   Prob > |T|
INTERCEP     1   -5.381451   3.96027205      -1.359        0.2073
DFRM         1    1.022350   0.36183435       2.825        0.0199
```

The SAS output reported in Example PR-4.2 indicates that computing the regression coefficients with SAS automatically yields the pair of standard errors (given rounding error) of

$$\left\{0.3621^2\,[(\frac{10}{11}) \times 5.2250^2 + 9.7455^2]\right\}^{\frac{1}{2}} = 3.95$$

(Note that the difference between 3.95 and 3.9602 reported in the results is rounding error) and

$$\left[\frac{7.7913^2 - 1.0224^2 \times 5.2250^2}{5.2250^2 \times (11-2)}\right]^{\frac{1}{2}} = 0.362$$

(see Example PR-12.3 in Griffith and Amrhein, 1991, p. 334).

4.2. Variable Selection and Conventional Diagnostics

Rarely will the MIN expression in (4.1) be close or equal to zero when a single geographic distribution is used to predict variable Y. The conceptual problem becomes one of devising the most meaningful and parsimonious $\hat{Y}$ by judiciously selecting a set of Xs to enter into the linear prediction equation that results in (4.1) being as close as possible to zero. Nevertheless, the error term, ϵ, remains the critical variable, indicating how $\hat{Y}$ deviates from Y and determining the foundation for statistical inference.[1] In classical statistics this term is to be constituted of only sampling error. The only class of serious calculation error that might enter regression analysis arises from rounding. Measurement error is assumed to be negligible, with its detailed treatment being beyond the scope of this book. Stochastic error refers to assumptions about the statistical behavior of the X variables that compose $\hat{Y}$. And, specification error can be of two types, namely, an incomplete set of Xs and nonadditive mathematical relationships between the Xs and Y (nonlinearity).

Selection of the set of Xs that specifies $\hat{Y}$ requires substantive and theoretical input. The linear model specification of classical statistics posits the additive mathematical structure

$$\hat{\mathbf{Y}} = \sum_{j=0}^{P} \mathbf{X}_j \beta_j + \epsilon \ , \qquad (4.5)$$

wwhere P denotes the number of geographic distributions (or variables) needed to predict Y. Recall that $X_0 = 1$. In addition, equation (4.4) is sufficiently general that it provides the solution for this general specification. Unfortunately, there are few problems in which the entire set of Xs is known.

EXAMPLE PR-4.3. Predict density of sugar cane production (DSGR) from other variables.

First the square root of DSGR is calculated (SQDSGR) to improve the symmetry of the frequency distribution for this variable [Shapiro-Wilk increases from 0.8147 to

[1]The Gauss-Markov Theorem states that given a linear equation of the form $\mathbf{Y} = \mathbf{X}\boldsymbol{\beta} + \epsilon$, where ϵ is statistically distributed with a mean of $\mathbf{0}$, a variance for each ϵ_i of σ^2, with the ϵ_is being pairwise uncorrelated, and matrix $\mathbf{X}$ consisting of mutually independent variables, then substitution of the least squares estimates for $\boldsymbol{\beta}$ in this linear equation leads to its minimum variance linear unbiased estimator.

0.8960 (p remains at 0.0001); Kolmogorov-Smirnov decreases from 0.1649 to 0.1362 (p increases from < 0.05 to ≈ 0.15)[2]] and to help stabilize its sample variance [Bartlett's test decreases from 17.77 (p = 0.0014) to 2.72 (p = 0.6858)]. Next, since sugar cane is consumptive of land, DSGR is expected to be inversely related to density of farms (DFRM) and directly related to the density of cultivated land (DCD). In the 1960s and 1970s, urbanization diffusing outward from the San Juan metropolitan area was displacing the use of land for sugar cane production, implying that distance from San Juan (DISTSJ) should be positively related to DSGR. Sizable production of rum, at that time, in Arecibo implies that distance from this urban area (DISTA) should be inversely related to DSGR. And, because some municipios have no sugar cane production, an indicator variable (I0) is introduced to distinguish between these two categories of municipio. Using the SAS procedure

```
PROC REG;
   MODEL SQDSGR = DFRM DCD DISTSJ DISTA I0;
RUN;
```

produced the SAS output

Dependent Variable: SQDSGR

Parameter Estimates

Variable	DF	Parameter Estimate	Standard Error	T for H0: Parameter=0	Prob > |T|
INTERCEP	1	-0.801023	0.60410655	-1.326	0.1894
DFRM	1	-0.089139	0.02661283	-3.349	0.0013
DCD	1	0.019063	0.01149570	1.658	0.1019
I0	1	1.496081	0.32684147	4.577	0.0001
DISTA	1	0.195598	0.09627586	2.032	0.0462
DISTSJ	1	0.335452	0.07491762	4.478	0.0001

These regression results confirm most of the conceptual expectations. Only the sign of DISTA is opposite of what was anticipated. The intercept term is not significant, as should be the case because when I0 is zero there is no sugar cane production. Finally, the anticipated relationship between density of cultivated land (DCD) and DSGR apparently fails to hold in the population.

One question that remains about these results, however, is whether they may be attributed to sampling error or one of the other sources of error. Of particular concern are specification and calculation error.

[2]Dutilleul and Legendre (1992, pp. 89–90) found "... that a high positive or negative [auto]correlation among the observations cannot be told apart from lack of normality. The numerical results ... for small and moderate sample sizes push the balance in favour of [the Kolmogorov-Smirnov statistic], whose properties are as good as or better than those of [the Shapiro-Wilk statistic], and counterbalance the slightly poorer power reported for [the Kolmogorov-Smirnov statistic] in the random sample case."

4.2.1. Statistical Distribution Properties of b

To draw inferences from **b**, about **ß**, properties of this statistics vector need to be established, leading to a third prominent question associated with regression analysis:

> *REG Question #3*: What are the statistical properties of linear model parameter estimators?

These properties also provide insights into assessment of the importance of variables in a regression equation, the next topic to be treated in this chapter.

One assumption of regression is that $E(\epsilon) = 0$. Because an intercept term, namely β_0, is

$$\begin{aligned} E(\mathbf{b}) &= E[(\mathbf{X}^T\mathbf{X})^{-1}\mathbf{X}^T\mathbf{Y}] \\ &= (\mathbf{X}^T\mathbf{X})^{-1}\mathbf{X}^T E(\mathbf{Y}) \\ &= (\mathbf{X}^T\mathbf{X})^{-1}\mathbf{X}^T[\mathbf{X}\beta + E(\epsilon)] = \beta + 0 , \end{aligned}$$

present in the equation, this assumption holds in the population because the intercept assumes the value of a conditional mean. This assumption holds in the sample when OLS is used with a linear model. Similarly, since its estimated conditional mean counterpart, b_0, appears in the estimated regression equation, this equality is forced to hold for the estimate of ϵ, usually denoted by **e**; $E(\mathbf{e}) = \bar{e}\mathbf{1} = \mathbf{0}$. One of the standard statistical properties emphasized in Griffith and Amrhein (1991, p. 11) is ***unbiasedness*** (the mean of the sampling distribution equals the corresponding population parameter). Here, using the conventional notation of E(☼) to denote an expected value, demonstrating that **b** is an unbiased estimator of **ß**. But this result depends upon matrix **X** being complete (i.e., there is no missing variables specification error present). Suppose that only a subset of the variables in matrix **X** is known, say $\mathbf{X}_I$, while the remainder of the variables, say $\mathbf{X}_{II}$, are unknown and therefore not included in the regression equation. Then

$$\begin{aligned} E(\mathbf{b}_I) &= E[(\mathbf{X}_I^T\mathbf{X}_I)^{-1}\mathbf{X}_I^T\mathbf{Y})] \\ &= (\mathbf{X}_I^T\mathbf{X}_I)^{-1}\mathbf{X}^T[X_I\beta_I + \mathbf{X}_{II}\beta_{II} + E(\epsilon)] \; . \\ &= \beta_I + (\mathbf{X}_I^T\mathbf{X}_I)^{-1}\mathbf{X}_I^T\mathbf{X}_{II}\beta_{II} \end{aligned}$$

Consequently, $\mathbf{b}_I$ is unbiased only in the unlikely event that $\mathbf{X}_I^T\mathbf{X}_{II} = \mathbf{0}$, which would signify that the two sets of variables are independent (or orthogonal). This result not only emphasizes how specification error can enter into a regression analysis but also highlights another gateway for computational error to enter into a regression analysis. Correlations between variables included in a model and those missing from this model complicate the computational solution. Because the $\mathbf{X}_{II}$ variables are missing, the term $\mathbf{X}_{II}\mathbf{ß}_{II}$ incorrectly becomes part of the error term, **e**. This misspecification is why explorations of relations between $\mathbf{X}_I$ and **e** can be illuminating.

The opposite situation occurs when the set of Xs is complete but additional variables, say $\mathbf{X}_{III}$, are included in a regression analysis. Because this extra set of Xs truly is not related to Y, then $\mathbf{ß}_{III} = \mathbf{0}$. Now working through the necessary algebra yields $E(\mathbf{b}) = \mathbf{ß}$ and $E(\mathbf{b}_{III}) = \mathbf{0}$, demonstrating that the estimated regression coefficients remain unbiased.

A second standard statistical property emphasized in Griffith and Amrhein (1991, p. 49) is ***efficiency*** (the standard deviation of an estimator's sampling distribution should be smaller than that of any other unbiased estimator of the parameter in question). Here, after working through the appropriate algebraic manipulations,

$$\mathrm{COV}(\mathbf{b}) = \mathrm{E}\{[(\mathbf{X}^T\mathbf{X})^{-1}\mathbf{X}^T\mathbf{Y} - \boldsymbol{\beta}][(\mathbf{X}^T\mathbf{X})^{-1}\mathbf{X}^T\mathbf{Y} - \boldsymbol{\beta}]^T\}$$
$$= (\mathbf{X}^T\mathbf{X})^{-1}\mathbf{X}^T\mathrm{E}(\epsilon^T\epsilon)\mathbf{X}(\mathbf{X}^T\mathbf{X})^{-1} \ .$$

Assuming that the errors are pairwise independent and have a constant variance renders $(\epsilon^T\epsilon) = \sigma^2\mathbf{I}$, yielding

$$\mathrm{COV}(\mathbf{b}) = \sigma^2(\mathbf{X}^T\mathbf{X})^{-1} \ ; \quad (4.6)$$

this result reveals yet another way in which calculation error can enter into a regression analysis.

Suppose, once again, that only the subset $\mathbf{X}_I$ of the variables constituting matrix $\mathbf{X}$ are known. Accordingly,

$$\mathrm{COV}(\mathbf{b}_I) = (\mathbf{X}_I^T\mathbf{X}_I)^{-1}\mathbf{X}_I^T\mathbf{X}_{II}\boldsymbol{\beta}_{II}\boldsymbol{\beta}_{II}^T\mathbf{X}_{II}^T\mathbf{X}_I(\mathbf{X}_I^T\mathbf{X}_I)^{-1} + \sigma^2(\mathbf{X}_I^T\mathbf{X}_I)^{-1} \ ,$$

which will have a minimum variance estimate only if $\mathbf{X}_I^T\mathbf{X}_{II} = \mathbf{0}$ *and* σ^2 are known. Because

$$\hat{\sigma}^2 = \frac{(\mathbf{Y} - \mathbf{Xb})^T(\mathbf{Y} - \mathbf{Xb})}{(n - P - 1)} \ , \quad (4.7)$$

if matrix $\mathbf{X}$ is incomplete, then $\hat{\sigma}^2$ generally is inflated even if $\mathbf{X}_I^T\mathbf{X}_{II} = \mathbf{0}$.

For the opposite situation of including extraneous variables in a regression analysis,

$$\mathrm{COV}(\mathbf{b}) = \sigma^2[\mathbf{X}^T\mathbf{X} - \mathbf{X}^T\mathbf{X}_{III}(\mathbf{X}_{III}^T\mathbf{X}_{III})\mathbf{X}_{III}^T\mathbf{X}]^{-1} \ ,$$

which will be correct only when $\mathbf{X}^T\mathbf{X}_{III} = \mathbf{0}$.

In summary,

OPERATIONAL FINDING 4.1: When matrix $\mathbf{X}$ is incomplete, $\mathbf{b}$ tends to be biased and $\hat{\sigma}^2$ tends to be a poor estimate, whereas when matrix $\mathbf{X}$ contains superfluous variables, $\mathbf{b}$ is unbiased but its covariance is inflated and is not based upon $(\mathbf{X}^T\mathbf{X})^{-1}$.

This finding acknowledges the seriousness of traditional specification error, underscores the importance of getting matrix $\mathbf{X}$ correct, and explicitly discloses the importance of having mutually independent Xs, a stationary process operating over a geographic landscape, and zero spatial autocorrelation.

4.2.2. Selection Criteria for the Matrix of Predictor Variables **X**

In the preceding section, the importance of specifying matrix **X** correctly was outlined. But since the complete set of variables in matrix **X** seldom is known, variable inclusion and exclusion decisions need to be made. Various criteria can be employed to support this decision-making process. Once more, these criteria will be complicated by computational error introduced through the presence of multicollinearity in matrix **X**.

Resorting to an assumption of normality for vector ϵ, the OLS estimate **b** is distributed as $N(\beta, \sigma^2(\mathbf{X}^T\mathbf{X})^{-1})$, **b** is distributed independently of **e**, and the unbiased $\hat{\sigma}^2 = \frac{(\mathbf{Y} - \mathbf{Xb})^T(\mathbf{Y} - \mathbf{Xb})}{(n - P - 1)}$ [equation (4.7)]. This result leads to the use of individual t-statistics to test the separate null hypotheses of H_0: $\beta_j = 0, j = 0, 1, 2, ..., P$. The appropriate standard error, s_{b_j}, is given by the square root of the j-th diagonal entry of $\hat{\sigma}^2(\mathbf{X}^T\mathbf{X})^{-1}$. Two features of this criterion merit emphasis. First, in the presence of multicollinearity, variation latent in X_h influences the computed standard error of X_j. Accordingly, if any variables are missing from matrix **X** or when multicollinearity introduces severe computational error, then the resulting t-tests based upon individual b_js will differ from those obtained with regression analyses using the complete matrix **X**. This is not the case when the X variables are mutually orthogonal. Second, even if the Xs were uncorrelated, appropriate probability levels depend upon the number of variables in matrix **X**. Hypothesis testing must be done simultaneously; in other words, each appropriate t-statistic and its accompanying probability are obtained for b_j only when the P remaining X variables already have been included in an estimated regression equation. As in univariate analysis, a researcher's level of significance should be declared before the regression analysis is undertaken.

EXAMPLE PR-4.3 (continued). Apply the t-test criterion to the prediction of DSGR from DFRM, DCD, I0, DISTA, and DISTSJ.

SAS reports each t-statistic together with the probability of t values greater than that statistic (in absolute value) occurring strictly due to sampling error. In this instance, if a researcher were to employ the customarily most generous 10% (5% in each tail of a two-tailed test) level of significance, the individual conclusions would be to fail to reject that $\beta_0 = 0$ and $\beta_{DCD} = 0$. This implies that the equation should be specified as

$$\sqrt{D\hat{S}GR} = \beta_{DFRM}DFRM + \beta_{I0}I0 + \beta_{DISTA}DISTA + \beta_{DISTSJ}DISTSJ \; .$$

This equation does not change if the level of significance is reduced to 5%. If it is reduced further to 1%, then one also would fail to reject that $\beta_{DISTA} = 0$. The new equation implication would become

$$\sqrt{D\hat{S}GR} = \beta_{DFRM}DFRM + \beta_{I0}I0 + \beta_{DISTSJ}DISTSJ \; .$$

This variable selection criterion may be summarized as follows:

SELECTION CRITERION 1: Exclude any variable from matrix **X** if its marginal t-statistic is not significant at the $\frac{\alpha}{2}$ level.

One must keep in mind with this criterion that dropping any single variable could, unfortunately, alter the t-statistics for other variables present in the regression equation, either through changes in standard errors attributable to multicollinearity or through changes in probabilities due to the ensuing modification of a simultaneous testing situation.

These fluctuations in decision-making support persuaded applied statisticians to seek ancillary variable selection criteria. A second criterion, which is a natural extension of the t-test idea, couches the problem in a sequential format and is called ***stepwise regression***. The basis of this selection standard is the notion of ***partial correlation*** (for Y and X_h given $\{\mathbf{X}_L\}$, the correlation between the residuals of Y regressed on $\mathbf{X}_L$ and the residuals of X_h regressed on $\mathbf{X}_L$).[3] The process can begin with no variables besides the intercept term in a regression equation (***forward selection***); the variables then are included sequentially, one by one, in a succession of no more than P steps. At each step, the variable selected is one of those not yet included in the regression equation but that has the largest absolute partial correlation with Y. For the first step, and for every step in the case of mutually orthogonal Xs, the partial correlation equals the bivariate Pearson's product moment correlation coefficient (see Griffith and Amrhein, 1991, p. 105). At step τ matrix $\mathbf{X}_\tau$ is expanded by one variable and a new regression solution, $\mathbf{b}_\tau = (\mathbf{X}_\tau^T\mathbf{X}_\tau)^{-1}\mathbf{X}_\tau^T\mathbf{Y}$, is computed, making this procedure numerically intensive. This procedure continues, at each step evaluating the (P-τ) X variables not yet included in the regression equation until no significant partial correlations remain. Because of differences between individual and simultaneous probabilities involved with this procedure, commercial software packages often use an extremely generous default level of significance (as large as 50%).

EXAMPLE PR-4.4. Execute a forward selection of variables multiple regression analysis for Example PR-4.3.

Using the SAS procedure

---

[3]Partial correlation coefficients, denoted $\mathbf{r}_{h\,\{L\}}$, may be obtained from a correlation matrix where r_{hj} is the simple linear correlation between X_h and X_j. It is defined as $\frac{-(-1)^{h+j}\ DET\ (\mathbf{R}_{hj})}{[(-1)^{h+h}\ DET\ (\mathbf{R}_{hh}) \times (-1)^{j+j}\ DET\ (\mathbf{R}_{jj})]^{\frac{1}{2}}}$,

where $\mathbf{R}_{hj}$ is obtained from the original correlation matrix **R** by removing its h-th row and j-th column. Thus, for p = 3,

$\mathbf{R} = \begin{bmatrix} 1 & r_{12} & r_{13} \\ r_{12} & 1 & r_{23} \\ r_{13} & r_{23} & 1 \end{bmatrix}$, and hence $r_{12.3} = \frac{(r_{12} - r_{23} \times r_{23})}{[(1-r_{23}^2)(1-r_{13}^2)]^{\frac{1}{2}}}$. As with simple linear correlation coefficients (see Griffith and Amrhein, 1991, p. 268), the test statistic is $t = \frac{r_{12.3}}{\left(\frac{(1-r_{12.3}^2)}{(n-3)}\right)^{\frac{1}{2}}}$, with df = 3, because s_1^2, s_2^2, and s_3^2 are estimates.

```
PROC STEPWISE;
   MODEL SQDSGR = DFRM DCD DISTSJ DISTA I0/FORWARD;
RUN;
```

produced the SAS output as shown on the following page.

SAS uses a 50% level of significance as its default value with the forward selection criterion. Even if a researcher believes that this level is too generous, one useful outcome of such an exercise is the ranking of variables from most important to least important, on the basis of partial correlations. One interesting result is that while b_{DISTA} is significant at the 5% level in the simultaneous evaluation of these five variables, if DCD is dropped from the analysis, then b_{DISTA} no longer is significant even at the 10% level.

Rather than beginning with no variables in a regression equation, the process can begin with all of the X variables in the equation (***backward elimination***). Then variables are deleted sequentially, one by one, in a succession of no more than P steps. At each step, the variable selected is one of those not yet excluded from the regression equation but that has the smallest absolute partial correlation with Y. For the first step, and for every step in the case of mutually orthogonal Xs, again

```
No other variable met the 0.5000 significance level for entry
into the model.

Summary of Forward Selection Procedure for Dependent Variable
SQDSGR
```

Step	Variable Entered	Number In	Partial R**2	Model R**2	C(p)	F	Prob>F
1	I0	1	0.4809	0.4809	27.5871	65.7862	0.0001
2	DISTSJ	2	0.0733	0.5543	15.9448	11.5132	0.0011
3	DFRM	3	0.0559	0.6101	7.5487	9.8876	0.0025
4	DISTA	4	0.0150	0.6252	6.7500	2.7285	0.1032
5	DCD	5	0.0148	0.6399	6.0000	2.7500	0.1019

the partial correlation equals the bivariate Pearson's product moment correlation coefficient. At step τ, matrix $\mathbf{X}_\tau$ is contracted by one variable and a new regression solution, $\mathbf{b}_\tau = (\mathbf{X}_\tau^T\mathbf{X}_\tau)^{-1}\mathbf{X}_\tau^T\mathbf{Y}$, is computed, making this procedure numerically intensive, too. This procedure continues, at each step evaluating the (P-τ) X variables not yet excluded from the regression equation, until only significant partial correlations remain. Because of differences between individual and simultaneous probabilities involved with this procedure, commercial software packages often use a generous default level of significance (as large as 10%).

EXAMPLE PR-4.4 (continued). Execute a backward elimination of variables multiple regression analysis for Example PR-4.3.

Using the SAS procedure

```
PROC STEPWISE;
   MODEL SQDSGR = DFRM DCD DISTSJ DISTA I0/BACKWARD;
RUN;
```

produced the SAS output

```
All variables left in the model are significant  at the 0.1000 level.

Summary  of  Backward  Elimination  Procedure  for  Dependent  Variable
SQDSGR

      Variable  Number   Partial   Model
Step  Removed       In      R**2    R**2       C(p)        F      Prob>F

1     DCD            4    0.0148   0.6252   6.7500   2.7500   0.1019
2     DISTA          3    0.0150   0.6101   7.5487   2.7285   0.1032
```

SAS uses a 10% level of significance as its default value with the backward elimination criterion. These results corroborate those obtained with the forward selection criterion. In an ideal situation, the variables would be deleted in exactly the reverse order they were included with the *forward* procedure.

A third sequential process combines these two procedures, beginning with forward selection, and at each step τ following the entry of a new variable into matrix $\mathbf{X}_\tau$ by a backward elimination evaluation. Accordingly, more than P steps are possible with this procedure, because variables have the possibility of repeatedly entering and exiting matrix $\mathbf{X}_\tau$.

EXAMPLE PR-4.4 (continued). Execute a stepwise selection of variables multiple regression analysis for Example PR-4.3.

Using the SAS procedure

```
PROC STEPWISE;
   MODEL SQDSGR = DFRM DCD DISTSJ DISTA I0/STEPWISE
                             SLE=0.15 SLS=0.10;
RUN;
```

produced the SAS output presented on the next page.

These results illustrate that forward selection criterion can introduce a variable that the backward elimination criterion later removes (e.g., DISTA).

This set of variable selection criteria may be summarized as follows:

```
All variables left in the model are significant  at the 0.1000 level.
No other variable met the 0.1500 significance level for entry into
the model.

Summary of Stepwise Procedure for Dependent Variable SQDSGR
 Variable     Number  Partial  Model
Step  Entered Removed   In        R**2    R**2    C(p)    F    Prob>F

1     IO                1     0.4809  0.4809  27.5871  65.7862  0.0001
2     DISTSJ            2     0.0733  0.5543  15.9448  11.5132  0.0011
3     DFRM              3     0.0559  0.6101   7.5487   9.8876  0.0025
4     DISTA             4     0.0150  0.6252   6.7500   2.7285  0.1032
5     DISTA             3     0.0150  0.6101   7.5487   2.7285  0.1032
```

> SELECTION CRITERION 2: Order all of the variables considered to be in matrix **X** in accordance with their individual sequential partial correlations, in terms of entering into the regression equation, being removed from the regression equation, or both. Next, select those variables in this ordering whose partial correlations satisfy some predetermined threshold level of significance.

The forward selection, backward elimination, and stepwise procedures are guaranteed to yield equivalent results (reverse orderings of variables between forward and backward) only when the Xs are mutually orthogonal and the levels of significance are commensurate. Furthermore, because of the step-by-step procedure involved, there is no guarantee that these automatic selection methods will uncover the overall best equation.

A third basis for assessing the importance of X variables exploits statistical distribution properties of the regression technique. A trade-off is sought between maximizing the relative prediction ability of the regression equation being composed and minimizing the bias introduced by excluding variables that might be part of matrix **X**. This criterion is implemented through a numerically intensive procedure called ***all possible regressions***. With P X variables there are (P+1) terms in a regression equation, including the intercept; an intercept term virtually always should be included in the equation. Assuming that the matrix **X** is complete, there are (2^P-1) subset regression equations; a variable X_j can be either entered into or kept out of the equation. Since the order of variables in an equation is unimportant, determining the number of possibilities is a combinatorial problem. Recall the combinatorial analysis of introductory statistics (see Griffith and Amrhein, 1991, pp. 153–55). The single equation including none of the X variables, C(P,0) = 1, refers to computing the mean of Y as a regression problem (Example PR-4.1) and usually is not evaluated in this procedure (hence the subtraction of 1 when counting the number of possibilities). Of course, the single equation including all of the X variables, C(P,P) = 1, is believed to be the correctly specified one. There are C(P,1) = P equations in which a single X_j variable appears, namely one for each X in turn. There are $C(P,2) = \frac{P(P-1)}{2}$ equations in which two of the X variables appear. In general, then, there are $C(C,p) = \frac{P!}{(p!)[(P-p)!]}$ equations in which p of the X variables appear. Summing these possibilities gives

$$\sum_{p=0}^{P} \frac{(P!)}{(P!)[(P-p)!]} = 2^P .$$

For each of these equations, denoting the selected set of X variables by $\mathbf{X}_p$, $(\mathbf{X}_p^T\mathbf{X}_p)^{-1}\mathbf{X}_p^T\mathbf{Y}$ must be calculated. To repeat, the *all possible regressions* procedure is numerically intensive. Because it is not a sequential procedure, unlike the stepwise ones, it enables a researcher to uncover overall best equations. A prominent motivation for developing this procedure involved high data collection costs. Applied statisticians sought a way of avoiding the collection of expensive data for some variable X_h, even though they knew this variable should be in the model (that is, was known to belong in matrix **X**). The idea is to determine what happens if X_h is not included in the model in order to gauge the benefit of incurring the collection costs.

Mallow's C_p index is popular and was initially formulated to gauge the degree of bias entering a linear regression equation when it is misspecified due to missing variables. This index is defined as

$$C_p = \frac{\text{SSE}(p)}{\hat{\sigma}^2} - n + 2(p+1) ,$$

where SSE(p) is the error sum-of-squares calculated for a subset of p of the X variables in the equation, and $\hat{\sigma}^2$ is the mean squared error with all P of the X variables in the equation. Using the calculus of expectations,

$$E(C_p|\text{bias} = 0) = p + 1 .$$

Accordingly, bias due to missing variables specification error can be assessed by plotting C_p versus p. Subset equations with substantial bias will fall above the expected value line, which is a straight line sloping upwards and to the right at a 45° angle and passing through the vertical axis at 1. Nearly unbiased equations typically will have (p, C_p) coordinates close to the 45° line.

Meanwhile, since the complete set of variables constituting matrix **X** is unknown, entering more and more variables into this matrix improves the chances of embracing the complete set of Xs while tending to enhance its prediction capabilities. Two statistics in particular often are employed to index potential improvement in the predictive power of $\hat{Y}$. The first is the multiple correlation coefficient, denoted by R, which generalizes the notion of correlation to one of measuring the strength (or degree) of the relation between Y and a set of Xs. Since $\hat{Y}$ is attempting to predict Y, R always is non-negative, so that the nature of this relationship is never in question; otherwise $\hat{Y}$ simply would be multiplied by -1 changing the signs on all of the regression coefficients. R is the product-moment correlation between Y and $\hat{Y}$. R^2 also equals $\frac{\mathbf{b}^T\mathbf{X}^T\mathbf{Y}}{\mathbf{Y}^T\mathbf{Y}}$, which defines it as the percentage of variance accounted for in Y by the set of X variables included in a regression equation (see Griffith and Amrhein, 1991, pp. 325–29). As an aside, sometimes a partial R^2 value of 0.05 is used, rather than significance levels, as the critical threshold value in the preceding stepwise procedures.

> *RULE OF THUMB 4.0*: A variable should not be considered important unless it can account for at least an additional 5% of the variance in Y beyond that accounted for by variables already entered into a regression equation.

This criterion is expressed again in *Rule of Thumb 6.2.* The second statistic used to measure the predictive power of $\hat{Y}$ is the *mean squared error* (MSE), which equals $\frac{SSE}{(n-P-1)}$ for the true equation; SSE is the quantity being minimized in expression (4.1).

Recall that a strategy in which **X** is augmented without restriction results in vector **b** remaining unbiased, while the COV(**b**) tends to be inflated. Combining these consequences with the number of degrees of freedom, (n-P-1), moves toward zero as P increases.

> *RULE OF THUMB 4.1*: To minimize the chances of focusing on sample specific rather than population information, there should be at least 20 degrees of freedom for each regression coefficient estimated.

These drawbacks motivated modifications to help avoid the careless use of R^2 and MSE. The modified version of the multiple correlation coefficient is known as the adjusted-R^2, denoted by R_p^2, and is defined as $\left[\frac{1-(n-1)}{(n-p-1)}\right](1-R^2)$. This revised quantity does not necessarily increase as additional X variables are entered into the regression equation. The modified version of MSE is a sequential, marginal decomposition, denoted by MSE_p, and is defined as $\frac{SSE_p}{(n-p-1)}$. The usual trajectory followed by MSE_p, as p increases from 1 to P, is as follows: initially it decreases, it stabilizes, and then eventually it may increase as p approaches P. In either case, the best predictive power is from the matrix $\mathbf{X}_p$ that maximizes R_p^2 while minimizing MSE_p and p.

EXAMPLE PR-4.5. Preform all possible regressions for Example PR-4.3.

Using the SAS procedure

```
PROC RSQUARE MSE ADJRSQ CP;
   MODEL SQDSGR = DFRM DCD DISTSJ DISTA I0;
RUN;
```

produced the SAS output as shown on the following page.

According to these results, no single- or two-variable subset regression equation is close to being unbiased. The three-variable equation based upon DFRM, I0, and DISTSJ is somewhat close, having a C_p value of 7.5 versus its expected value of 4. Two four-variable equations—one based upon DFRM, I0, DISTA, and DISTSJ and the other based upon DFRM, DCD, I0, and DISTSJ—have C_p values relatively close to 5. These three subsets of variables suggest that at least DFRM, I0, and DISTSJ should be included in the regression equation. Adding either or both of the remaining two variables only slightly increases either R^2 or the adjusted-R^2. The MSE_p plot appears

N = 73   Regression Models for Dependent Variable: SQDSGR

Number in Model	R-square	Adjusted R-square	C(p)	MSE	Variables in Model
1	0.480942	0.473631	27.5871	1.30640	I0
1	0.272479	0.262232	66.3782	1.83107	DISTSJ
1	0.106885	0.094306	97.1921	2.24784	DFRM
1	0.032549	0.018923	111.0	2.43494	DCD
1	0.007819	-.006155	115.6	2.49718	DISTA
2	0.554255	0.541520	15.9448	1.13790	I0 DISTSJ
2	0.497689	0.483338	26.4706	1.28231	DFRM I0
2	0.492929	0.478441	27.3564	1.29446	DCD I0
2	0.481341	0.466522	29.5128	1.32404	I0 DISTA
2	0.458408	0.442934	33.7802	1.38258	DFRM DISTSJ
2	0.358580	0.340254	52.3562	1.63742	DISTA DISTSJ
2	0.273224	0.252459	68.2395	1.85532	DCD DISTSJ
2	0.210505	0.187948	79.9104	2.01543	DFRM DCD
2	0.106885	0.081368	99.1921	2.27996	DFRM DISTA
2	0.061383	0.034565	107.7	2.39611	DCD DISTA
3	0.610124	0.593173	7.5487	1.00971	DFRM I0 DISTSJ
3	0.577275	0.558896	13.6612	1.09478	I0 DISTA DISTSJ
3	0.554284	0.534905	17.9395	1.15432	DCD I0 DISTSJ
3	0.529900	0.509461	22.4768	1.21747	DFRM DCD I0
3	0.499122	0.477344	28.2041	1.29718	DFRM DISTA DISTSJ
3	0.497873	0.476042	28.4365	1.30041	DFRM I0 DISTA
3	0.497656	0.475815	28.4769	1.30098	DCD I0 DISTA
3	0.472425	0.449487	33.1719	1.36632	DFRM DCD DISTSJ
3	0.361833	0.334087	53.7509	1.65273	DCD DISTA DISTSJ
3	0.221686	0.187846	79.8298	2.01569	DFRM DCD DISTA
4	0.625164	0.603115	6.7500	0.98503	DFRM I0 DISTA DISTSJ
4	0.617761	0.595276	8.1276	1.00448	DFRM DCD I0 DISTSJ
4	0.579652	0.554925	15.2190	1.10463	DCD I0 DISTA DISTSJ
4	0.532199	0.504682	24.0490	1.22933	DFRM DCD I0 DISTA
4	0.527344	0.499541	24.9525	1.24209	DFRM DCD DISTA DISTSJ
5	0.639942	0.613073	6.0000	0.96032	DFRM DCD I0 DISTA DISTSJ

in Figure 4.2[4]. It shows a dramatic decline as I0, DISTSJ, and DFRM sequentially are included in matrix **X**. But this graph levels off with the further inclusion of either DCD or DISTA, corroborating the findings based upon C_p and R^2.

This set of variable selection criteria may be summarized as follows:

SELECTION CRITERION 3: Compute the $(2^P - 1)$ possible subset regressions. Select that subset of p variables that simultaneously has a C_p as close as possible to (p+1), an R_p^2 as large as possible, and the smallest possible MSE_p and p combination.

---

[4]This is similar to the scree plots presented in Chapter 6.

Of course, the selection could be p = P.

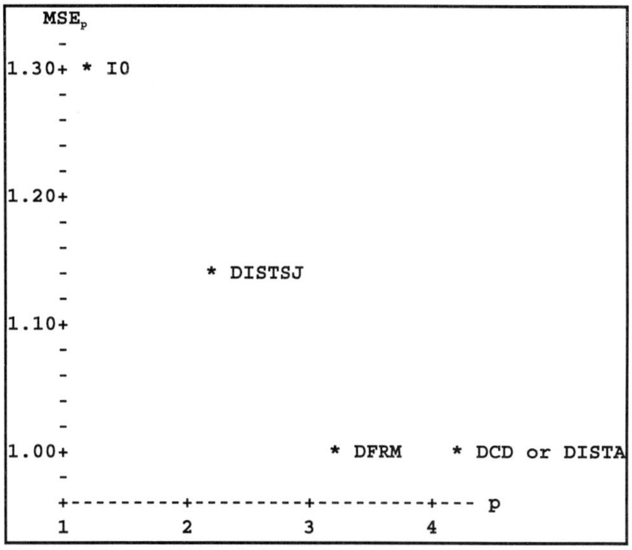

Figure 4.2. Incremental decrease in MSE with the addition of each variable for example PR-4.5.

4.2.3. Calculation Error Diagnostics

To achieve good statistical descriptions and to draw sound inferences, remedial action usually should be taken when assumptions associated with the regression technique are found not to hold in sample data. As mentioned, $E(e_i) = E(\epsilon_i) = 0$ by construction and ϵ_i should conform to a normal frequency distribution. The assumption of no measurement error in either Y or X_j is unlikely to hold and can be quite problematic[5]; nevertheless, it is assumed that the error present is due solely to sampling. Therefore, the remaining assumptions to be subjected to diagnostic evaluation are

(1) the X_js are mutually independent, implying that only negligible calculation error is present, so that computation of $(\mathbf{X}^T\mathbf{X})^{-1}$ is not numerically problematic;

(2) $VAR(\epsilon) = \sigma^2\mathbf{I}$, implying a homogeneous geographic landscape in which there is constant variance and the absence of nonzero spatial autocorrelation; and,

(3) relationships between the X_js and Y are linear.

The second assumption alludes to the problem of data anomalies, too.

As mentioned, even if matrix **X** is complete, it can serve as a gateway for computational error to enter into a regression analysis. Sometimes such calculation error results from cumulative rounding error, as is noticeable in Example PR-4.2. More problematic, however, is calculation error due to the presence of multicollinearity. This feature of data analysis allies with another important question connected to regression analysis.

REG Question #4: What is the impact of prevailing multicollinearity on a statistical analysis?

---

[5]Consider the case of simple linear regression. Suppose the X variable is measured without error, while the dependent variable is measured with error, becoming $(Y + \xi_1)$. Then, $E(\mathbf{b})$ is biased when $\mathbf{X}^T\xi_1 \neq 0$, $E(\xi_1) \neq \mathbf{0}$, or $\xi_1 \propto Y$. If ξ_1 is random, its elements are uncorrelated, and $E(\xi_1) = 0$, then no new complications are introduced into the regression analysis. Now suppose the regressor variable is measured with error, such that it is $(X + \xi_1)$, but Y is not. In this case the statistical distribution of **b** behaves in a way analogous to that for mis-specification due to missing variables in matrix **X**.

Recall that as multicollinearity increases in matrix **X**, $DET(\mathbf{X}^T\mathbf{X})$ approaches zero, causing $(\mathbf{X}^T\mathbf{X})^{-1}$ to be undefined. Two extreme types of multicollinearity can exist, namely ***perfect multicollinearity*** and ***complete multicollinearity***. Prior to assessing multicollinearity, each X variable must be converted to its z-score counterpart to remove variance differences in the X_js. If the usual transformation $[(x_i - \bar{x})/s]$ is used, then each Z_j has a variance of 1. Perfect multicollinearity occurs when at least one Z variable can be expressed as an exact linear function (defined by δ) of one or more of the other Z variables:

$$Z_j = \delta_{0,h,j} + Z_h\delta_{h,j} \text{ , or } Z_j = \delta_{0,j} + \sum_{h=1}^{P} Z_h\delta_{h,j} \text{ where } h \neq j \ (\xi_j = 0), \tag{4.8}$$

where $\delta_{h,j}$ is the slope coefficient (a conditional mean) for regression of Z_j on an entire set of predictor variables $\{Z_h; h = 1, 2, ..., P\}$. $\delta_{0,h,j}$ is the intercept term when this regression is a simple linear one involving a single predictor variable Z_h. In contrast, $\delta_{0,j}$ is the intercept term when the regression is a multiple regression. Complete multicollinearity occurs when

$$Z_j = \delta_{0,h,j} + Z_h\delta_{h,j} \text{ , } (\xi_{h,j} = 0) \tag{4.9}$$

for each of the (P-1) remaining Z variables. Each time condition (4.8) holds, one of the eigenvalues of matrix $(\mathbf{Z}^T\mathbf{Z})$ becomes zero, and the principal eigenvalue, λ_1, increases. When condition (4.9) holds, the principal eigenvalue of matrix $(\mathbf{Z}^T\mathbf{Z})$ becomes (P+1). But, conditions (4.8) and (4.9) rarely hold, and the usual case is

$$Z_j = \delta_{0,h,j} + Z_h\delta_{h,j} + \xi_{h,j} \ . \tag{4.10}$$

In contrast, when all of the $\delta_{h,j}$ coefficients are zero, the Zs are mutually orthogonal, and all of the eigenvalues of $(\mathbf{Z}^T\mathbf{Z})$ equal 1 (property P11, Table 3.1). Therefore, inspection of the eigenvalues of matrix $(\mathbf{Z}^T\mathbf{Z})$, to see whether λ_1 is close to (P+1) and/or any of the remaining λ_js are close to zero—as opposed to a more uniform distribution of values—indicates the extent and seriousness of computational error. One measure used to index this source of calculation error is the ***condition number***, defined (in keeping with SAS) as $\sqrt{\frac{\lambda_1}{\lambda_j}}$.

> *RULE OF THUMB 4.2*: Condition number values, $\sqrt{\frac{\lambda_1}{\lambda_j}}$, in excess of 10 are indicative of the presence of multicollinearity that can introduce serious calculation error into a regression analysis. Values between 10 and 32 imply moderate to strong multicollinearity, whereas values exceeding 32 imply severe multicollinearity.

As $DET(\mathbf{X}^T\mathbf{X})$ approaches zero, $\mathbf{X}^T\mathbf{X}$ becomes increasingly more difficult to invert, and the diagonal entries of $(\mathbf{X}^T\mathbf{X})^{-1}$ increase, inflating the variance estimate of at least some of the s_{b_j}. This inflation is measured by the variance inflation factor (VIF), which indexes how much the variances of estimated regression coefficients are inflated as compared to when the Xs are not linearly related. VIF_j follows equation (4.8), regressing each X_j on the remaining (P-1) X variables, such that

$$X_j = \delta_{0,j} + \sum_{h=1}^{p} X_h \delta_{h,j} + \xi_{h,j} . \tag{4.11}$$

Its definition utilizes the R_j^2 value accompanying equation (4.11); VIF_j is defined as $\frac{1}{1 - R_j^2}$. When the Xs are mutually orthogonal, these R_j^2s equal 0, and hence the VIF_js equal 1.

> *RULE OF THUMB 4.3*: A maximum VIF_j value, $\frac{1}{1 - R_j^2}$, in excess of 10 indicates that multicollinearity may be unduly influencing the OLS estimates, **b**, as well as their corresponding t-statistics. Values between 5 and 10 indicate questionably bothersome multicollinearity, whereas values between 1 and 5 indicate the presence of inconsequential calculation error due to multicollinearity.

Equations (4.10) and (4.11) show why the inspection of pairwise correlations among the X variables may not disclose the extent of multicollinearity problems. For example, a pairwise comparison will not reveal the correlation between a single variable and a linear combination of other variables. In contrast, the correlation matrix for the parameter estimates can be quite revealing. When $|r_{b_h b_j}| > 0.90$, then the parameters β_h and β_j often are redundant, suggesting that either X_h or X_j does not belong in matrix **X**.

> *RULE OF THUMB 4.4*: When the correlation between b_j and b_h exceeds 0.9, in absolute value, variables X_j and X_h may be introducing serious multicollinearity into the regression analysis.

EXAMPLE PR-4.6. Calculate the multicollinearity diagnostics for Example PR-4.3.

Using the SAS procedure

```
PROC REG;
   MODEL SQDSGR = DFRM DCD DISTSJ DISTA I0/
                          VIF COLLIN CORRB;
RUN;
```

produced the SAS output for COLLIN that includes

Number	Eigenvalue	Condition Number
1	5.14804	1.00000
2	0.34980	3.83627
3	0.30534	4.10610
4	0.09141	7.50467
5	0.07908	8.06824
6	0.02633	13.98325

There are several indications in this eigenvalue analysis that the matrix **X** for this regression analysis contains seriously troublesome multicollinearity. First, λ_1 is close to its maximum of 6. Second, each of the six eigenvalues is markedly different from 1. And third, the final condition number exceeds 10.

EXAMPLE 4.6 (continued). Calculate the variance inflation factor (VIF).

The SAS output for VIF is

Variable	DF	Variance Inflation
INTERCEP	1	0.00000000
DFRM	1	1.38966149
DCD	1	1.41695014
IO	1	1.50859400
DISTA	1	1.34408257
DISTSJ	1	1.62043826

Although the eigenvalue diagnostic results are symptomatic of disturbing multicollinearity, the VIF_js suggest that it is not unduly influencing the OLS estimates, **b**, or their variances used in computing the t-statistics. These estimates should not be poor strictly due to the presence of multicollinearity.

EXAMPLE 4.6 (continued). Calculate the parameter estimate correlations for Example PR-4.3.

The SAS output for CORRB is

Correlation of Estimates

CORRB	INTERCEP	DFRM	DCD	I0	DIST A	DISTSJ
INTERCEP	1.0000	-0.2805	-0.5612	-0.1642	-0.70 08	-0.1554
DFRM	-0.2805	1.0000	-0.3061	0.3865	0.04 15	-0.1924
DCD	-0.5612	-0.3061	1.0000	-0.1339	0.24 78	-0.1844
I0	-0.1642	0.3865	-0.1339	1.0000	-0.23 50	-0.4732
DISTA	-0.7008	0.0415	0.2478	-0.2350	1.00 00	0.3151
DISTSJ	-0.1554	-0.1924	-0.1844	-0.4732	0.31 51	1.0000

The largest $|r_{b_h b_j}|$ value in this correlation matrix is 0.7008, ignoring self-correlations. This value is considerably smaller than 0.9, suggesting that none of the variables in the regression equation are near perfectly redundant.

In conclusion, potentially bothersome multicollinearity is present in the matrix **X** for Example PR-4.3. Fortunately, diagnostics suggest that it is not introducing serious calculation error into the OLS regression analysis results.

4.2.4. Other Useful Diagnostics

Once a regression equation has been specified and any necessary corrective action taken so that it is not infected with serious calculation error, other standard diagnostic analyses should be performed. Especially because the t-test variable selection criterion assumes normality, the frequency distribution for the estimated error term, **e**, should be evaluated. This diagnostic should be accompanied by one for variance homogeneity. Such analyses performed on variable Y are informative (see Example PR-4.3); however, the relevant inferential basis in regression analysis is established by the error term, ϵ. Further, as pointed out in Chapter 1, because spatial autocorrelation can play havoc with statistical diagnostics, both the Shapiro-Wilk and the Kolmogorov-Smirnov tests should be performed on the frequency distribution of **e**. For the most part, these concerns can be dismissed by using a very large random sample, a situation that rarely occurs in geographic analyses (but a situation that may well change as GIS technology and adoption move forward).

EXAMPLE PR-4.7. Calculate a battery of diagnostics for the error term in Example PR-4.3.

Using the SAS procedure

```
PROC REG;
   MODEL SQDSGR = DFRM DCD DISTSJ DISTA I0/
                  SPEC INFLUENCE R;
                  OUTPUT OUT=TEMP3 R=YRESID;
                  PLOT RESIDUAL.*PREDICTED.;
```

RUN;

the SAS output for SPEC and PLOT is illustrated in the following output:

```
            Test of First and Second Moment Specification
DF:     19    Chisq Value: 22.145633741    Prob>Chisq:    0.2771
```

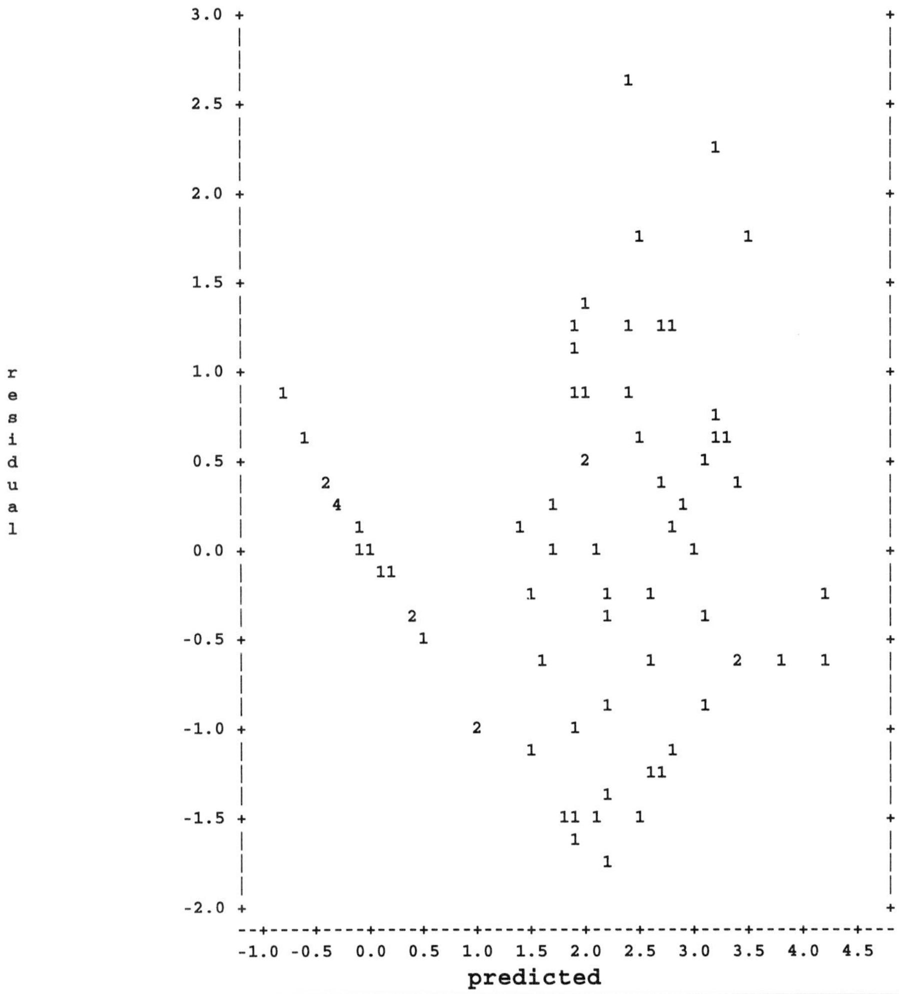

The SAS option SPEC invokes White's formal test for heteroscedasticity, found not to be significant in this case. This finding, however, is contradicted by the visual finding. The scatter diagram portrays a nonconstant variance for the error terms; the linear relation occurring in the ranges of (-1, 1), coupled with too few high predicted values with negative residuals, is bothersome. The graphical indication is that some remedial action is needed to strengthen the inferential basis. This inconsistency highlights a well-known dilemma arising from the use of multiple statistics for a single diagnostic. Since each statistic is tuned to subtly different idiosyncrasies and nuances in the data, conflicting implications are not uncommon. Nevertheless, it is sensible to examine the data in as many different ways as possible.

EXAMPLE PR-4.7 (continued). Calculate normality and spatial variance homogeneity diagnostics for the error term in Example PR-4.3.

Using the SAS procedure

```
PROC UNIVARIATE DATA=TEMP3 FREQ NORMAL;
   VAR YRESID;
RUN;
PROC DISCRIM POOL=TEST;
   CLASS ADM;
   VAR YRESID;
RUN;
```

produced the SAS output as shown on the following page.

Thus, according to the S-W statistic, there is no evidence to indicate that ϵ is not normally distributed in the population. Tabulating the frequency information as

z-score	-0.84	-0.25	0.25	0.84	∞
empirical distribution	0.219	0.411	0.562	0.822	1
normal distribution	0.2045	0.4012	0.5987	0.7955	1

yields a Kolmogorov-Smirnov statistic that corroborates this finding with the S-W statistic (its disaggregate value is 0.0499). Meanwhile, Bartlett's test for variance homogeneity across the agricultural administrative regions, obtained with PROC DISCRIM POOL=TEST, reveals that there is no evidence to indicate that ϵ is not homogeneously distributed across the geographic landscape of Puerto Rico in the corresponding statistical population.

Besides addressing the more focused question of constant variance, researchers often are concerned about whether the sample was drawn from more than one population. In other words, do any of the observations appear to be aberrations? An answer to this question refers to the notions of *outlier* and *influential point*, or some small subset of the data that, when minimizing equation (4.1), exerts a disproportionate influence on the regression parameter estimates. An ***outlier*** is a suspected extreme value of either Y or the **X** matrix and is defined as an observation whose y_i value differs from its fitted $\hat{y}_i$ value by far more than do values for the other observations. An ***influential point*** is an outlier that has a considerable effect on the OLS estimation of the regression equation. This point may produce serious distortions in the statistical description (i.e., predicted values and the usual summary statistics) rendered by, or the inferences that are drawn from, a regression analysis. A battery of statistics has been devised to expose outliers; some of the more popular ones are Cook's D_i statistic, the d_i* studentized deleted residual, the h_{ii} leverage statistic, and the standard influence of observations on predicted values ($DFFITS_i$) statistic, $D_i = \frac{(\mathbf{b} - \mathbf{b}_i)^T\mathbf{X}^T\mathbf{X}(\mathbf{b} - {}_i\mathbf{b})}{(P \times MSE)}$ where vector ${}_i\mathbf{b}$ is the regression coefficients with observation i deleted from the OLS estimation. Even though this statistic does not follow an F distribution, relating its value to F(1-α, P, n-P), α = .8 or .9 has proven useful in ascertaining whether the i-th observation has little apparent influence on the

```
W:Normal    0.974205   Prob<W         0.3708

                         Frequency Table
                      Percents                        Percents
    Value Count   Cell    Cum         Value Count   Cell    Cum
 -1.72019     1    1.4    1.4      0.051807     1    1.4   52.1
 -1.60393     1    1.4    2.7      0.062699     1    1.4   53.4
 -1.54167     1    1.4    4.1       0.06823     1    1.4   54.8
 -1.51252     1    1.4    5.5      0.097078     1    1.4   56.2
 -1.48008     1    1.4    6.8       0.25405     1    1.4   57.5
 -1.46487     1    1.4    8.2      0.255368     1    1.4   58.9
 -1.35395     1    1.4    9.6      0.264204     1    1.4   60.3
 -1.26875     1    1.4   11.0      0.264527     1    1.4   61.6
 -1.23347     1    1.4   12.3      0.299525     1    1.4   63.0
  -1.1392     1    1.4   13.7        0.3062     1    1.4   64.4
 -1.09058     1    1.4   15.1      0.350333     1    1.4   65.8
 -1.00617     1    1.4   16.4      0.370612     1    1.4   67.1
 -0.99548     1    1.4   17.8      0.380131     1    1.4   68.5
 -0.96227     1    1.4   19.2      0.405238     1    1.4   69.9
 -0.92637     1    1.4   20.5      0.449821     1    1.4   71.2
  -0.9052     1    1.4   21.9       0.48309     1    1.4   72.6
 -0.67579     1    1.4   23.3      0.513091     1    1.4   74.0
 -0.65815     1    1.4   24.7      0.562667     1    1.4   75.3
 -0.64073     1    1.4   26.0      0.580772     1    1.4   76.7
 -0.63333     1    1.4   27.4      0.600018     1    1.4   78.1
 -0.61557     1    1.4   28.8      0.637218     1    1.4   79.5
 -0.56502     1    1.4   30.1      0.733704     1    1.4   80.8
 -0.45403     1    1.4   31.5      0.813466     1    1.4   82.2
 -0.41454     1    1.4   32.9      0.844688     1    1.4   83.6
  -0.3965     1    1.4   34.2      0.848223     1    1.4   84.9
 -0.35038     1    1.4   35.6      0.888456     1    1.4   86.3
 -0.32283     1    1.4   37.0       1.11647     1    1.4   87.7
 -0.31173     1    1.4   38.4      1.194894     1    1.4   89.0
 -0.30552     1    1.4   39.7       1.23314     1    1.4   90.4
 -0.25774     1    1.4   41.1      1.280265     1    1.4   91.8
 -0.21759     1    1.4   42.5      1.290555     1    1.4   93.2
 -0.15842     1    1.4   43.8      1.344005     1    1.4   94.5
 -0.07555     1    1.4   45.2      1.717567     1    1.4   95.9
 -0.02487     1    1.4   46.6      1.803942     1    1.4   97.3
 -0.02206     1    1.4   47.9      2.234402     1    1.4   98.6
 -0.02145     1    1.4   49.3      2.676335     1    1.4  100.0
 0.049683     1    1.4   50.7

 Test Chi-Square Value =      5.637615
 with       4 DF       Prob > Chi-Sq = 0.2279
```

fitted regression equation. D_i assists in the identification of observations with unusual X values. The quantity $d_i^* = e_i \sqrt{\dfrac{(n-P-1)}{[SSE \times (1-h_{ij}) - e_i^2]}}$, which is distributed as $t_{\alpha/2,n-P-1}$ and helps identify outlying Y observations. This statistic assists in the identification of observations with unusual Y values. Both h_{ii} and $DFFITS_i$ combine these two notions. $h_{ii} = \mathbf{X}_i(\mathbf{X}^T\mathbf{X})^{-1}\mathbf{X}_i^T$, where $\mathbf{X}_i$ is the i-th row of matrix $\mathbf{X}$. The critical threshold value here is $\frac{2P}{n}$ or $\frac{3P}{n}$. Finally, $DFFITS_i = \left[\frac{h_{ii}}{1 - h_{ii}}\right]\left[\frac{e_i}{MSE + VAR(\hat{y}_i)}\right]$. The critical threshold value for this statistic is 1 or 2.

RULE OF THUMB 4.5: If observation i has a Cook's D_i value $< F(.1, P, n\text{-}P)$, a $|d_i^*| < t_{\frac{\alpha}{2}, n-P-1}$, and $h_{ii} < \frac{2P}{n}$, *and* a $DFFITS_i < 2\sqrt{\frac{P}{n}}$, then there is no evidence that it is an influential outlier, and a researcher has some assurance that the regression analysis is not being misled by it. In contrast, if observation i has a Cook's D_i value $> F(0.2, P, n\text{-}P)$, a $|d_i^*| > t_{\frac{\alpha}{2}, n-P-1}$, and $h_{ii} > \frac{3P}{n}$, *and* a $DFFITS_i > 2\sqrt{\frac{P}{n}}$, then there is strong evidence that it is an influential outlier and deserves special attention. Otherwise, the status of the observation is debatable.

In the case of areal units whose status is debatable but leaning toward being classified as anomalies, often the goal is to further explore features of these areal units in order to resolve their respective status.

EXAMPLE PR-4.7 (continued). Identify outliers and influential points for Example PR-4.3.

The SAS output for the INFLUENCE and R options is voluminous and will be only summarized here. The largest D_i value is 0.264 (Aguada), which is notably less than $F(.1, 6, 67) = 0.3626$. The largest h_{ii} value is 0.1626 (San Juan/Rio Piedras), which is slightly less than $\frac{2(6)}{73} = 0.1644$. Both Aguada and Santa Isabela have $|d_i^*|$ values that exceed $t_{0.975, 73\text{-}6\text{-}1} = 1.9966$. And Aguada has a $DFFITS_i$ value that exceeds 1 but not 2. Therefore, there is little evidence to indicate that any of the 73 observations deserve special attention as influential outliers; the controversial exception perhaps is Aguada.

One commonality running across the members of this group of four statistics is that each assesses what happens to the regression estimation with and without each individual observation. The presence of outliers *hints* at the possibility of sampling having been from more than one population. However, in practice the outlier often is interpreted as a data recording error or some other mistake having been made. If this latter interpretation causes the outlier to be dropped, additional serious error may enter the analysis.

Assumption (2) of §4.2.3 also requires that the errors be uncorrelated. With regard to spatial autocorrelation, the most powerful test statistic for this purpose is the Moran Coefficient (MC), although corroboration of its implication by the Geary Ratio is desirable. Recall from §2.3.1 that MC can be computed with results obtained from two simple linear regressions. In the first regression, the spatially lagged error vector **Ce** (recall that **C** is the binary connectivity matrix) is regressed on the error vector **e**, producing regression coefficient b_{eCe}. In the second regression, the row sum of connections **C1** is regressed on the vector **1**, producing regression coefficients b_{1C1}. Combining these results produces $MC = \frac{b_{eCe}}{b_{1C1}}$. The regression analyses executed should be for ***no-intercept*** (option NOINT in SAS) equations. The expected value of MC in this general case is defined as

$$E(MC) = \frac{-n \times TR[(\mathbf{X}^T\mathbf{X})^{-1}\mathbf{X}^T\mathbf{C}\mathbf{X}]}{[(n-P-1)\times \mathbf{1}^T\mathbf{C}\mathbf{1}]} \tag{4.12}$$

(recall that *TR* denotes the matrix trace operator), which for a problem like Example PR-4.1, where $\mathbf{X} = \mathbf{1}$, reduces to the standard univariate result of $\frac{-n[(\frac{1}{n})\mathbf{1}^T\mathbf{C}\mathbf{1}]}{(n-0-1)\mathbf{1}^T\mathbf{C}\mathbf{1}} = \frac{-1}{(n-1)}$. If the Xs are mutually orthogonal, then equation (4.12) reduces to $-\frac{(1+\sum_{j+1}^{P} MC)}{(n-P-1)}$. A routine multivariate technique introduced in Chapter 6 will be used to simplify its calculation by exploiting this feature. Until then, when diagnostics indicate that serious calculation error is not being introduced by multicollinearity, then this sum of MCs furnishes a good approximation to E(MC). Meanwhile, the variance of MC in this general case is defined as

$$VAR(MC) = \left(\frac{(2\times n^2)}{(\mathbf{1}^T\mathbf{C}\mathbf{1})^2(n-P-1)(n-P+1)}\right) \times \left(\mathbf{1}^T\mathbf{C}\mathbf{1} + TR[(\mathbf{X}^T\mathbf{X})^{-1}\mathbf{X}^T\mathbf{C}\mathbf{X}(\mathbf{X}^T\mathbf{X})^{-1}\mathbf{X}^T\mathbf{C}\mathbf{X}] - 2TR[(\mathbf{X}^T\mathbf{X})^{-1}\mathbf{X}^T\mathbf{C}^2\mathbf{X}] - \frac{TR[(\mathbf{X}^T\mathbf{X})^{-1}\mathbf{X}^T\mathbf{C}\mathbf{X}^2}{(n-P-1)}\right), \tag{4.13}$$

which for a problem like Example PR-4.1, where $\mathbf{X} = \mathbf{1}$, reduces to

$$\frac{2(n-1)}{(n-1)^2(n+1)} + \frac{2n^2}{(\mathbf{1}^T\mathbf{C}\mathbf{1})(n-1)(n+1)} - \frac{4n(\mathbf{1}^T\mathbf{C}^2\mathbf{1})}{(\mathbf{1}^T\mathbf{C}\mathbf{1})^2(n-1)(n+1)} .$$

Asymptotically, this value roughly equals $\frac{2}{\mathbf{1}^T\mathbf{C}\mathbf{1}}$ (see Griffith and Amrhein, 1991, p. 330). This approximation may be computed as $\frac{2}{n\times b_{1C1}}$.

EXAMPLE PR-4.8. Calculate MC for the residuals from the regression analysis executed in Example PR-4.3 (consult the SAS code presented on pp. 44-5).

Let CYRESID denote **Ce**, CSUM denote **C1**, and X0 denote **1**. Using the SAS procedure

```
PROC REG;
  MODEL CYRESID = YRESID/NOINT;
RUN;
```

produced SAS output that includes

Variable	DF	Parameter Estimate	Standard Error	T for H0: Parameter=0	Prob > |T|
YRESID	1	0.693951	0.26500613	2.619	0.0108

Meanwhile, using the SAS procedure

```
PROC REG;
  MODEL CSUM = X0/NOINT;
RUN;
```

produced SAS output that includes

Variable	DF	Parameter Estimate	Standard Error	T for H0: Parameter=0	Prob > |T|
X0	1	4.712329	0.18555057	25.396	0.0001

Thus, MC $= \frac{0.693951}{4.712329} = 0.147263$. As an aside, the corresponding Geary Ratio value is 0.774455. The approximate expected value of MC, given by $-\frac{(1+\sum_{j=1}^{5} MC_j)}{67}$, is -0.051186 (recall that n=73, P=5 and 73-5-1=67). The exact value is -0.050980. The approximate standard error of MC equals $\frac{\sqrt{2}}{(73 \times 4.712329)} \approx 0.076249$. Therefore, $z_{MC} \approx 2.6026$. Note that the exact z_{MC} equals $\frac{(0.147263 + 0.050980)}{0.067387} = 2.9419$. Significant spatial autocorrelation is being displayed in the residuals.

After all of the results from PR-4.3 to PR-4.8 have been mastered, what has been learned? The relationship between DSGR and five independent variables has been explored more completely than usually found in the literature. We now can state with far more confidence that the regression equation for Example PR-4.3 (1) accounts for roughly 60% of the variation in Y, (2) perhaps should be simplified by removing variables DCD and DISTA, which apparently would not introduce severe regression parameter estimation bias, (3) has errors that seem not to deviate markedly from normally distributed ones having constant variance but looks to have errors that are spatially autocorrelated in the population, and (4) does not appear to have complicating multicollinearity or influential outliers.

4.3. Exploring Nonlinearities: Trend Surface Models and Polynomial Regression

The third critical assumption listed in §4.2.3 concerns linearity. Standard classical statistical models are specified as being linear in nature. If this assumption fails to hold, serious specification error can be introduced into an analysis. If all of the X variables are nonlinear in their relationship with Y, then the regression equation may well be of the form

$$\hat{Y} = \beta_0 \times \prod_{j=1}^{p} X^{\beta_j}. \tag{4.14}$$

Equation (4.14) can be treated as an OLS regression problem if its error term is assumed to be multiplicative, such that

$$\hat{Y} = \beta_0 \times \prod_{j=1}^{p} X^{\beta_j} \times \epsilon \quad .$$

This specification becomes linear in form when natural logarithms, *LN*(☼), are taken of both sides of the equation, yielding

$$LN(Y) = LN(\beta_0) + \sum_{j=1}^{P} \ln(X_j)\,\beta_j + \ln(\epsilon) \quad .$$

Unfortunately, this type of specification violates the assumption of $E(\epsilon) = 0$, creating additional problems.

A second quasi-nonlinear situation occurs when the linear regression equation ($Y = \hat{Y} + \epsilon$) is a correct specification, but powers and cross-products of some of the Xs also are present in the complete matrix **X**. The appropriate specification in this case is called polynomial regression. For a third-order model (a model with three X variables), polynomial regression differs from equation (4.14) and is written

$$\hat{Y} = \beta_0 + \sum_{g=1}^{P} \sum_{h=1}^{P} \sum_{j=1}^{P} X_g X_h X_j \beta_{ghj}.$$

Misspecification of this form generates outliers that do not arise from some extraneous cause (such as a coding mistake). Rather, an outlier arises because the regression equation should have, say, two of its X variables multiplied, instead of added together. Two concerns must be addressed as the order of the polynomial increases: (1) the number of degrees of freedom begins to decrease rapidly, and (2) increasingly severe multicollinearity is introduced into the equation. As a result, in practice polynomials beyond the third order rarely are used.

EXAMPLE PR-4.9. Formulate a second-order polynomial regression equation counterpart to Example PR-4.3.

Using the SAS procedure

```
PROC RSREG;
   MODEL SQDSGR = DFRM DCD I0 DISTA DISTSJ;
RUN;
```

produced SAS output that includes

Parameter	Degrees of Freedom	Parameter Estimate	Standard Error	T for H0: Parameter=0	Prob > |T|
INTERCEPT	1	4.930960	2.178704	2.263	0.0277
DFRM	1	-0.451108	0.188019	-2.399	0.0200
DCD	1	0.009804	0.082325	0.119	0.9057
I0	1	-1.467492	2.101166	-0.698	0.4880
DISTA	1	-1.270837	0.683061	-1.861	0.0684
DISTSJ	1	-0.730160	0.522423	-1.398	0.1680
DFRM*DFRM	1	0.002259	0.005014	0.451	0.6541
DCD*DFRM	1	0.004023	0.003446	1.168	0.2482
DCD*DCD	1	-0.001151	0.000937	-1.229	0.2247
I0*DFRM	1	-0.255150	0.090747	-2.812	0.0069
I0*DCD	1	0.060600	0.037964	1.596	0.1164
I0*I0	0	0	.	.	.
DISTA*DFRM	1	0.052274	0.027275	1.917	0.0607
DISTA*DCD	1	0.012981	0.011824	1.098	0.2772
DISTA*I0	1	0.521078	0.402745	1.294	0.2013
DISTA*DISTA	1	0.019488	0.069820	0.279	0.7812
DISTSJ*DFRM	1	0.057994	0.017797	3.259	0.0020

A peculiarity of this output relates to I0*I0. Since I0 is a binary, 0-1, variable, it is equal to its square (that is, $1^2 = 1$), introducing perfect multicollinearity into the polynomial regression equation. The default action taken in SAS upon detection of perfect multicollinearity is to automatically drop the troublesome variable from the regression analysis, which is equivalent to equating its parameter estimate to zero. In comparison with the output reported in Example PR-4.3, this polynomial regression has 14 fewer degrees of freedom. The individual t-statistics imply that a number of the cross-product terms involving I0 are important. This feature of the data could account for the disturbing graphic portrayal appearing in Example PR-4.7.

Trend surface analysis is a particular geo-referenced data situation in which this polynomial form often is used. The geographical problem is one of predicting y_i from a polynomial function of its Cartesian coordinates (u_i, v_i). Of course, other X variables also might be in the complete matrix **X**. To reduce spurious multicollinearity complications in this problem, the coordinate system

(U, V) should be centered at zero and perhaps rescaled to a smaller range. Such a modification does not affect the substantive part of an analysis because geo-referencing scales are at the interval and not the ratio level of measurement (Griffith and Amrhein, 1991, pp. 36–41).

EXAMPLE PR-4.9 (continued). Formulate a second-order polynomial regression trend surface model counterpart to Example PR-4.3.

Using the SAS procedures

```
PROC STANDARD MEAN=0 STDEV=1 OUT=TEMP4;
   VAR U V;
RUN;
PROC RSREG DATA=TEMP4;
   MODEL SQDSGR = U V;
RUN;
```

produced SAS output that includes

Parameter	Degrees of Freedom	Parameter Estimate	Standard Error	T for H0: Parameter=0	Prob > |T|
INTERCEPT	1	0.117588	0.247849	0.474	0.6367
U	1	-0.273610	0.126563	-2.162	0.0342
V	1	-0.195788	0.128065	-1.529	0.1310
U*U	1	0.826317	0.141741	5.830	0.0000
V*U	1	-0.314338	0.133131	-2.361	0.0211
V*V	1	1.002215	0.148713	6.739	0.0000

The t-statistics for this analysis imply that the geographic surface for SQDSGR could be described as bowl shaped, with the highest densities tending to occur along the coastal lowlands, and the lowest densities tending to occur in the central, interior highlands.

Therefore, in a fairly restricted sense, polynomial regression can help supply evidence that the assumption of linearity is being violated. It assumes that the exponents on X variables are known and are integers. A more detailed approach is to inspect pairwise plots of Y versus modified Xs. The straightforward task of constructing scatter diagrams of Y versus each X_j can be misleading, and in a number of situations they are almost useless. Modifications are necessary because of the presence of multicollinearity in matrix **X** and the simultaneous influence of combinations of Xs on Y. Not surprisingly, these modifications are reminiscent of partial correlation (see footnote 2 of this chapter). The relevant notion is called *added variable plots* in the exploratory data analysis literature and involves the following steps:

Step 1: regress **Y** on matrix **X***, which is the complete matrix **X** from which a single variable X_j has been removed, and obtain the vector of residuals $\mathbf{e}_{Y|X^*}$;

Step 2: regress $\mathbf{X}_j$ on matrix **X***, which does not contain $\mathbf{X}_j$, and obtain the vector of residuals $\mathbf{e}_{X_j|X^*}$; and,

Step 3: plot the residuals $\mathbf{e}_{Y|X^*}$ against $\mathbf{e}_{X_j|X^*}$, which is the added variable plot.

This plot of the two sets of residuals shows the relationship between variables Y and X_j, with the effects of the (P - 1) other Xs partialed out. It is this plot that is assumed to be linear. Haining (1990) outlines the use of this type of scatter diagram for selecting an appropriate spatial statistical model for geo-referenced data.

4.4. Analysis of Variance (*ANOVA*) as a Regression Problem

Example PR-4.1 suggests, with some subtlety, that most any statistical technique can be cast in terms of the regression model. This feature of regression highlights the general relevance of the four preceding *REG* questions to much of statistical analysis and implies the following question:

> *REG Question #5*: How can a given statistical technique be rewritten as a regression analysis?

By doing so, residuals can be obtained for diagnostic analysis. ANOVA will be used here to illustrate this conversion.

ANOVA is a statistical technique used to decide whether or not the means, calculated for some variable Y, of K different mutually exclusive and collectively exhaustive groups of observations are equal in the population (Griffith and Amrhein, 1991, pp. 305–10). In geography, the groups usually are regions, while the observations are areal units. With regard to measurement, Y is at the interval/ratio scale, and when the data are observational (the usual case in geography) the groupings are for classification purposes,[6] and hence a numerical label for each would be at the nominal scale (Griffith and Amrhein, 1991, pp. 36–37, 101–2). Thus, this classification can be converted from a qualitative to a quantitative gauge by using a set of K 0-1 binary variables, each indicating whether an areal unit belongs to region k; accordingly, these are called indicator variables, and may be denoted by I_k. The use of a single classification variable leads to one-way ANOVA. If more than one classification scheme is involved, then multiple sets of indicator variables are needed, one set for each grouping variable. This leads to N-way ANOVA. The crucial question then is whether the means of the cross-classification groups differ in the population.

ANOVA deviates from the preceding multiple regression approach in that it drops the third assumption of linearity listed in §4.2.3. Classification leading to indicator variables may be either qualitatively (there is no underlying natural scale of measurement) or quantitatively (a well-defined scale of measurement exists) based. The assumption of normality becomes more restrictive when the classification is quantitative. Rather than assuming only that ϵ conforms to a normal distribution in the population, ANOVA assumes that each of the K regional sets of ϵ also conforms to a normal distribution in its respective population.

---

[6]Classification occurs when the groupings are not under the control of the investigator, as is the case with census data. If observations are assigned at random to the groupings, then the investigation is experimental in nature, as in agricultural field plot studies. Experimental data provide firmer foundations for drawing conclusions than do observational data.

4.4.1. Basic Experimental Design

In conventional ANOVA terminology, the classification variables are referred to as factors. One-way ANOVA constitutes a single-factor study; N-way ANOVA constitutes a multifactor study. When the classification, or factor, categories are chosen because of intrinsic interest in them, such as regions, and they are not considered to be a random sample from a population of possible categories, then the ANOVA model involves a fixed effect. Conclusions will pertain to just those categories of the classification variable that have been included in a study. In contrast, if categories constitute a sample from some population of all possible categories, then the ANOVA model involves a random effect. Conclusions are sought for the population of categories from which the sample has been drawn. In one-way ANOVA, the nature of the classification variable has no effect on the numerical results. In N-way ANOVA, not only does a distinction between the nature of the classification variables render different numerical results, but also some classification variables can be fixed and others can be random, yielding mixed effects. Regardless of the nature of effects, a probability distribution for Y will correspond to each category, and for inferential purposes this probability distribution is assumed to be normal.

The K indicator variables constructed to designate category membership can be organized into the matrix **X** of multiple regression.[7] Griffith and Amrhein (1991, p. 371) review the relationship between regression results and an ANOVA table. Some modifications are necessary. Since $\sum_{j=1}^{K} I_k = 1$ jointly these K indicator variables will have perfect multicollinearity with X_0, the vector **1** in matrix **X** [as in expression (4.8)]. Any one of three strategies may be employed to circumvent this problem. First, vector **1** can be removed from matrix **X**; the no-intercept regression will not estimate an overall mean and will yield certain statistics, such as R^2, that will be quite difficult to interpret. Second, any one of the K indicator variables, *arbitrarily* selected, can be removed from matrix **X**; the intercept will estimate both an overall mean and the mean for the arbitrarily ignored region, the t-statistic for the regression coefficient b_k will be for a null hypothesis of H_0: $\beta_k = 0$, and tests for differences in regional means will not be straightforward. Third, any one of the K indicator variables, say k*, being *arbitrarily* selected, can be removed from matrix **X**, as well as subtracted from the remaining (K-1) indicator variables; the intercept will estimate the overall mean, and the t-statistic for the regression coefficient b_k will test the null hypothesis H_0: μ_k - μ_{k*}. Each of these specifications of the regression equation produces exactly the same results for the ANOVA null hypothesis; the choice of implementation often is a matter of preference and sometimes a matter of convenience.

EXAMPLE PR-4.10. Replicate Example PR-11.9 (and PR-13.3) of Griffith and Amrhein (1991, pp. 308–9, 371), concerning the difference between the five regional means of DCD in Puerto Rico.

[7]In fact, as Example PR-4.1 alludes to as well, most statistical techniques can be rewritten in terms of a regression equation.

From the previous book, the five regional variances of DCD were deemed equal (Griffith and Amrhein, 1991, pp. 292–93, Example PR-11.3), DCD was found to be normally distributed (Griffith and Amrhein, 1991, pp. 253–54, Example PR-10.2), and the one-way ANOVA yielded F = 6.902. Using the SAS procedure

```
PROC ANOVA;
   CLASS ADM;
   MODEL DCD = ADM;
   MEANS ADM/TUKEY LINES;
RUN;
```

produced the SAS output

```
                    Analysis of Variance Procedure
Dependent Variable: DCD
                              Sum of           Mean
Source              DF       Squares         Square  F Value     Pr > F
Model                4     2973.2607       743.3152     6.90     0.0001
Error               68     7323.4399       107.6976
```

Indicator variables can be created in the data statement with the SAS statements

```
IF ADM=1 THEN ISJ=1; ELSE ISJ=0;
IF ADM=2 THEN IA=1; ELSE IA=0;
IF ADM=3 THEN IM=1; ELSE IM=0;
IF ADM=4 THEN IP=1; ELSE IP=0;
IF ADM=5 THEN IC=1; ELSE IC=0;
```

Using the SAS procedure

```
PROC REG;
   MODEL DCD = IA IM IP IC;
RUN;
```

produced this SAS output

```
                         Analysis of Variance
                       Sum of          Mean
Source        DF      Squares        Square        F Value         Prob>F
Model          4   2973.26066     743.31517          6.902         0.0001
Error         68   7323.43988     107.69765
```

Here the indicator variable for the San Juan region (ISJ) has been arbitrarily selected for deletion. Next, differenced indicator variables are created in the data statement, arbitrarily selecting the indicator variable for the Caguas region (IC) for deletion and to subtract, with the SAS statements

```
DISJ = ISJ - IC;
DIA = IA - IC;
DIM = IM - IC;
DIP = IP - IC;
```

Using the SAS procedure

```
PROC REG;
   MODEL DCD = DISJ DIA DIM DIP;
RUN;
```

exactly reproduced the preceding SAS output.

Two steps are involved in one-way ANOVA. In the first step, the following null hypothesis about equality of means is tested, as is done in Example PR-4.10:

$$H_0\colon\ \mu_1 = \mu_2 = \ldots = \mu_K\ .$$

If this null hypothesis is rejected, as in Example PR-4.10, the second step is to ascertain how the means differ and what implications stem from the detected differences. In other words, the alternate hypothesis of

$$H_a\colon\ \mu_h \neq \mu_k\ ,\ h \neq k$$

states that at least one mean differs from the remaining (K-1) means. But all the means might differ, or subsets of equal means could differ. Moreover, there are many specific alternate hypotheses, the number involving m subsets being given by

$$\frac{1}{m!}\sum_{j=1}^{m} (-1)^{m-j} C\,(m,j) \times j^{K}\ .^{8} \qquad (4.15)$$

Thus, for the (K =) five agricultural administrative regions of Puerto Rico, there are 51 possible alternative hypotheses (see Griffith and Amrhein, 1991, pp. 153–55, for the background combinatorial framework underlying this type of counting). The single null hypothesis H_0 is for m = 1. When m = 2, expression (4.15) yields 15. There are 5 possibilities of 4 regional means being equal, while only 1 differs [i.e., C(5,4)×C(1,1)], such as

---

[8]This expression counts the number of ways K objects can be sorted into m nonempty groups and will be referenced again in Chapter 8, when cluster analysis is discussed.

$$H_a: \mu_1 = \mu_2 = \mu_3 = \mu_4 \neq \mu_5 \ ;$$

and there are 10 possibilities of 3 regional means being equal, while differing from the remaining 2 regional means, which are equal [i.e., C(5,3)×C(2,2)], such as

$$H_a: \mu_1 = \mu_2 = \mu_3 \neq \mu_4 = \mu_5 \ .$$

When m = 3, expression (4.15) yields 25. There are 10 possibilities of 3 regional means being equal, while differing from the remaining 2 means, each of which also differs [i.e., $\frac{C(5,3)\times C(2,1)\times C(1,1)}{2!}$, where the denominator indicates that the order of the 2 distinct means in the alternative hypothesis is unimportant], such as

$$H_a: \mu_1 = \mu_2 = \mu_3 \neq \mu_4 \neq \mu_5 \ ;$$

and, there are 15 possibilities of 2 sets of 2 equal regional means, these sets differing from each other as well as the remaining single regional mean [i.e., $\frac{C(5,2)\times C(3,2)\times C(1,1)}{2!}$], such as

$$H_a: \mu_1 = \mu_2 \neq \mu_3 = \mu_4 \neq \mu_5 \ ;$$

When m = 4, expression (4.15) yields 10. There are 10 possibilities of 2 regional means being equal, while differing from the remaining 3 distinct means [i.e., $\frac{C(5,2)\times C(3,1)\times C(2,1)\times C(1,1)}{3!}$], such as

$$H_a: \mu_1 = \mu_2 \neq \mu_3 \neq \mu_4 \neq \mu_5 \ .$$

Finally, When m = 5, expression (4.15) yields 1 [i.e., C(5,5)]; all of the means differ, rendering

$$H_a: \mu_1 \neq \mu_2 \neq \mu_3 \neq \mu_4 \neq \mu_5 \ .$$

The second step in one-way ANOVA is concerned with reducing this set of possibilities to no more than a few and, it is hoped, only one. The statistical approach involved in this analysis requires simultaneous rather than sequential or independent treatment of individual regional means.

Discriminating between these various alternate hypothesis possibilities involves conducting C(K,2) difference of means tests latent in a one-way ANOVA. It is difficult, however, to establish equivalent Type I error probability levels across all of these pairwise tests so that results are comparable. After all, even when H_0 is not rejected, some of these multiple pairwise comparisons might differ simply by chance. Thus, given K regions, the problem becomes one of determining the probability of a particular alternative hypothesis being appropriate. If the inferences were independent (but they are not, since they are based upon the same sample of data, instead of separate pairs of not-yet-used samples for each test made), this probability would be given by the binomial theorem (Griffith and Amrhein, 1991, p.171). If the confidence interval in the one-way ANOVA is $(1 - \alpha)$, where α is the probability of a Type I error in the ANOVA, then the problem is to find

α^*, the pairwise comparison probability of a Type I error. This quantity may be obtained as follows:

$$[(1-\alpha^*)+\alpha^*]^K = (1-\alpha^*)^K + \sum_{k=1}^{K} C(K,k)\times(\alpha^*)^k\times(1-\alpha^*)^{K-k} , \tag{4.16}$$

such that $(1 - \alpha^*)^K = (1 - \alpha)$. Work pioneered by Bonferroni, Duncan, Fisher, Scheffé, and Tukey, among others (Neter, Wasserman and Kutner, 1990), shows that independent, separate confidence intervals yield $\alpha^* \approx \frac{\alpha}{K}$ ($\frac{\alpha}{K}$ is referred to as the *Bonferroni* adjustment throughout this book). Thus, when $\alpha = 0.05$ for a one-way ANOVA problem involving the Puerto Rico agricultural administrative regions, $\alpha^* \approx \frac{0.05}{5}$ or 0.01; $(1 - 0.01)^5 = 0.95110$. When $\alpha = 0.01$ for such Puerto Rico problems, $\alpha^* \approx \frac{0.01}{5}$ or 0.002; $(1 - 0.002)^5 = 0.99004$. Of course, when two-tailed tests are involved, $\alpha^* \approx \frac{\alpha}{2K}$. These values of α^* constitute a lower bound.

The other extreme would occur when the separate pairwise comparisons are perfectly correlated, in which case once an arbitrarily selected pairwise test is made, all remaining pairwise tests would render exactly the same null hypothesis decision. In this extreme case, if $\alpha^* = \frac{\alpha}{K}$ for the first pairwise test conducted, the conditional probabilities (see Griffith and Amrhein, 1991, pp. 158–61) for subsequent tests would become 1 rather than $(1-\alpha^*)$ or α^*.

The first term in the right-hand side of equation (4.16) gives the probability of all K confidence intervals, one for each of the individual μ_Ks, being correct, and relates to H_0. The remaining terms in this right-hand side relate to the Type I errors that could suggest any one of the 51 possible alternate hypotheses in the Puerto Rico study, and provide the probabilities for (K-1), or (K-2), or (K-3), ... , or 1, or 0 of the confidence intervals being correct. Thus, for the Puerto Rico example, the probability of exactly one incorrect confidence interval when the null hypothesis is true, of which there are five possibilities, is

$$C(5,1)\left(\frac{\alpha}{5}\right)^1\left(1-\frac{\alpha}{5}\right)^4.$$

Meanwhile, the probability of exactly two incorrect confidence intervals in this example, which may or may not overlap (hence there are 20 possibilities), is $C(5,2)\left(\frac{\alpha}{5}\right)^2\left(1-\frac{\alpha}{5}\right)^3$.

Although a number of multiple comparisons tests are available, some of these methods do not lend themselves to *data snooping* (i.e., studying effects suggested by the data). This type of exploration can be a valuable exercise when H_0 is rejected, because the natural next step, and the second one in one-way ANOVA, is to examine the nature of detected differences. Multiple comparison tests seek to reveal the most likely H_a, so that the means can be ordered, if possible, only when the pairwise confidence intervals do not overlap. Statistics developed to achieve this goal include Tukey's multiple range test, defined as

$$(\bar{y}_1 - \bar{y}_2) \pm \left[\frac{q(1-\alpha;K,n-K)}{\sqrt{2}}\right] s_{\bar{y}_1-\bar{y}_2},$$

where q(1-α; K, n-K) follows a studentized range distribution.[9] Tukey's test statistic has been selected here because it has the advantage of being applicable to situations with unequal sample sizes, a common characteristic of geographic landscapes and certainly of the Puerto Rican study.

EXAMPLE PR-4.10 (continued) Identify the most likely alternate hypotheses for Example PR-11.9 of Griffith and Amrhein (1991, pp. 308–9).

The SAS statement

```
MEANS ADM/TUKEY LINES;
```

appearing in the ANOVA code produces the following output:

Tukey Grouping		Mean	N	ADM
	A	44.286	14	4
	A			
	A	39.636	11	2
	A			
	A	38.825	16	3
	A			
B	A	35.477	13	5
B				
B		26.437	19	1

These multiple comparison results suggest that the best alternate hypothesis for the one-way ANOVA for DCD is either H_a: $\mu_1 \neq \mu_2 = \mu_3 = \mu_4 = \mu_5$ or H_a: $\mu_1 = \mu_5 \neq \mu_2 = \mu_3 = \mu_4$. According to the sample data, the other 49 possible alternate hypotheses are not probable, under H_0.

Applying the multiple comparisons method to the test for normality, which always is one-tailed for S-W, means $\alpha^* = \frac{\alpha}{5}$; thus, for $\alpha = 0.05$ the level of significance becomes 0.01. Individual, regional S-W statistics may be obtained by including a BY ADM statement with the PROC UNIVARIATE code, which produced this SAS output:

[9] Following the construction of the Student's t distribution, where $t = \frac{(\bar{x} - \mu)}{\frac{s}{\sqrt{n}}}$, the studentized range distribution is constructed as $\frac{[MAX(X) - MIN(X)]}{s}$.

```
------------------------------ ADM=1 ------------------------------
              W:Normal    0.962501   Prob<W        0.6189
------------------------------ ADM=2 ------------------------------
              W:Normal    0.944624   Prob<W        0.5541
------------------------------ ADM=3 ------------------------------
              W:Normal    0.967994   Prob<W        0.7747
------------------------------ ADM=4 ------------------------------
              W:Normal    0.969169   Prob<W        0.8223
----------------------------- ADM=5 ------------------------------
              W:Normal    0.856785   Prob<W        0.0336
```

By applying a multiple comparison test criterion, the S-W statistic for the Caguas region (ADM=5) switches from being noticeably significant at the 0.05 level to noticeably nonsignificant at the 0.01 level. Given results for n = 20 reported in Dutilleul and Legendre (1992, p. 85), coupled with the modest but significant autocorrelation that has been detected here, these normality test results should be reasonably sound.

The ANOVA model presents serious restrictions to spatial analysts. Two prominent restrictions are overcome by replacing a conventional ANOVA technique with the regression equation form of ANOVA. First, a researcher no longer is restricted to those special cases covered by conventional ANOVA, such as equal numbers of observations in all cross-classifications for N-way ANOVA; this feature will be exploited in Chapter 5. In addition, and of considerable importance for geo-referenced data, the residuals can be tested for the presence of spatial autocorrelation. For DCD regression residuals in Example PR-4.10, MC = 0.155301 (GR = 0.881215); the approximate test statistic is $z = \frac{(0.155301 + 0.051932)}{0.076249} = 2.718$. Thus, since 2.718 > 1.96, there is evidence to indicate that the one-way ANOVA regression residuals for DCD are spatially autocorrelated in the population.

4.4.2. N-way Balanced Designs

For decades theoretical statistics and the state of computer technology constrained multifactor ANOVA studies to very special cases, the most common of which required that each cross-classification group had the same number of observations, which will be denoted here by $n_{\bullet}$ using the subscript dot ($_{\bullet}$) to indicate that this is a constant cross-classification sample size. Such setups are called ***balanced designs***, exhibit several appealing properties (especially the component sums-of-squares adding up to the total sum-of-squares), and govern virtually all of the standard ANOVA software that is available. Because geographical studies rarely have the luxury of dictating the number of areal units into which a landscape is partitioned, this binding constraint either discouraged geographers from employing ***N-way ANOVA***—the usual reaction—or caused somewhat artificial analyses to be undertaken in which a subset of areal units was selected to satisfy this constraint. The trade-off was between feasibility of implementation (a balanced design), on the one hand, and the increased efficiency that multifactor studies afforded (as n increased) over an approach that manipulated only one classification effect at a time, keeping all other conditions constant, on the other hand. The chief benefit of a multifactor ANOVA is that it requires a smaller sample to provide knowledge about effects of the different classifications involved; it also enables evaluation of interactions between classification schemes.

Interaction effects may be defined as differential differences between the means of groupings based on one classification variable across the groupings defined by another classification variable; in other words, the classification variables are not independent. With regard to indicator variables, interaction effects are represented by multiplying the indicator variables constituting one set of groupings by those indicator variables constituting another set. Because of the presence of these interaction effects, the null hypothesis must be stated, for a two-way ANOVA involving two classification variables respectively defining K_1 and K_2 groups, as

$$H_0: \mu_{11} = \mu_{12} = K = \mu_{1K_1} = \mu_{21} = \mu_{22} = K = \mu_{2K_1} = K = \mu_{K_21} = \mu_{K_22} = K = \mu_{K_2K_1} .$$

The critical question is whether *all* interactions are zero in theory, or negligible and unimportant in practice. If interaction effects are inconsequential, then the problem is "simple" in that each classification scheme can be studied separately, and the multiple-classification effects become additive. Accordingly, the N-way composite null hypothesis decomposes into a set of N individual null hypotheses, one for each classification variable. For a two-way ANOVA, the pair of hypotheses in this case is

$$H_0: \mu_{\bullet 1} = \mu_{\bullet 2} = \ldots = \mu_{\bullet K_1} \text{ , and}$$
$$H_0: \mu_{1\bullet} = \mu_{2\bullet} = \ldots = \mu_{K_2\bullet} \text{ ,}$$

where the $\bullet$ indicates that groupings of the opposite classification are being pooled. Because of this N-way ANOVA feature, *interaction effects always should be the first ones tested.* The test statistic used is

$$F = \frac{\text{interaction mean squared error}}{\text{residual mean squared error}} = \frac{MSAB}{MSE} ,$$

which is distributed with $(K_1-1)(K_2-1)$ degrees of freedom in the numerator and $K_1K_2(n_\bullet-1)$ degrees of freedom in the denominator for a balanced design involving $n_\bullet$ cross-classification observations. Note that MSAB is the mean squared error of interaction effects, and MSE is the mean squared error of the residuals. If the null hypothesis of no interaction is rejected, then only the composite null hypothesis can be tested. But if a researcher fails to reject the null hypothesis of no interaction, then the composite null hypothesis can be decomposed into the N individual classification null hypotheses, and each can be tested separately.

After decomposition, proper testing of the individual classification null hypotheses requires that each variable be declared as either a fixed effects or a random effects variable. The test statistic used for a fixed effects variable, which is the most prevalent in geography, defining K_1 groupings, is

$$F = \frac{\text{classification mean squared error}}{\text{residual mean squared error}} = \frac{MSA}{MSE}$$

and is distributed with (K_1-1) degrees of freedom in the numerator and $K_1K_2(n_\bullet-1)$ degrees of freedom in the denominator for a balanced design. Note that MSA is the mean squared error of classification variable A. In contrast, the test statistic used for a random effects variable, defining K_1 groupings, is

$$F = \frac{\text{classification mean squared error}}{\text{interaction mean squared error}} = \frac{MSA}{MSAB}$$

and is distributed with (K_1-1) degrees of freedom in the numerator and (K_1-1)(K_2-1) degrees of freedom in the denominator.

EXAMPLE PR-4.11. Execute a two-way ANOVA of DCD using the agricultural administrative regions (ADM) and the coastal lowlands/interior highlands (LND) classification variables, for a stratified random sample, without replacement,[10] where $n_\bullet = 2$ (Appendix C enumerates the sample areal units).

A 0-1 indicator variable was created for the stratified random sample (1 denotes membership in the sample) that was drawn, and then multiplied by the municipio number variable to identify the sample areal units. This second step was achieved using the SAS code

```
RSWOR = RSWOR*MUNNUM3;
```

to extract the sample from the data set, RSWOR is used in a data statement as follows:

```
DATA STEP2;
   SET STEP1;
      IF RSWOR > 0;
```

Using the SAS procedure

```
PROC ANOVA DATA=STEP2;
   CLASS ADM LND;
```

[10]The total number of possible samples in this case is given by the product of the number of possible samples of size 2 that can be drawn, without replacement, from each cross-classification set of areal units:

$$C(12,2)\times C(7,2)\times C(7,2)\times C(4,2)\times C(9,2)\times C(7,2)\times C(10,2)\times C(4,2)\times C(4,2)\times C(9,2) \approx 7.699680\times 10^{12}.$$

```
    MODEL DCD = ADM LND ADM*LND;
RUN;
```

produced the SAS output

```
              Analysis of Variance Procedure
Dependent Variable: DCD
                              Sum of       Mean
Source              DF     Squares      Square F Value   Pr > F
Model                9   1345.6180   149.5131    1.92    0.1617
Error               10    778.1500    77.8150
Corrected Total     19   2123.7680

Source    DF    Anova SS    Mean Square   F Value   Pr > F
ADM        4   492.75300    123.18825       1.58    0.2529
LND        1   328.05000    328.05000       4.22    0.0671
ADM*LND    4   524.81500    131.20375       1.69    0.2290
```

The default F-ratios are for a fixed effects classification. In fact, the agricultural administrative regions and the terrain dichotomy for Puerto Rico are this type of classification variable. Accordingly, one would fail to reject the null hypothesis of no interaction effects, based upon the calculated F-ratio of 1.69, allowing each classification effect to be evaluated separately. If $\alpha = 0.05$, then one would fail to reject the null hypothesis for each of the individual classification variables, too, based respectively upon the calculated F-ratios of 1.58 and 4.22. These individual outcomes are in keeping with the composite null hypothesis result, which is reported in the first part of the table and which has a nonsignificant F-ratio of 1.92.

The sums-of-squares additivity property can be demonstrated here by noting that

$$492.75300 + 328.05000 + 524.81500 = 1345.6180 \ .$$

This additivity is further shown by explicitly treating this two-way ANOVA within the context of multiple regression. This additivity property is lost when the design becomes unbalanced (Chapter 5).

EXAMPLE PR-4.11 (continued).

DATA STEP2 also includes the following indicator variable calculations, which utilize the choice of subtracting an arbitrarily selected indicator variable from the remaining ones established for a given classification variable (consult p. 114):

```
ILND = 2*LND - 1;
DISJILND = DISJ*ILND;
DIAILND = DIA*ILND;
DIMILND = DIM*ILND;
DIPILND = DIP*ILND;
```

Using the SAS procedure

```
PROC REG DATA=STEP2;
   MODEL DCD = DISJ DIA DIM DIP ILND DISJILND DIAILND
                  DIMILND DIPILND/SS1;
RUN:
```

produced SAS output that includes

Variable	DF	Type I SS
INTERCEP	1	27736
DISJ	1	73.205000
DIA	1	68.681667
DIM	1	28.060208
DIP	1	322.806125
ILND	1	328.050000
DISJILND	1	208.080000
DIAILND	1	1.306667
DIMILND	1	9.275208
DIPILND	1	306.153125

The option SS1 instructs SAS to print these sequential sums-of-squares. Because the design is balanced, when compared with the preceding tabulated two-way ANOVA results, the additivity property again can be shown:

$$73.205000 + 68.681667 + 28.060208 + 322.806125 = 492.75300$$
$$328.050000 = 328.05000$$
$$208.080000 + 1.306667 + 9.275208 + 306.153125 = 524.81500.$$

Moreover, the total of the sequential sums-of-squares for the four differenced ADM indicator variable regression terms equals the ADM sum-of-squares, and the total of the sequential sums-of-squares for the four differenced indicator variable product regression terms equals the interaction sum-of-squares, ADM*LND. Because of the balance in the matrix **X** when all cross-classification sample sizes are equal, these totals of the sums-of-squares are the same regardless of the order of the variables in the regression equation.

EXAMPLE PR-4.12. Execute a three-way ANOVA of DCD using aggregated agricultural administrative regions (ADM2[11]), the coastal lowlands/interior highlands (LND), and the rural-urban (RU) classification variables, for a stratified random

[11]The San Juan and Arecibo agricultural administrative regions were combined, as were the Ponce and Caguas regions, yielding a northern region, a western region (Mayaguez), and a southern region.

sample, with replacement,[12] where $n_\bullet = 2$ (Appendix C enumerates the sample areal units).

As in Example PR-4.11, an indicator variable (RSWR) was created for the stratified random sample (1 denotes membership in the sample) that was drawn and then multiplied by the municipio number variable to identify the sample areal units. A frequency variable (RSWRFREQ) also was constructed, since sampling with replacement allows individual areal units to appear more than once in a sample. This second step was achieved using the SAS code

```
RSWRFREQ = RSWR*MUNNUM3;
```

which then was used in the following data statement to extract the sample from the data set:

```
DATA STEP3;
   SET STEP1;
      IF RSWR > 0;
IF ADM=1 OR ADM=2 THEN ADM2=1; ELSE ADM2=3;
IF ADM=3 THEN ADM2=2;
```

Using the SAS procedure

```
PROC ANOVA DATA=STEP3;
   CLASS ADM2 LND RU;
   MODEL DCD = ADM2 LND RU ADM2*LND
               ADM2*RU LND*RU ADM2*LND*RU;
   FREQ RSWRFREQ;
RUN;
```

produced SAS output that includes

[12]The total number of possible samples in this case is given by the product of the number of possible samples of size 2 that can be drawn, with replacement, from each cross-classification set of areal units (that is, ADM2×LND×RU = 3×2×2 = 12), or $15^2 \times 2^2 \times 4^2 \times 9^2 \times 3^2 \times 1^2 \times 6^2 \times 6^2 \times 3^2 \times 5^2 \times 11^2 \times 8^2 = 2.370508 \times 10^{16}$.

```
                  Analysis of Variance Procedure
Dependent Variable: DCD
Frequency:          RSWRFREQ
                         Sum of        Mean
Source            DF    Squares       Square    F Value   Pr > F
Model             11  1660.1746     150.9250      3.84    0.0146
Error             12   472.1050      39.3421
Corrected Total   23  2132.2796

Source            DF     Anova SS   Mean Square  F Value     Pr > F
ADM2               2    601.05333    300.52667      7.64     0.0072
LND                1      2.10042      2.10042      0.05     0.8212
RU                 1    486.90042    486.90042     12.38     0.0042
ADM2*LND           2     20.58333     10.29167      0.26     0.7741
ADM2*RU            2     18.50333      9.25167      0.24     0.7940
LND*RU             1    152.51042    152.51042      3.88     0.0725
ADM2*LND*RU        2    378.52333    189.26167      4.81     0.0292
```

The null hypothesis that *must* be tested first in Example PR-4.12 is for the triple interaction term, ADM2*LND*RU. If $\alpha = 0.01$, then there is no evidence [Pr(>F) = 0.0292 > 0.0100] to indicate that the three classification variables do not interact jointly in the population (hence, they are considered independent). The next set of null hypotheses that *must* be tested in this example are for the three pairwise interaction terms. Again, using $\alpha = 0.01$, one would fail to reject each of these null hypotheses. Now that the interactions have been dispensed with, the following three null hypotheses obtained through decomposition, for the classification variables of interest, can be evaluated:

$$H_0\colon\ \mu_{SJ/A\bullet\bullet} = \mu_{M\bullet\bullet} = \mu_{P/C\bullet\bullet}\ ,$$
$$H_0\colon\ \mu_{\bullet CL\bullet} = \mu_{\bullet IH\bullet}\ ,$$
$$H_0\colon\ \mu_{\bullet\bullet R} = \mu_{\bullet\bullet U}\ .$$

Using $\alpha = 0.01$ once more, (1) there is evidence to indicate that the regional means of DCD are not equal across the three aggregate agricultural administrative regions [Pr(>F) = 0.0072 < 0.0100], (2) there is no evidence to indicate that the regional means of DCD are not equal between the coastal lowlands and interior highlands regions [Pr(>F) = 0.8212 > 0.0100], and (3) there is evidence to indicate that the regional means of DCD are not equal between urban and rural municipios [Pr(>F) = 0.0042 < 0.0100], in the population.

These last results assume that the rural-urban dichotomy variable is a fixed effect. But, municipios can change their status in the classification scheme. In addition, a percentage other than 50% could be used for classification purposes. Both facts argue for RU to be treated as a random effect. Invoking this assumption retains the F-ratio for the triple interaction effect, ADM2*LND*RU, while causing the following F-ratio changes:

Source	F-ratio	Pr > F
ADM2	32.484	0.0299
LND	0.014	0.9256
RU	negative	
ADM2*LND	0.054	0.9484
ADM2*RU	0.049	0.9534
LND*RU	0.806	0.4641

Each of the pairwise interaction F-ratios is

$$\frac{\text{pairwise interaction mean squared error}}{\text{triple interaction mean squared error}}.$$

For example, the F-ratio for ADM2*LND is $0.054378 = \frac{10.29167}{189.26167}$. The individual classification variable mean squared errors have rather complicated denominators, which, as RU indicates, are not guaranteed to be nonnegative.

Returning to the results from the ANOVA procedure, if $\alpha = 0.05$ rather than 0.01, then there is evidence to indicate that the three classification variables interact jointly in the population [Pr(>F) = 0.0292 < 0.0500]. Consequently, the null hypothesis can not be decomposed.

EXAMPLE PR-4.12 (continued) Execute a three-way ANOVA of CDC as a one-way ANOVA.

An indicator variable was created for the cross-classification groups using the following SAS code:

```
IF ADM2=1 AND LND=0 AND RU=0 THEN GRP=1; ELSE GRP=12;
IF ADM2=1 AND LND=0 AND RU=1 THEN GRP=2;
IF ADM2=1 AND LND=1 AND RU=0 THEN GRP=3;
IF ADM2=1 AND LND=1 AND RU=1 THEN GRP=4;
IF ADM2=2 AND LND=0 AND RU=0 THEN GRP=5;
IF ADM2=2 AND LND=0 AND RU=1 THEN GRP=6;
IF ADM2=2 AND LND=1 AND RU=0 THEN GRP=7;
IF ADM2=2 AND LND=1 AND RU=1 THEN GRP=8;
IF ADM2=3 AND LND=0 AND RU=0 THEN GRP=9;
IF ADM2=3 AND LND=0 AND RU=1 THEN GRP=10;
IF ADM2=3 AND LND=1 AND RU=0 THEN GRP=11;
IF ADM2=3 AND LND=1 AND RU=1 THEN GRP=12;
```

Using the SAS procedure

```
PROC ANOVA DATA=STEP3;
```

```
    CLASS GRP;
    MODEL DCD = GRP;
    FREQ RSWRFREQ;
RUN;
```

produced SAS output that includes

```
                    Analysis of Variance Procedure
Dependent Variable: DCD
Frequency:          RSWRFREQ
                       Sum of       Mean
Source            DF   Squares      Square     F Value   Pr > F
Model             11   1660.1746   150.9250    3.84      0.0146
Error             12    472.1050    39.3421
Corrected Total   23   2132.2796
```

These are exactly the results reported earlier for the composite null hypothesis. Being cast as a one-way ANOVA enables multiple comparison tests to be performed and regression residuals to be evaluated.

Unfortunately, because of the artificiality of the sampling procedure employed in these N-way ANOVA examples, latent spatial autocorrelation in geo-referenced data cannot easily be diagnosed. Chapter 5 discusses N-way ANOVA with unbalanced designs, which circumvents such flaws.

4.5. Analysis of Covariance ($P \geq 2$, $K > 3$)

Analysis of covariance (ANCOVA) is a technique that conceptually combines features of ANOVA as a regression problem and multiple regression (see Griffith and Amrhein, 1991, pp. 392–402). The goal may be viewed as augmenting an ANOVA with interval/ratio variables to reduce the error variance, making an analysis more precise, and hence making the comparison of classification means more powerful. As in the ANOVA procedure, ANCOVA can be written as a regression equation.

By combining quantitative and indicator variables in a regression equation, a researcher can ask either or both of the following questions:

(1) Is the conditional mean (regression intercept) the same in all regions or cross-classifications?

(2) Are the slopes attached to the quantitative variables the same in all regions or cross-classifications?

Traditionally these questions are answered after two data adjustments have been made. First, each variable in matrix **X** is centered (has its mean subtracted from it). This alteration shifts the effect of each mean to the intercept term, β_0. Second, the differenced form of indicator variables is calculated. This modification allows a direct test of the difference of regional means. These two conventions help to simplify interpretation of the regression analysis results and directly supply the sums-of-squares that go into a conventional ANCOVA table.

EXAMPLE PR-4.13. Execute an ANCOVA of SQDSGR (Example PR-4.9 revisited).

The binary 0-1 indicator variable, I0, was converted to its differenced version with the SAS code

```
DI0 = 2*I0 - 1;
```

The relevant variables were centered using the SAS procedure

```
PROC STANDARD MEAN=0 OUT=TEMP5;
     VAR DFRM DISTSJ DISTA;
```

Using the SAS procedure

```
PROC GLM DATA=TEMP5;
  MODEL SQDSGR = DFRM DISTSJ DISTA DI0 DI0*DFRM DI0*DISTSJ;
RUN;
```

produced SAS output that includes

Parameter	Estimate	T for H0: Parameter=0	Pr > |T|	Std Error of Estimate
INTERCEPT	1.172490486	5.27	0.0001	0.22263779
DI0	1.078845324	4.80	0.0001	0.22476654
DFRM	-0.042248278	-1.31	0.1945	0.03223567
DISTSJ	0.228518287	2.16	0.0345	0.10582214
DISTA	0.154069730	1.65	0.1042	0.09351509
DI0*DFRM	-0.049438926	-1.54	0.1277	0.03204718
DI0*DISTSJ	0.172174988	1.70	0.0942	0.10138002

The variables are selected for this analysis because of their prominence in the preceding quadratic nonlinear analysis. As expected, shifting most of the mean effect to b_0 causes it to become markedly significant. There is a significant difference between the cluster of municipios with nonzero density of sugar cane production and those with zero density, which certainly is not surprising. The cross-product terms, DI0*DFRM and DI0*DISTSJ, imply that the regression slopes in these two classes of municipios are the same. Therefore, little has changed from the analysis reported in Example PR-4.4 that does not seem to be attributable to multicollinearity (e.g., DFRM going from a highly significant to a clearly nonsignificant variable).

4.6. Effects of Spatial Autocorrelation

Conventional linear regression assumes that any spatial autocorrelation present in geo-referenced data can be accounted for by variables included in matrix **X**. But this assumption fails to hold for most geographic data, resulting in the presence of spatial autocorrelation in the residuals produced by a multiple linear regression analysis.

Significant spatial autocorrelation has been detected repeatedly in the illustrative analyses reported in this chapter. Impacts on tests of normality already have been noted. The general seriousness of this complication is portrayed graphically (see Dutilleul and Legendre, 1992, p. 89) in Figure 4.3. Using selected theoretical levels of spatial auto-correlation associated with a 10-by-10 regular square lattice geographic landscape and standardizing the three popular normality tests reveal the following: (1) as spatial autocorrelation increases, especially in the positive direction, all three of the test statistics become seriously compromised; and, (2) the Kolmogorov-Smirnov test statistic apparently suffers the least from the presence of moderate levels of spatial autocorrelation in geo-referenced data. Going beyond diagnostics, a deeper understanding of what this detection means can be attained by returning to the statistical distribution properties of **b** (see §4.2.1).

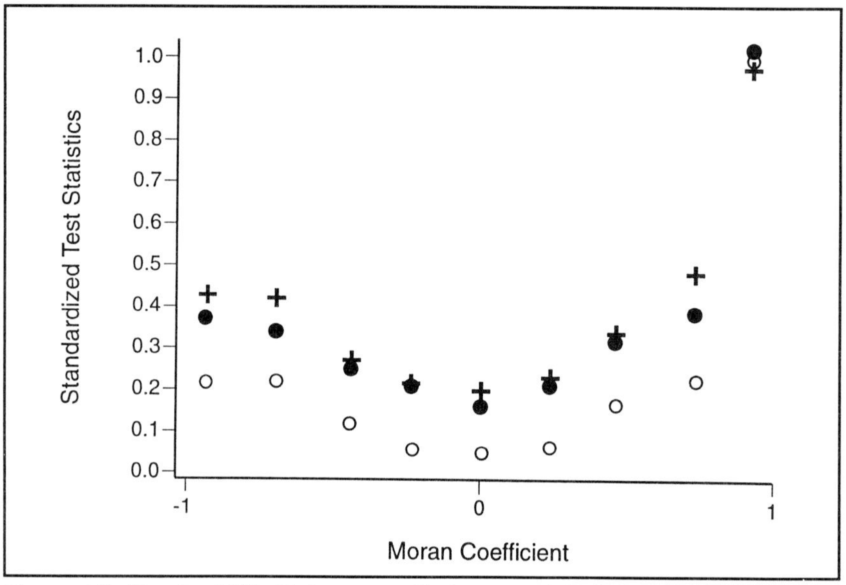

Figure 4.3. Shapiro-Wilk, Kolmogorov-Smirnov, and Anderson-Darling tests of normality (from top to bottom).

When regression residuals display spatial autocorrelation, the standard practice in spatial statistics is to modify the regression specification to that of a simultaneous autoregressive equation. Salient features as well as estimation procedures for this equation are presented in Chapter 12. The useful characteristic to recognize here is that the error term of a regression equation is viewed as being of the form

$$\epsilon = (\mathbf{I} - \rho\mathbf{W})^{-1}\xi \ , \qquad (4.17)$$

where ρ is the spatial autocorrelation parameter, **W** is the geographic weights matrix (Chapter 3), and ξ is an error vector that is more than just blindly assumed to be distributed as $N(\mathbf{0}, \sigma^2\mathbf{I})$. Accordingly, the regression equation employed in geo-referenced data analyses, which frequently is diagnosed as having spatially autocorrelated residuals, is of the form

$$\mathbf{Y} = \mathbf{X}\boldsymbol{\beta} + (\mathbf{I} - \rho\mathbf{W})^{-1}\xi \ . \qquad (4.18)$$

Multiplying both sides of equation (4.18) by $(\mathbf{I} - \rho\mathbf{W})$ yields

$$(\mathbf{I} - \rho\mathbf{W})\mathbf{Y} = (\mathbf{I} - \rho\mathbf{W})\mathbf{X}\boldsymbol{\beta} + \xi \text{ , or}$$

$$\mathbf{Y} = \mathbf{X}\boldsymbol{\beta} + [(\rho\mathbf{WY} - \rho\mathbf{WX}\boldsymbol{\beta}) + \xi].$$

The expression $(\rho\mathbf{WY} - \rho\mathbf{WX}\boldsymbol{\beta})$ involving spatially lagged variables that appears as part of the residuals in OLS is the source of the spatial autocorrelation detected with an MC.

Returning to §4.2.1, the expected value of **b** for equation (4.18) is

$$E\{(\mathbf{X}^T\mathbf{X})^{-1}\mathbf{X}^T[\mathbf{X}\boldsymbol{\beta} + (\mathbf{I} - \rho\mathbf{W})^{-1}\xi]\} = \boldsymbol{\beta} \ .$$

Therefore, the presence of spatial autocorrelation in this simultaneous autoregressive specification form *does not introduce bias into the regression parameter estimates.* Meanwhile, the variance of the parameter estimates comes from

$$E[\{(\mathbf{X}^T\mathbf{X})^{-1}\mathbf{X}^T[\mathbf{X}\beta + (\mathbf{I} - \rho\mathbf{W})^{-1}\xi] - \beta\}\{(\mathbf{X}^T\mathbf{X})^{-1}\mathbf{X}^T[\mathbf{X}\beta + (\mathbf{I} - \rho\mathbf{W})^{-1}\xi] - \beta\}^T]$$
$$= \sigma^2(\mathbf{X}^T\mathbf{X})^{-1}\mathbf{X}^T(\mathbf{I} - \rho\mathbf{W})^{-1}(\mathbf{I} - \rho\mathbf{W}^T)^{-1}\mathbf{X}(\mathbf{X}^T\mathbf{X})^{-1} \ .$$

This expression reduces to $\sigma^2(\mathbf{X}^T\mathbf{X})^{-1}$ only when $\rho = 0$, as reported in equation (4.6). As ρ increases toward 1, the diagonal elements of $(\mathbf{I} - \rho\mathbf{W})^{-1}$ increase from unity, causing the standard errors of the individual b_js of **b** to increase. Therefore, the presence of spatial autocorrelation in this simultaneous autoregressive specification causes OLS standard error estimates to be less efficient (Griffith and Amrhein, 1991, p. 49). This bias, in turn, will tend to cause Type I error probabilities to be wrong; latent positive spatial autocorrelation will modify standard errors so that they tend to inflate the significance of the regression while making certain covariates appear to be more highly significant than they actually are, whereas latent negative spatial autocorrelation will tend to suppress the regression results and suggest that certain covariates are less significant than they actually are. In other words, the presence of non-

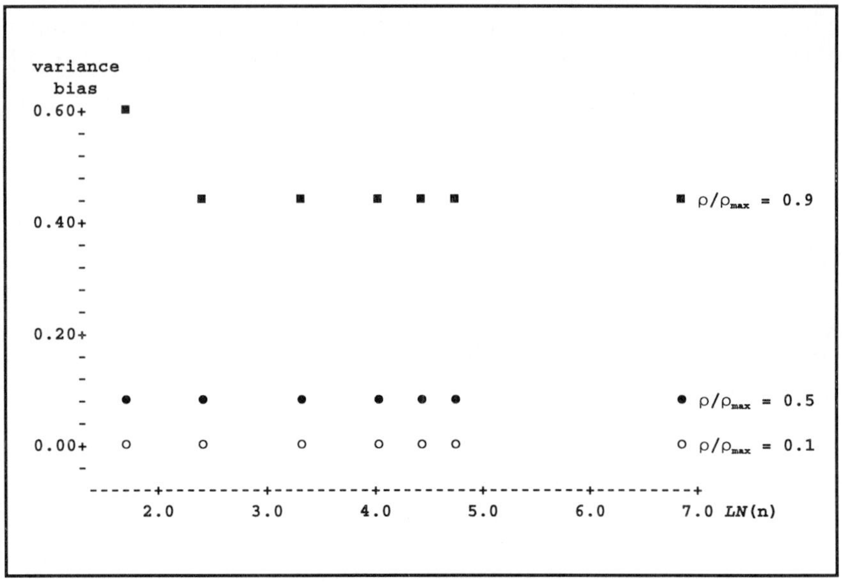

Figure 4.4. Efficiency of OLS estimates for geo-referenced data.

zero spatial autocorrelation tends to cause conventional confidence intervals for regression coefficients and the corresponding assessment of the significance of any of the covariates in matrix **X** to be invalid. The general seriousness of this complication is portrayed graphically (after Cordy and Griffith, 1993) in Figure 4.4.

A third problem area has to do with heteroscedasticity (nonconstant variance). It is unlikely that the variance of regression errors will be constant across a geographic landscape, especially one partitioned into areal units of different sizes. This is one of the fundamental concerns of spatial econometrics and is beyond the scope of this book (see Anselin 1988). Caution should be exercised to not confuse nonconstant variance with spatial dependence.

In conclusion, a principal consequence of the presence of nonzero spatial autocorrelation in geo-referenced data is the poor standard errors that can be produced by OLS. OLS tends to be reasonably robust to a departure from normality, as long as it is not too extreme; kurtosis seems to be more onerous than does skewness. Accordingly, unless spatial autocorrelation is quite marked, results of the S-W statistic should remain valid. In addition, nonconstant variance tends not to induce pronounced effects as long as the ratio of the larger to the smaller variance in the population is not unusually large (e.g., exceeds 2). This situation tends to degenerate in ANOVA as a design becomes increasingly unbalanced. Furthermore, Bartlett's equality of variance test is highly sensitive to slight differences between regional variances and is unable to disentangle nonnormality and nonconstant variance. In general, the consequences of nonzero spatial autocorrelation tend to be much less severe than those of heteroscedasticity, but far more severe than those of nonnormality.

4.7. Problems

P-4.1. Given the *Old Maps* (Table 10.1 in Andrews and Herzberg, 1985) data set, construct two regression equations, and perform the battery of diagnostics on them; regress the actual latitude (this is **Y**) on the 11 map latitudes (e.g., latitude from Cornelli, 1688; these are **X**) and regress the actual longitude on the 11 map longitudes.

P-4.2. Given the *Identifying Groundwater Populations* (Table 17.1 in Andrews and Herzberg, 1985) data set, construct a trend surface regression equation that describes the spatial distribution of PH across the northwest Texas region.

4.8. References

Andrews, D. and H. Herzberg. 1985. *Data: A Collection of Problems from Many Fields for the Student and Research Worker.* Berlin: Springer-Verlag.

Anselin, L. 1988. *Spatial Econometrics: Methods and Models.* Dordrecht: Kluwer Academic Publishers.

Bailey, T. and A. Gatrell. 1995. *Interactive Spatial Data Analysis.* Harlow, U.K.: Longman Scientific Publishing.

Cordy, C. and D. Griffith. 1993. "Efficiency of Least Squares Estimators in the Presence of Spatial Autocorrelation." *Communications in Statistics*, Series B 22: 1161–79.

Dutilleul, P. and P. Legendre. 1992. "Lack of Robustness in Two Tests of Normality Against Autocorrelation in Sample Data." *Journal of Statistical Computation and Simulation* 42: 79–91.

Griffith, D. and C. Amrhein. 1991. *Statistical Analysis for Geographers.* Englewood Cliffs, N.J.: Prentice Hall.

Haining, R. 1990. *Spatial Data Analysis in the Social and Environmental Sciences.* Cambridge, U.K.: Cambridge University Press.

Hair Jr., J., R. Anderson, R. Tatham, and W. Black. 1995. *Multivariate Data Analysis.* 4th ed. Englewood Cliffs, N.J.: Prentice Hall.

Neter, J., W. Wasserman, and M. Kutner. 1990. *Applied Linear Statistical Models.* Homewood, Il: Irwin.

PART II

INSTANCES OF THE GENERAL LINEAR MODEL

CHAPTER 5

MULTIVARIATE ANALYSIS OF VARIANCE

Chapter Preview. Multivariate analysis of variance (often implemented as a balanced design MANOVA) is an extension of the classic analysis of variance technique (often implemented as a balanced design ANOVA). In the multivariate world, two or more categorical variables are analyzed for their influence on a set of two or more interval/ratio variables. As with many of the computationally intensive techniques explored in this book, general MANOVA has become practical only with the availability of large amounts of inexpensive computing power.

Extending ANOVA, we first derive multivariate analysis of variance from its roots in analysis of variance before turning to its relationship with regression analysis.

As with analysis of variance, the key research question facing analysts in MANOVA is whether the sample data set is balanced (has the same number of observations in each cross classification defined in the problem). Because geographical problems often involve unbalanced subsamples derived from fieldwork (e.g., surveys in human geography, field studies in physical geography), we treat both balanced and unbalanced designs here, with fixed and/or random effects.

The value of MANOVA-type analysis is well established today. Increasingly, as geo-referenced data sets that were unimaginable ten years ago become more common (e.g., remotely sensed data, public use microsamples, and scanner data) and more readily available, spatial analysts should find themselves frequently turning to MANOVA. For this reason, we give high profile to this technique as the second one presented.

5.0. Introduction

Analysis of variance (ANOVA) is a technique that commonly is treated toward the end of an introductory statistics course. In its simplest form, it generalizes a two-sample difference of means test to K samples, under the assumption of homogeneity of variance. This K-samples problem can be solved by writing the ANOVA model as a regression (see §4.4) model with sample membership for each observation identified by a binary indicator variable, and then any one of these indicator variables is chosen *arbitrarily* and either deleted from the analysis or subtracted from the remaining K-1 indicator variables. In addition, an n-by-1 vector of ones, denoted **1**, is introduced as the variable associated with the intercept term. This formulation is for ***one-way*** ANOVA, and is the basis of the general linear model (***GLM***).

The **X** matrix for the regression model then consists of the vector **1**, and the K-1 differenced indicator variables. In this context, it is referred to as a design matrix because of its close association with experimental designs and because it is used to partition the total sum-of-squares into various components. The sum-of-squares computed for the regression model is equivalent to the traditional ANOVA between regions sum-of-squares, while the sum-of-squares computed for the regression residuals is equivalent to the traditional ANOVA within regions sum of squares. The vector of regression coefficients, **b**, allows the following marginal null hypotheses to be tested:

$$H_0: \ \beta_j = \mu_j - \mu_K = 0, j = 2, 3, ..., K\text{-}1 \ .$$

The regression model itself allows the following composite null hypothesis to be tested:

$$H_0: \ \beta = \mathbf{0} \Rightarrow \mu_1 = \mu_2 = \mu_3 = ... = \mu_K \ .$$

The equivalence of these two situations facilitates diagnostic treatment of the statistical model.

EXAMPLE PR-5.1 (Example PR-4.10 revisited). Test the differences of agricultural administrative regional (ADM) means for the density of cultivated farm land (DCD) in Puerto Rico.

Using the SAS code in the initial data statement, where ADM denotes the agricultural administrative region label variable,

```
IF ADM=1 THEN IR1=1; ELSE IR1=0;
IF ADM=2 THEN IR2=1; ELSE IR2=0;
IF ADM=3 THEN IR3=1; ELSE IR3=0;
IF ADM=4 THEN IR4=1; ELSE IR4=0;
IF ADM=5 THEN IR5=1; ELSE IR5=0;
IR1=IR1-IR5;
IR2=IR2-IR5;
IR3=IR3-IR5;
IR4=IR4-IR5;
```

and then using the SAS procedure

```
PROC REG;
   MODEL DCD = IR1-IR4;
RUN;
```

produced SAS output that includes

```
Model: MODEL1
Dependent Variable: DCD

                        Analysis of Variance

                         Sum of          Mean
Source          DF      Squares         Square       F Value       Prob>F

Model            4   2973.26066      743.31517         6.902       0.0001
Error           68   7323.43988      107.69765
C Total         72  10296.70055

                  Analysis of Variance Procedure

Dependent Variable: DCD
                                 Sum of          Mean
Source                DF        Squares         Square  F Value      Pr > F
Model                  4      2973.2607       743.3152     6.90      0.0001
Error                 68      7323.4399       107.6976
Corrected Total       72     10296.7005
```

Numerical results reported in these two tables differ only by rounding error. Therefore, the null hypothesis is rejected at the 5% level. Hence, there is evidence to indicate that at least two of the agricultural administrative regional means for DCD are different in the population. And, nearly 0% of the time one would expect to obtain an F-ratio value of this magnitude strictly due to sampling error.

The diagnostic statistics that accompany the preceding ANOVA tables follow:

```
                    Within Regions Normality Test
      REGION          San Juan   Arecibo   Mayaguez    Ponce    Caguas
      SHAPIRO-WILK     0.9625    0.9446     0.9680    0.9692    0.8568
      PROBABILITY      0.6189    0.5541     0.7747    0.8223    0.0336

    Homogeneity of Variance Test         Spatial Autocorrelation Test
    OBSERVED CHI SQUARE    6.8615        MORAN COEFFICIENT        0.1553
    PROBABILITY            0.1434        Z TEST STATISTIC         3.0604
```

These diagnostics suggest that (1) there is no evidence to indicate that the individual within-regions frequency distributions do not conform to a normal distribution in their respective populations (the null hypothesis for Caguas would be rejected at the 5% level of significance but not at the corresponding Bonferroni 1% level), (2) the variance of DCD is constant across the five regions in the population, and (3) there is evidence to indicate that DCD is weakly and positively spatially autocorrelated across Puerto Rico in the population. These first two findings suggest that the F-ratio of 6.90 should be relatively free of specification error with regard to ANOVA. This third finding could account for the possibility that the frequency distribution of DCD in Caguas does not conform to a normal distribution in the population. It also suggests that perhaps one or more variables are missing from the regression model equation. This missing variable possibility is supported by the change in the Moran Coefficient (MC) between the raw DCD values and the ANOVA errors; the ANOVA differenced indicator variables reduced the MC from 0.3610 ($z = 5.0868$, $p = 0.000$) to 0.1469, the value reported.

Treating ANOVA as a multiple regression problem reveals that it actually is a multivariate technique. Including additional relevant predictor variables moves ANOVA even deeper into the domain of multivariate analysis. If these additional predictors are categorical variables, then the problem becomes one of multiple classification ANOVA, known as N-way ANOVA. Rarely will N exceed 3, meaning that most studies employing ANOVA involve one-way, two-way, or three-way designs.

5.1. N-way Unbalanced Designs

For decades, computational restrictions on ANOVA models involving multiple classification variables forced all cross-classification groups to have the same number of observations (known as a balanced design, see Chapter 4), or, as a special case, a number of observations proportional to the marginal totals. Most standard statistical software and textbook discussions of two- and three-way ANOVA assume a balanced design. This framework is fine if a researcher has control over the assignment of observations to groups, as often is the case in educational, medical, and psychological experiments. But, areal units for geo-referenced data sets tend to be fixed in number and location, almost always defying balanced design tabulations. One result of the incompatibility between ANOVA and geo-referenced data sets is that virtually all geographic ANOVA applications to date are one-way.

Advances in computing technology coupled with development of the GLM in statistics and dissemination of its technology through standard statistical software packages allow this restriction to be circumvented. The resulting designs, in which different numbers of observations occur in the cross-classification groups, are known as unbalanced designs.

> *RULE OF THUMB 5.1*: If cross-classification sample sizes are unequal, use the general linear model (GLM) formulation to implement analysis of variance.

One of the principal costs of moving from a balanced design to an unbalanced design is a reduction in statistical power; unbalanced designs tend to have more difficulty distinguishing between the null hypothesis and alternate hypotheses. This loss of power increases as differences in group sample sizes increase.

The critical point for interpretative purposes in an N-way ANOVA concerns whether interaction terms are significant (see §4.4.2 for a detailed discussion). To recapitulate, the null hypothesis states that for a two-way scheme the cross-classification means are equal and may be stated as

$$H_0: \mu_{11} = \mu_{12} = \mathrm{K} = \mu_{1K_1} = \mu_{21} = \mu_{22} = \mathrm{K} = \mu_{2K_1} = \mathrm{K} = \mu_{K_21} = \mu_{K_22} = \mathrm{K} = \mu_{K_2K} \; .$$

If interaction effects are not significant, then this composite null hypothesis can be restated as the following pair of individual null hypotheses:

$$H_0\colon \mu_{\bullet 1} = \mu_{\bullet 2} = K = \mu_{\bullet K_1}$$
$$H_0\colon \mu_{\bullet 1} = \mu_{\bullet 2} = K = \mu_{K_1 \bullet}$$

indicating that effects of the individual classification variables can be interpreted separately. In the presence of significant interaction, the composite null hypothesis *cannot* be simplified in this fashion. Recall that this feature of N-way ANOVA is why interaction effects *must* be tested first.

Larger numbers of classification variables become even more problematic because the number of interaction terms goes up at an increasing rate as N becomes larger. For instance, three-way designs have four interaction terms (i.e., $C_1 \times C_2$, $C_1 \times C_3$, $C_2 \times C_3$, $C_1 \times C_2 \times C_3$), and four-way designs have eleven interaction terms. Counting the number of interaction terms is a combinatorial problem with the total number of terms equal to $2^N - C(N,0) - C(N,1) = 2^N - 1 - N$. Because so many interaction terms emerge as N increases, just by chance some will be significant. This complication is a basic reason why many studies are restricted to no more than a three-way design. An additional complication that tends to be of particular relevance for geo-referenced data concerns the loss of degrees of freedom. N classification variables will cost $K_1 \times K_2 \times \ldots \times K_N$ degrees of freedom forcing n to need to become very large as N increases, an unlikely event for most geo-referenced data.

5.1.1. A Two-way Unbalanced Design: The Fixed Effects Model

The first step in employing an ANOVA model is to decide whether each classification variable is a fixed effect or a random effect. In the case of the Puerto Rico data set, the agricultural administrative regions (ADM) will not change through time or replication, meaning that this classification will be constant across data analyses. This is a fixed effect. Similarly, classifying municipios as being in the coastal lowlands or the interior highlands (LND) will not change across data analyses; this, too, is a fixed effect. Therefore, each of the F-ratios for a two-way ANOVA involving these two classification variables will have the mean square error in its denominator. The cross-classification of municipios for these two variables yields the unbalanced design sample sizes, say n_{jk}, of

Cross-classification Sample Sizes

REGION	San Juan	Arecibo	Mayaguez	Ponce	Caguas
COASTAL LOWLANDS	12	7	9	10	4
INTERIOR HIGHLANDS	7	4	7	4	9

The range of n_{jk} is such that the maximum is three times the minimum value.

A key distinction between one-way and two-way ANOVA is the presence of interaction terms in N-way designs. If the two classification variables act independently, then the single interaction term for a two-way design will be zero in the population. Interaction terms are created by taking the set of differenced indicator variables for the first classification variable, say C_1, and the set of differenced indicator variables for the second classification variable, say C_2, and multiplying them. If there are K_1 categories for C_1 and K_2 categories for C_2, then there will be $(K_1-1)(K_2-1)$ interaction indicator variables. Assuming independence of the classification variables is equivalent to assuming that all of the regression coefficients for the product indicator variables are zero. One result is an additive model, having each classification variable interpretable on its own, since C_1 does not have differential effects across C_2, and vice versa. A second result is that ignoring existing interaction between C_1 and C_2 will result in specification error, causing vector **b** to be biased (§4.2.1). Consequently, in most situations researchers include interaction effects and hope that these effects will be statistically insignificant.

EXAMPLE PR-5.2. Test the differences of ADM and LND regional means for the density of milk production (DMLK) in Puerto Rico.

Using the SAS procedure

```
PROC GLM;
   CLASS ADM LND;
   MODEL DMLK = ADM LND ADM*LND;
RUN;
```

produced SAS output that includes

Dependent Variable: DMLK

Source	DF	Sum of Squares	Mean Square	F Value	Pr > F
Model	9	55742.607	6193.623	3.17	0.0032
Error	63	123037.959	1952.983		
Corrected Total	72	178780.566			

Source	DF	Type III SS	Mean Square	F Value	Pr > F
ADM	4	22687.214	5671.803	2.90	0.0286
LND	1	3124.661	3124.661	1.60	0.2106
ADM*LND	4	17618.472	4404.618	2.26	0.0730

Consider a 5% level of significance. Because the null hypothesis of no interaction is not rejected, separate statements can be made about the difference of means for classification variables ADM and LND. Moreover, there is evidence to indicate that at least two of the agricultural administrative milk production density regional means are different in the population, while there is no evidence to indicate that the average density of milk production differs between coastal lowlands and interior highlands in the population.

Since a fixed effects model is used in this example, the F-ratios have been computed as follows:

$$\frac{\text{mean square for interaction}}{\text{mean square for within-cross-classification groups}} = \frac{4404.6}{1953.0} = 2.26$$

$$\frac{\text{mean square between ADM regions}}{\text{mean square for within-cross-classification groups}} = \frac{5671.8}{1953.0} = 2.90$$

$$\frac{\text{mean square between LND regions}}{\text{mean square for within-cross-classification groups}} = \frac{3124.7}{1953.0} = 1.60$$

These are usually the default calculations produced by standard statistical software packages.

The diagnostic statistics that accompany the preceding ANOVA table are:

Within Cross-classification Groups Normality Test

REGION	COASTAL LOWLANDS SHAPIRO-WILK	PROBA-BILITY	INTERIOR HIGHLANDS SHAPIRO-WILK	PROBA-BILITY
San Juan	0.8984	0.1446	0.7313	0.0078
Arecibo	0.5978	0.0003	0.7081	0.0154
Mayaguez	0.7153	0.0025	0.7951	0.0364
Ponce	0.9033	0.2274	0.9466	0.6662
Caguas	0.8187	0.9087	0.8748	0.1361

Homogeneity of Variance Test

OBSERVED CHI SQUARE	113.1247
PROBABILITY	0.0001

Spatial Autocorrelation Test

MORAN COEFFICIENT	-0.0387
Z TEST STATISTIC	0.3847
PROBABILITY	0.3502

According to the normality tests, there is evidence to indicate that five of the cross-classification frequency distributions do not conform to a normal distribution in their respective populations. But using the equivalent Bonferroni level of 0.005 identifies only two of these regional deviants. This finding casts some doubt on the homogeneity of variance test statistics, which in turn casts serious doubt on the validity of the ANOVA F-ratios. Furthermore, these violations of assumptions cast doubt on the inference implied by MC, namely that the regression residuals are not spatially autocorrelated in the population.

Two corrective actions can be taken at this point in the analysis. On the one hand, a nonparametric ANOVA can be employed (see Griffith and Amrhein, 1991, p. 310–14). But ordinal data do not permit interaction to be handled, forcing the assumption that interaction is irrelevant and risking the introduction of another type of specification error. While this approach might seem reasonable for this particular example, the documented presence of serious specification error indicates that an insignificant interaction F-ratio may not be due strictly to sampling error. An alternative is to employ a data transformation. One finding reported in Chapter 1 is that transforming DMLK to TDMLK = LN(DMLK + 1) stabilizes its variance across the agricultural administrative regions.

EXAMPLE PR-5.3. (Example PR-5.2 revisited). Test the differences of ADM and LND regional means for the transformed density of milk production (TDMLK) in Puerto Rico.

Dependent Variable: TDMLK

Source	DF	Sum of Squares	Mean Square	F Value	Pr > F
Model	9	84.573147	9.397016	6.78	0.0001
Error	63	87.341405	1.386372		
Corrected Total	72	171.914553			

Source	DF	Type III SS	Mean Square	F Value	Pr > F
ADM	4	45.009105	11.252276	8.12	0.0001
LND	1	3.461904	3.461904	2.50	0.1191
ADM*LND	4	19.228963	4.807241	3.47	0.0127

Therefore, because the null hypothesis of no interaction is rejected at the 5% level, separate statements cannot be made about the difference of means for classification variables ADM and LND.

The diagnostic statistics that accompany the preceding ANOVA table are:

Within Cross-classification Groups Normality Test

REGION	COASTAL LOWLANDS SHAPIRO-WILK	PROBA-BILITY	INTERIOR HIGHLANDS SHAPIRO-WILK	PROBA-BILITY
San Juan	0.9320	0.3761	0.9083	0.3995
Arecibo	0.8071	0.0481	0.9636	0.7686
Mayaguez	0.9021	0.2600	0.8635	0.1679
Ponce	0.8571	0.0677	0.9853	0.9087
Caguas	0.9020	0.4346	0.9527	0.7131

Homogeneity of Variance Test

OBSERVED CHI SQUARE	7.4882
PROBABILITY	0.5864

Spatial Autocorrelation Test

MORAN COEFFICIENT	0.1313
Z TEST STATISTIC	2.9296
PROBABILITY	0.0017

Clearly the variable transformation resulted in a failure to reject the null hypothesis of a normal distribution in the parent population for each of the within-cross-classification frequency distributions, especially when viewed in terms of Bonferroni levels. In turn, the variance is stabilized across the cross-classification groups. But now the interaction effect is significant, as is the regression residuals' MC index value. Moreover, specification error present when analyzing the raw data causes a Type II error to be made here for both interaction effects and spatial autocorrelation.

5.1.2. A Two-way Unbalanced Design: The Mixed Effects Model

In the case of the Puerto Rico data set, the rural-urban (R-U) classification is based upon the percentage of urban population; once a municipio crosses a certain threshold, its classification changes. In other words, this classification mechanism may result in a different category assignment for a given municipio depending upon the census year for which data were acquired or the threshold definition (a randomly selected single threshold value is taken from a population of possible threshold values). This is a random effect. Including this classification variable with ADM results in a mixed effects ANOVA. Therefore, the F-ratio for interaction uses the cross-classification mean square in its denominator, whereas the F-ratio for ADM has the interaction mean square in its denominator. Calculation of the expected mean square for the denominator of a random effect factor is quite complicated in an unbalanced design. Recall that in the balanced design situation, the expected mean square equals the mean square for within-cross-classification groups. In the unbalanced design situation the expected mean square equals the weighted sum

δ_1(mean square for interaction) + δ_2(mean square for within cross-classification groups),

where δ_j are functions of the cross-classification sample sizes n_{jk}, and the marginal sample totals $n_{j\bullet}$ and $n_{\bullet k}$. The cross-classification of municipios for these two variables produces the unbalanced design sample sizes of

	Cross-classification Sample Sizes				
	San Juan	Arecibo	Mayaguez	Ponce	Caguas
RURAL	6	7	12	12	7
URBAN	13	4	4	2	6

Now the range of n_{jk} is such that the maximum is six and one-half times the minimum value. The geographic distribution of this cross-classification scheme appears in Map 5.1.

As emphasized in §4.2.1., if a variable is missing from the set of predictors, then the estimated regression coefficients will be biased, unless this variable is uncorrelated with those already in the regression model. Often, MC will detect such a missing variable for geo-referenced data with the absence of the missing variable(s), causing a significant MC index value, whereas the presence of the missing variable(s) causes an insignificant or drastically reduced MC value.

EXAMPLE PR-5.4. (Example PR-5.1 revisited). Test the difference of ADM regional and rural-urban (R-U) classification means for the density of cultivated land (DCD) in Puerto Rico.

The significant MC value obtained in a one-way ANOVA suggests that perhaps a variable is missing from the analysis. Since rural environments should have more land available for farming than do urban environments, one might expect the density of cultivated land to differ between these two areas.

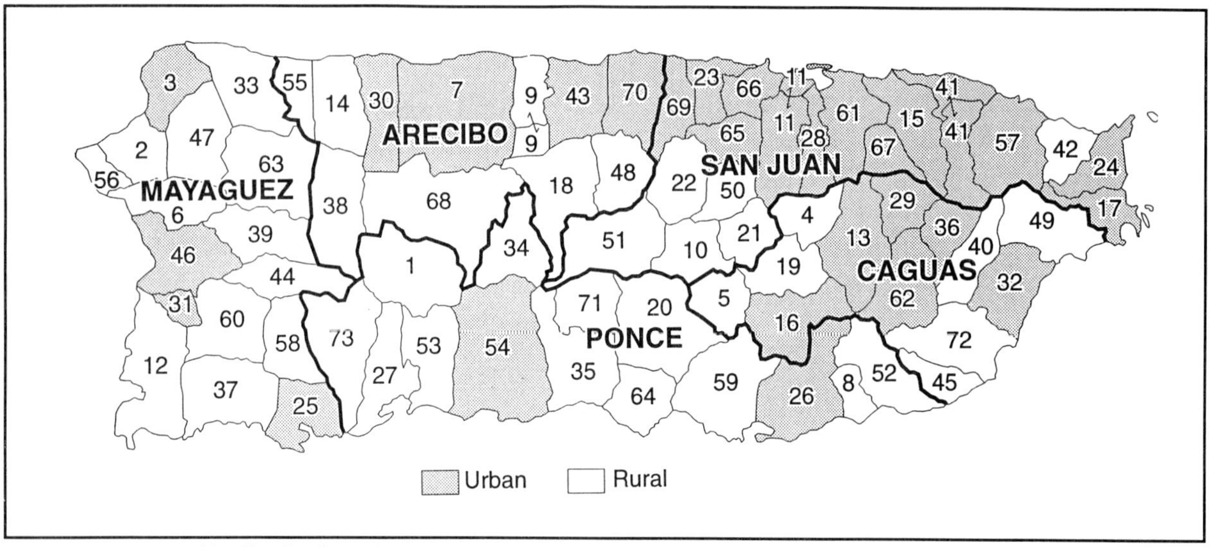

Map 5.1. Geographic distribution of the rural-urban dichotomy binary variable, with the ADMs superimposed on the island.

Using the SAS procedure

```
PROC GLM;
   CLASS ADM RU;
   MODEL DCD = ADM RU ADM*RU;
   RANDOM RU ADM*RU / TEST;
RUN;
```

produced SAS output as shown on the following page.

Because the null hypothesis of no interaction is rejected at the 5% level, separate statements cannot be made about the difference of means for classification variables ADM and R-U.

Since a mixed effects model is used in this example, the F-ratios are computed as follows:

$$\frac{\text{mean square for interaction}}{\text{mean square for within-cross-classification groups}} = \frac{218.1}{84.7} = 2.57$$

$$\frac{\text{mean square ADM regions}}{\text{mean square for interaction}} = \frac{214.6}{218.1} = 0.98$$

$$\frac{\text{mean square for R-U areas}}{0.9366 \times (\text{mean square for interaction}) + 0.0634 \times (\text{mean square for within-cross-classification groups})}$$

$$= \frac{827.1}{209.6} = 3.95 \ .$$

```
Dependent Variable: DCD
                                          Sum of           Mean
Source                        DF         Squares         Square   F Value     Pr > F

Model                          9       4958.5706       550.9523      6.50     0.0001
Error                         63       5338.1299        84.7322
Corrected Total               72      10296.7005

Source: ADM
Error: MS(ADM*RU)
                              Denominator     Denominator
     DF     Type III MS                DF              MS     F Value     Pr > F
      4    214.58612927                 4    218.07258728       0.984     0.5060

Source: RU
Error: 0.9366*MS(ADM*RU) + 0.0634*MS(Error)

                              Denominator     Denominator
     DF     Type III MS                DF              MS     F Value     Pr > F
      1    827.10980842              4.21    209.62097476       3.946     0.1144

Source: ADM*RU
Error: MS(Error)
                              Denominator     Denominator
     DF     Type III MS                DF              MS     F Value     Pr > F
      4    218.07258728                63     84.73222135       2.574     0.0461
```

The diagnostic statistics that accompany the preceding ANOVA table are

```
              Within Cross-classification Groups Normality Test
     REGION     COASTAL RURAL     PROBA-      INTERIOR URBAN     PROBA-
                 SHAPIRO-WILK     BILITY       SHAPIRO-ADM       BILITY
     San Juan       0.9386        0.6657         0.9506          0.5681
     Arecibo        0.9230        0.5115         0.8977          0.4140
     Mayaguez       0.9677        0.8318         0.8973          0.4121
     Ponce          0.9673        0.8276         1.0000          1.0000
     Caguas         0.9208        0.4935         0.7845          0.0406

  Homogeneity of Variance Test          Spatial Autocorrelation Test
  OBSERVED CHI SQUARE    9.5016         MORAN COEFFICIENT         0.0200
  PROBABILITY            0.3923         Z TEST STATISTIC          1.1416
                                        PROBABILITY               0.1268
```

These diagnostics suggest that each cross-classification frequency distribution conforms to a normal distribution in its respective population, especially when compared with Bonferroni levels; the variance across the cross-classification groups is constant in the population; and the regression residuals are not spatially autocorrelated in the population. Unfortunately, the analysis implies that there is a differential effect on the density of cultivated land across agricultural administrative regions according to whether a municipio is rural or urban.

5.1.3. Two-way ANOVA Interpretation When Interaction Is Significant

The composite null hypothesis must be tested when significant interaction is detected in a two-way design. In other words, the cross-classifications become the groupings of interest. Because these cross-classifications are what normality and homogeneity of variance diagnostics have been computed for, further assessment of these two properties is unnecessary. The resulting one-way ANOVA relates to the overall GLM regression for the two-way design, ignoring the individual clusters of differenced indicator variables. Accordingly, the number of degrees of freedom involved is $K_1K_2 - 1$, because there are K_1K_2 cross-classifications and any one of these indicator variables is *arbitrarily* chosen and subtracted from the remaining indicator variables. The quantity $(K_1 - 1) + (K_2 - 1) + (K_1 - 1)(K_2 - 1)$ equals the number of degrees of freedom involved in the two-way scheme and is the sum of the degrees of freedom for C_1, C_2, and the interaction of these two classificatory variables.

EXAMPLE PR-5.5. (Example PR-5.3 extended). Test the differences of ADM and LND regional means for the transformed density of milk production (TDMLK) in Puerto Rico, using a single composite cross-classification variable (ADM×LND).

Let the regional groups of interest be the coastal lowlands subgroup of the five agricultural administrative regions and the interior highlands subgroup of the five agricultural administrative regions, producing a total of ten cross-classification regions.

Dependent Variable: TDMLK

Source	DF	Sum of Squares	Mean Square	F Value	Pr > F
Model	9	84.573147	9.397016	6.78	0.0001
Error	63	87.341405	1.386372		
Corrected Total	72	171.914553			

The null hypothesis is rejected at the 5% level, and there is evidence to indicate that at least two of the cross-classification regional means for DMLK are different in the population. Nearly 0% (actually 0.0001) of the time one would expect to obtain an F-ratio value of this magnitude strictly from sampling error.

The diagnostic statistics for this example are identical to those reported in Example PR-5.3.

Because one-way ANOVA does not distinguish between fixed effects and variable effects classification variables, differentiating these perspectives becomes irrelevant when dealing with the composite null hypothesis. Furthermore, because one-way ANOVA is being used now, causing interaction effects to be taken into account implicitly rather than explicitly, a nonparametric alternative becomes feasible. For instance, results reported in Example PR-5.5 are consistent with the Kruskal-Wallis nonparametric test result, which yields $\chi^2 = 34.737$ ($p = 0.0001$).

5.2. Multicollinearity and Simultaneous Testing: Multivariate Analysis of Variance (*MANOVA*)

Consider the geographic distribution of the density of banana/plantain production (DBN). A researcher may wish to determine whether this productivity differs across the five agricultural administrative regions. The ANOVA table for this situation, incorporating an appropriate data transformation [where DBN has been transformed to TDBN = LN(DBN + 0.025)], is

Dependent Variable: TDBN

Source	DF	Sum of Squares	Mean Square	F Value	Pr > F
Model	4	8.0700564	2.0175141	0.79	0.5325
Error	68	172.5687061	2.5377751		
Corrected Total	72	180.6387625			

The null hypothesis is not rejected at the 5% level indicating no evidence that the agricultural administrative regional means for TDBN are different in the population.

The diagnostic statistics that accompany this ANOVA table are

Within Regions Normality Test

REGION	San Juan	Arecibo	Mayaguez	Ponce	Caguas
SHAPIRO-WILK	0.9377	0.9058	0.9678	0.8833	0.9574
PROBABILITY	0.2469	0.2077	0.7711	0.0647	0.6666

Homogeneity of Variance Test

OBSERVED CHI SQUARE	3.9358
PROBABILITY	0.4148

Spatial Autocorrelation Test

MORAN COEFFICIENT	0.3580
Z TEST STATISTIC	6.0457
PROBABILITY	0.0000

These diagnostics suggest that (1) there is no evidence to indicate that the individual within regions frequency distributions do not conform to a normal distribution in their respective populations, (2) the variance of LN(DBN + 0.025) is constant across the five regions in the population, and (3) there is evidence to indicate that LN(DBN + 0.025) is spatially autocorrelated across Puerto Rico in the population. The first two findings suggest that the F-ratio of 0.79 should be relatively free of specification error with regard to ANOVA. The third finding suggests that perhaps one or more variables are missing from the analysis, possibly explaining why this variable required a transformation in the first place.

But what if a researcher wants to know whether or not agricultural production in general differs across the five agricultural administrative regions? A statistical decision based upon DCD suggests that the answer to this question is "yes," whereas a statistical decision based upon TDBN suggests that the answer to this question is "no."

> *MANOVA Question #1*: Given some classification(s) and many response variables, do the overall classification means for these variables differ jointly rather than individually?

Of course, there would be no dilemma if both statistical decisions produced the same answer. The frequency of such conflicting findings, merely by chance, can be treated as a binomial distribution *if the individual variables are independent.* By choosing a 5% level of significance, $p = 0.95$ and $(1\text{-}p) = 0.05$. Hence for P variables,

$$\text{probability of agreement} = p^{P} + (1\text{-}p)^{P} \ ,$$

$$\text{probability of disagreement} = \sum_{x=1}^{m-1} C(P,x)p^{x}(1-p)^{P-x}.$$

Consequently, as the number of variables being analyzed increases, the chances of observing some disagreement between individual ANOVA results merely by happenstance dramatically increases. One strategy for handling this problem is to treat the P variables simultaneously and thus undertake what commonly is known as a MANOVA. This technique differs from N-way classifications in that it expands the dependent variable Y from a vector into matrix **Y** rather than expanding the set of classificatory indicator variables **X**. MANOVA allows a statistical decision to be made taking into account the possibility that some individual variables may imply rejection of the null hypothesis whereas others may imply a failure to reject the null hypothesis. Roughly speaking, *if the individual variables are independent*, then a 5% level of significance for MANOVA will translate into a (5/P)% level for each individual ANOVA, where m is the number of categories in the analysis. This concept, recalling the Bonferroni technique (§4.4.1), relates to the notion of multiple comparisons (Chapter 4). Consequently, a key point in this discussion of simultaneous testing is whether or not the variables composing matrix **Y** are independent, or orthogonal. If they are perfectly correlated, meaning that all the differences of means information across the P variables is completely redundant, then all individual ANOVA results will agree, and

$$\text{probability of agreement} = 1 \ ,$$
$$\text{probability of disagreement} = 0 \ .$$

Naturally most researchers will find that their data fall somewhere between these two extreme cases of mutually independent and perfectly correlated variables; one of the crucial concerns of multivariate analysis is properly accounting for such multicollinearity in a simultaneous fashion.

Another salient reason for disagreement between univariate and multivariate results involves the relationship between the number of dependent variables, P, constituting the matrix **Y**, and the within groups sample sizes. It is necessary to have $n_j > P$ in a one-way design or $n_{jk} > P$ in a two-way design.

RULE OF THUMB 5.2: Only undertake a MANOVA when $n_{jk} > P$ for *all* cross-classification categories.

As P increases, this requirement becomes burdensome, especially because the total sample size (n) for geo-referenced data sets almost always is fixed. When this requirement is not satisfied, the power of MANOVA is reduced, which in turn can render a nonsignificant MANOVA F-ratio even though many of the ANOVA F-ratios are significant.

5.2.1. The MANOVA Format

MANOVA posits a null hypothesis claiming that all K regional means are equal for each of a set of P variables. This null hypothesis is a vector generalization of the scalar one appearing in §4.4.1 and is denoted by

$$H_0: \boldsymbol{\mu}_1 = \boldsymbol{\mu}_2 = \boldsymbol{\mu}_3 = \ldots = \boldsymbol{\mu}_K \ ,$$

where $\boldsymbol{\mu}_k$ is a P-by-1 vector of P population variable means for region K. This joint hypothesis replaces the following set of individual variable ANOVA hypotheses:

$$\begin{aligned} H_0&: \mu_{11} = \mu_{21} = \mu_{31} = \ldots = \mu_{K1} \\ H_0&: \mu_{12} = \mu_{22} = \mu_{32} = \ldots = \mu_{K2} \\ &\quad . \quad . \quad . \quad \ldots \quad . \\ &\quad . \quad . \quad . \quad \ldots \quad . \\ &\quad . \quad . \quad . \quad \ldots \quad . \\ H_0&: \mu_{1P} = \mu_{2P} = \mu_{3P} = \ldots = \mu_{KP} \end{aligned}$$

The primary goal is to synthesize the inferences drawn from each of these univariate null hypotheses.

The test statistic for MANOVA is a generalization of $(\frac{ss_B}{ss_W})(\frac{df_w}{df_B})$, the statistic used in ANOVA. First, the sum-of-squares between regions is generalized to matrix **B**, which is diagonal only if the variables are mutually orthogonal (nonzero off-diagonal entries mean that multicollinearity is present). This matrix may be defined as follows:

$$\mathbf{B} = \sum_{k=1}^{K} n_k(\overline{\mathbf{X}}_k - \overline{\mathbf{X}}_T)(\overline{\mathbf{X}}_k - \overline{\mathbf{X}}_T)^T,$$

where n_k is the sample size for region k, $\overline{\mathbf{X}}_k$ is a P-by-1 vector of sample variable means for region k, and $\overline{\mathbf{X}}_T$ is a P-by-1 vector of sample variable means for all regions. Next, the sum-of-squares within regions is generalized to matrix **W**, which also is diagonal only if the variables are mutually orthogonal. This matrix may be defined as follows:

$$\mathbf{W} = \sum_{k=1}^{K} \sum_{i=1}^{n_k} (\mathbf{X}_{ik} - \overline{\mathbf{X}}_k)(\mathbf{X}_{ik} - \overline{\mathbf{X}}_k)^T,$$

where $\mathbf{X}_{ik}$ is a P-by-1 vector of variable values for observation i in region k. Both of these matrices are symmetric, because $COV(X_j,X_k) = COV(X_k,X_j)$. In univariate analysis, a case is made linking reductions in degrees of freedom to data analysis constraints. In some sense, this argument can be carried over to MANOVA by inspecting these definitions of matrices **B** and **W**, although multicollinearity and simultaneity features of multivariate analysis can result in fractional degrees of freedom. One set of constraints in MANOVA that is a straightforward generalization of ANOVA constraints consists of the individual regional variable means and the overall variable means. There are P variables, with K-1 individual regional means plus one overall mean for each variable.

Now the test statistic becomes $\mathbf{W}^{-1}\mathbf{B}$, and is analogous to the ANOVA sum-of-squares ratio; recall that the order of matrix multiplication is important, whereas that of scalar multiplication is not. The ANOVA test statistic follows an F distribution, with df_B and df_W degrees of freedom. MANOVA offers several test statistics, of which Wilks' Λ is the most popular. This statistic is defined as follows:

$$\frac{1}{\Lambda} = DET\,|\mathbf{I} + \mathbf{W}^{-1}\mathbf{B}\,| \rightarrow \Lambda = \frac{DET\,|\mathbf{W}\,|}{DET\,|\mathbf{W} + \mathbf{B}\,|} = \frac{DET\,|\mathbf{W}\,|}{DET\,|\mathbf{T}\,|} = \prod_{j=1}^{G} (1 + \lambda_j),$$

where matrix **T** is the total sum-of-squares matrix, G = *MIN*(P, K-1), and λ_j are the eigenvalues of matrix $\mathbf{W}^{-1}\mathbf{B}$ (Property P6, Table 3.1). The previously mentioned constraints result in there being *at most MIN*(P, K-1) nonzero eigenvalues for the matrix product $\mathbf{W}^{-1}\mathbf{B}$ and is the origin of the definition of G. Note that 1 is added to each eigenvalue of $\mathbf{W}^{-1}\mathbf{B}$ because the identity matrix **I** is being added to this matrix (property P4, Table 3.1). The determinant of a covariance matrix is called the ***generalized variance***, and geometrically is the square of the P-dimensional "volume" in n-dimensional observational space. The generalized variance equals zero when at least one instance of perfect multicollinearity is present somewhere in the data. When P = 1, the generalized variance relates to the univariate F-ratio, since *DET*|**W**| reduces to ss_W and *DET*|**B**| reduces to ss_B. A transformation of Wilks' lambda (Λ) follows an F distribution:

$$\left[\frac{(1 - \Lambda^{\frac{1}{s}})}{\Lambda^{\frac{1}{s}}}\right] \frac{\left\{\left[n - 1 - \frac{(P+K)}{2}\right]s - \frac{P(K-1)}{2} + 1\right\}}{P(K-1)},$$

where $s = \sqrt{\frac{P^2(K-1)^2 - 4}{P^2 + (K-1)^2 - 5}}$. The ratio contained in the large square brackets, [ ], of this expression is analogous to the ANOVA degrees of freedom ratio $(\frac{df_w}{df_B})$ and exactly equals $(\frac{df_w}{df_B})$ when P = 1.

Three other commonly reported test statistics are the Hotelling-Lawley's trace, Pillai's trace, and Roy's maximum root. Each of these statistics also is computed from the eigenvalues of matrix $\mathbf{W}^{-1}\mathbf{B}$, with Wilks' lambda and Hotelling-Lawley's trace being the most powerful, displaying about the same degree of statistical power. The Hotelling-Lawley trace follows an F distribution, and may be defined as

$$TR(\mathbf{W}^{-1}\mathbf{B}) = \sum_{j=1}^{G} \lambda_j .$$

(property P#, Table 3.1). Because *MIN*(P, K-1) defines the maximum number of nonzero eigenvalues and all eigenvalues of matrix $\mathbf{W}^{-1}\mathbf{B}$ must be nonnegative, then even if G = 1, calculating $TR(\mathbf{W}^{-1}\mathbf{B})$ involves adding a number of zeroes to a single nonzero scalar. G = 1 when P = 1, K = 2, or perfect and complete multicollinearity exists in the data. Pillai's trace also follows an F distribution and may be defined as

$$TR(\mathbf{I} + \mathbf{W}^{-1}\mathbf{B}) = \sum_{j=1}^{G} \frac{\lambda_j}{(1 + \lambda_j)} .$$

Each of the three preceding statistics uses all of the eigenvalues of matrix $\mathbf{W}^{-1}\mathbf{B}$. In contrast, Roy's maximum root uses only the principal eigenvalue of matrix $\mathbf{W}^{-1}\mathbf{B}$, say λ_1. This statistic follows an F distribution, too, and may be defined as

$$\frac{\lambda_1}{(1 + \lambda_1)} .$$

Since Roy's maximum root uses only the largest eigenvalue, it is consistent with Wilks' lambda and Hotelling-Lawley's trace in the rare situation of perfect and complete multicollinearity or when P = 1 and K = 2.

The summary table for MANOVA looks similar to that for ANOVA, with the ANOVA sum-of-squares quantities replaced by the generalized variances and the ANOVA mean square quantities replaced by Wilks' lambda.

EXAMPLE PR-5.6. Execute one-way joint ANOVA of DCD and TDBN.

Using the SAS procedure (note: ANOVA also could be used here)

```
PROC GLM;
   CLASS ADM;
   MODEL DCD TDBN=ADM;
   MANOVA H=ADM/PRINTE PRINTH;

RUN;
```

produced SAS output that includes

	E = Error SS&CP Matrix		H = Anova SS&CP Matrix for ADM	
	DCD	TDBN	DCD	TDBN
DCD	7323.4398849	375.51950882	2973.2606631	125.16382574
TDBN	375.51950882	172.56870608	125.16382574	8.0700564448

Statistic	Value	F	Num DF	Den DF	Pr > F
Wilks' Lambda	0.69768348	3.30329	8	134	0.0018

The null hypothesis is rejected at the 5% level. Hence, there is evidence to indicate that at least some of the variable regional means for agricultural productivity, based upon DCD and TDBN, are different in the population.

The following conventional ANOVA-format MANOVA table can be constructed from this preceding information:

The "between regions" generalized variance is the determinant of the "error SS&CP matrix" and is computed as $(7323.4398849)(172.56870608) - 375.51950882^2$. Similarly, the "Within regions" generalized variance is the determinant of the "ANOVA SS&CP matrix for ADM" and is computed as $(2973.2606631)(8.0700564448) - 125.16382574^2$.

Selected diagnostic statistics that accompany this MANOVA table are:

Homogeneity of Variance Test		*Partial Correlations from W*	
OBSERVED CHI SQUARE	14.4743		LN(DBN+0.025)
PROBABILITY	0.2715	DCD	0.3340

One-way MANOVA for Puerto Rican Agricultural Production

Source	Degrees of Freedom	Generalized Variance	Wilks' Lamdbda	F-Ratio	p
Between regions	8	1122781.6	0.6977	3.3	0.0018
Within regions	134	8328.4			

For this example, the homogeneity of variance test suggests that the five 2-by-2 regional covariance matrices are not different in the population. The nonzero eigenvalues of $\mathbf{W}^{-1}\mathbf{B}$ are 0.4077 and 0.0182, which respectively account for 95.73% and 4.27% of the difference of means information content of the two variables. The partial correlation reported here signifies that a weak relationship exists between DCD and LN(DBN + 0.025), whereas the eigenvalues indicate that most of the information content of these two variables supports a finding of a difference of regional means. Consequently, the conflicting inferences drawn from each of the univariate ANOVAs is resolved here by the single inference drawn from the MANOVA results.

5.2.2. Selected Diagnostics for MANOVA

Two diagnostic issues are of concern at this time. First, violation of the assumption of multivariate normality can lead to serious specification error in testing for homogeneity of the covariance matrices across regions. Second, violation of observational independence (i.e., the presence of nonzero spatial autocorrelation) can introduce serious specification error especially into the significance test results.

The relevant question is whether the two variables in Example PR-5.6 conform to bivariate normal (in general, multivariate normal) distributions in their respective regional populations. With regard to Example PR-5.6, the elements of vector **L** presented in Chapter 11, together with the resulting S-W statistic and its associated probability, for each agricultural administrative region, are

Within Regions Multivariate Normality Test

REGION	COEFFICIENTS		SHAPIRO-	PROBA-
	DCD	LN(DBN+0.025)	WILK	BILITY
San Juan	0.0100000	-0.1274560	0.9136	0.0888
Arecibo	0.0265127	-0.0829209	0.9699	0.8709
Mayaguez	0.0146631	-0.0421325	0.9417	0.3646
Ponce	0.0063680	-0.1058300	0.8400	0.0157
Caguas	-0.0300455	-0.0206221	0.8496	0.0268

Consequently, although the individual ANOVA S-W statistics support the assumption of normality for the individual marginal distributions, this joint test casts some doubt concerning multivariate normality for Ponce and Caguas in their respective populations. This outcome emphasizes the failure for individual marginal normal distributions to ensure joint multivariate normality. As a result, caution must be exercised when interpreting the homogeneity of covariance matrices test statistic reported earlier. This finding also suggests that a joint transformation of the data may yield results that better mimic the target multivariate normal distribution.

A second troublesome assumption violation in this example concerns the presence of spatial autocorrelation among the MANOVA residuals (obtained by treating it as a GLM problem, but they actually are the individual ANOVA residuals). Corrective techniques for dealing with this problem are presented in Chapter 12. Again in reference to Example PR-5.6, weak to moderate levels of positive spatial autocorrelation have been detected in residuals from a MANOVA based upon DCD and LN(DBN + 0.025), which may well be the source of a significant F-ratio rather than a "true" difference of mean vectors in the population. Reduction of redundant attribute information, by constructing DF1 (a linear combination of DCB and TDBN using coefficients given by the first eigenvector of matrix $\mathbf{W}^{-1}\mathbf{B}$; see Chapter 7), fails to correct this situation given that the ANOVA F-ratio of 6.93 and the MC of 0.1434 still are significant. Spatial dependence also may be the reason why certain regional bivariate frequency distributions deviate from a multivariate normal distribution. In summary, the detection of positive spatial autocorrelation suggests that caution should be exercised when interpreting especially significance test calculations for the MANOVA results reported in Example PR-5.6.

5.3. Two-way MANOVA

Like two-way ANOVA, two-way MANOVA has the **X** matrix expanded to include additional classification variables, as well as cross-products of the accompanying sets of differenced indicator variables. But now the **Y** vector is expanded into an n-by-P matrix, as was done in one-way MANOVA. Again, one of the primary questions to address is whether each classificatory variable represents a fixed or a random effect. Similar to ANOVA, regardless of the type of effect, interaction is tested first using the within-cross-classifications matrix **W**.

> *MANOVA Question #2*: Given simultaneous classifications and many response variables, is the behavior of these variables across the categories constituting a selected classification scheme identical for, or differential between categories of a remaining classification scheme?

The interaction effects matrix $\mathbf{S}_{12}$ is used in testing fixed effects. As before, the matrix for testing random effects is not straightforward.

Suppose matrix $\mathbf{B}_1$ denotes the between-regions covariance matrix for the first classification variable, $\mathbf{B}_2$ denotes the between regions covariance matrix for the second classification variable, and $\mathbf{S}_{12}$ denotes the interaction covariance matrix. The relevant matrix products for different effects become

for a model with interaction effects, $\mathbf{W}^{-1}\mathbf{S}_{12}$,
for a fixed effects model, $\mathbf{W}^{-1}\mathbf{B}_j$,
for a random effect model, $\mathbf{S}_{12}^{-1}\mathbf{B}_j$,and
for a mixed effects model, $\mathbf{S}_{12}^{-1}\mathbf{B}_j$ (fixed effect) or $\mathbf{Q}^{-1}\mathbf{B}_j$ (random effect).

where matrix **Q** should be a direct generalization of the weighted sums-of-squares and cross-products matrices for its ANOVA counterpart. Based upon the preceding univariate problem in which R-U is declared to be a random effect, $\mathbf{Q} \approx \mathbf{S}_{12}$. Accordingly, this approximation is used in the two-way MANOVA results reported in this section. Use of the wrong inverse matrix may result in an incorrect statistical decision because these are the matrices from which eigenvalues are extracted (and on which decisions are based). These inverses are constructed within the context of the GLM for unbalanced designs.

Diagonal entries of matrix **W** refer to the individual two-way ANOVA error terms. Accordingly, as found with one-way MANOVA, spatial autocorrelation detected in the individual ANOVAs and redundant attribute information reflect upon this MANOVA model.

5.3.1. Fixed Effects MANOVA

Consider the geographic distribution of DCD and TDBN, with a researcher wishing to determine whether agricultural productivity differs across the five agricultural administrative regions as well as the coastal lowlands/interior highlands regions. Both of these classification schemes already are labeled as fixed effects. The individual ANOVAs for this problem suggest that, in the population, (1) there is no interaction effect, (2) there is a difference in mean vectors between the coastal lowlands and interior highlands, and (3) there is conflicting information regarding the difference of mean vectors among administrative regions.

EXAMPLE PR-5.7. Execute two-way (joint analysis) MANOVA of DCD and TDBN.

Statistic	Value	F	Num DF	Den DF	Pr > F
Wilks' Lambda	0.67367815	3.3845	8	124	0.0015
Wilks' Lambda	0.62098639	18.9206	2	62	0.0001
Wilks' Lambda	0.8040113	1.78624	8	124	0.0858

Therefore, the null hypothesis with regard to interaction effects is not rejected (at the 5% level with 0.0858 > 0.0500), allowing each classification variable to be interpreted separately. There is evidence to indicate that the regional mean vectors based upon DCD and TDBN are different in the population across the five ADM regions, as well as across the two LND regions (0.0015 and 0.0001 are both < 0.0500).

The following conventional ANOVA-format MANOVA table can be constructed from output obtained with SAS for this preceding problem, too:

TWO-WAY FIXED EFFECTS MANOVA
FOR PUERTO RICAN AGRICULTURAL PRODUCTION

SOURCE	DEGREES OF FREEDOM	GENERALIZED VARIANCE	WILKS' LAMBDA	F-RATIO	p
Between ADM regions	8, 124	11289.7	0.6737	3.4	0.0015
Between LND regions	2, 62	0.0	0.6210	18.9	0.0001
Interaction	8, 124	6751.7	0.8040	1.8	0.0858
Within cross-classifications		569910.5			

The univariate conflict affiliated with ADM has been resolved by moving from an ANOVA format to a MANOVA format.

The diagnostic statistics that accompany this preceding MANOVA table are as follows:

Within Cross-classification Groups Normality Test

REGION/	COEFFICIENTS		SHAPIRO-	PROBA-
COASTAL LOWLANDS	DCD	LN(DBN+0.025)	WILK	BILITY
San Juan	.002132	-.226136	0.8587	0.0443
Arecibo	.021572	-.648087	0.8501	0.1263
Mayaguez	.020068	.002184	0.9499	0.6825
Ponce	.003166	-.1404900	0.7993	0.0145
Caguas	-.089320	1.218780	0.9570	0.7277
REGION/				
INTERIOR HIGHLANDS				
San Juan	.046424	-.331560	0.8851	0.2600
Arecibo	-.072608	-.120272	0.7506	0.0395
Mayaguez	-.030643	-.092173	0.7457	0.0112
Ponce	.053622	.057771	0.8039	0.1055
Caguas	.070928	.187983	0.6949	0.0015

Homogeneity of Variance Test

OBSERVED CHI SQUARE	38.1609
PROBABILITY	0.0754

Partial Correlations from W

	LN(DBN+0.025)
DCD	0.1140

For this example, the homogeneity of variance test suggests that the ten 2-by-2 regional covariance matrices are not different in the population. The partial correlation reported here signifies that a weak relationship exists between DCD and LN(DBN + 0.025) but that this correlation coefficient is not significant. Meanwhile, the bivariate normality test casts doubt on the validity of the homogeneity of covariance matrices test when a 5% level of significance is used but not when a 1% level is used. As a result, caution should be exercised when interpreting this test statistic, particularly because it is close to being significant. This finding also suggests that a joint transformation of the data may yield results that better mimic the target multivariate normal distribution.

In addition, two-way ANOVA results for DCD reveal that the GLM error term renders MC = -0.0227, implying insignificant spatial autocorrelation in the population. Two-way ANOVA results for LN(DBN + 0.025) reveal that the GLM error term yields MC = 0.1822, which is significant. And, the cross-MC = -0.0738, which suggests a very weak relationship (as noted, the correlation between these two error terms is 0.1140). Consequently, spatial autocorrelation may be a complicating factor here, suggesting caution when interpreting the two-way MANOVA outcomes.

5.3.2. Mixed Effects MANOVA

Consider the geographic distribution of DCD and TDBN, with a researcher wishing to determine whether agricultural productivity differs across the five agricultural administrative regions as well as the rural-urban dichotomy classes. The second of these classification schemes already has been labeled as representing a random effect. The individual ANOVAs for this problem suggest that in the population (1) there possibly is conflicting information regarding an interaction effect, (2) there is no difference between rural and urban area mean vectors, and (3) there is possibly conflicting information regarding the difference of administrative region mean vectors. Both of these potential conflicts barely occur at a 5% level of significance.

EXAMPLE PR-5.8. A two-way joint analysis of DCD and TDBN.

TWO-WAY MIXED EFFECTS MANOVA
FOR PUERTO RICAN AGRICULTURAL PRODUCTION

SOURCE	DEGREES OF FREEDOM	GENERALIZED VARIANCE	WILKS' LAMBDA	F-RATIO	p
Between ADM regions	8, 6	4690.4	0.2039	0.9	0.5615
Between R-U areas	2, 3	0.0	0.3582	2.7	0.2144
Interaction	8, 124	5075.2	0.8065	1.7	0.0913
Within cross-classifications		686365.2			

Therefore, the null hypothesis with regard to interaction effects is not rejected, allowing each classification variable to be interpreted separately and resolving the univariate conflict. There is no evidence to indicate that the regional mean vectors based upon DCD and TDBN are different in the population across either the five ADM regions, resolving the other univariate conflict, or across the two R-U areas.

In Example PR-5.8, if R-U is treated as a fixed effect, the resulting univariate statistical decisions are consistent and suggest that the null hypothesis of equality of means across R-U should be rejected. These inferences are the reverse of those for a random effect with the respective probabilities increasing from 0.0027 to 0.1233 and from 0.0023 to 0.0555. This change in conclusion probably is attributable to specification error. Since one cross-classification group has $n_{jk} = 2$, the power of MANOVA is compromised and may be the source of any inconsistencies.

The diagnostic statistics that accompany this preceding MANOVA table are

Within Cross-classification Groups Normality Test

REGION/ COASTAL LOWLANDS	COEFFICIENTS DCD	LN(DBN+0.025)	SHAPIRO-WILK	PROBA-BILITY
San Juan	-.005824	-.247886	0.8184	0.0786
Arecibo	-.043588	.173216	0.9655	0.8704
Mayaguez	-.017553	-.082152	0.8841	0.0947
Ponce	.005889	-.106760	0.8574	0.0425
Caguas	-.053524	.462656	0.7723	0.0212
REGION/ INTERIOR HIGHLANDS				
San Juan	.003578	-.213732	0.8589	0.0359
Arecibo	.029349	.228032	0.7768	0.0653
Mayaguez	.008695	-.195450	0.9791	0.8680
Ponce	.500000	.500000	1.0000	1.0000
Caguas	-.045200	.012574	0.7752	0.0339

Homogeneity of Variance Test

OBSERVED CHI SQUARE	36.2954
PROBABILITY	0.1090

Partial Correlations from W

	LN(DBN+0.025)
DCD	0.1466

For this example the homogeneity of variance test suggests that the ten 2-by-2 regional covariance matrices are not different in the population. The partial correlation reported is not significant but signifies that a weak relationship exists between DCD and LN(DBN + 0.025). Meanwhile, the bivariate normality test casts little doubt on the validity of the homogeneity of covariance matrices test when a 0.005 Bonferroni level of significance is used (which is equivalent to an overall level of 5%). Again, caution is warranted when interpreting this test statistic. This finding suggests that a joint transformation of the data may yield results that better mimic the target multivariate normal distribution.

Finally, two-way ANOVA results for DCD, reported with Example PR-5.4, reveal that the GLM error term renders MC = 0.0200, implying insignificant spatial autocorrelation in the population. Two-way ANOVA results for LN(DBN + 0.025) reveal that the GLM error term yields MC = 0.2345, which is significant. And the cross-MC = -0.0073, which appears negligible (as noted, the correlation between these two error terms is 0.1466). Consequently, spatial autocorrelation may be a complicating factor here, suggesting caution when interpreting the two-way MANOVA outcomes.

5.4. Problems

P-5.1. Given the *Uniform Trials: Variation and Correlation in the Yield of Wheat* (Table 6.1 in Andrews and Herzberg, 1985) data set, undertake a one-way MANOVA of the yield of grain and straw per plot in pounds (this is **Y**), using the horizontal row code (1–20) as the classification variable (this is **X**). Repeat this exercise using the vertical rows (there are 25) as the classification variable.

P-5.2. Given the *Maize Fertilizer Experiments on the Islands of St. Vincent and Antigua* (Table 58.1 in Andrews and Herzberg, 1985) data set, undertake a two-way MANOVA of number of ears and weight of harvest, using "location" and "treatment" as the two classification variables.

5.5. References

Andrews, D. and H. Herzberg. 1985. *Data: A Collection of Problems from Many Fields for the Student and Research Worker*. Berlin: Springer Verlag.

Griffith, D. and C. Amrhein. 1991. *Statistical Analysis for Geographers*. Englewood Cliffs, N.J.: Prentice Hall.

CHAPTER 6

PRINCIPAL COMPONENTS AND FACTOR ANALYSIS

Chapter Preview. In this chapter, we treat principal components analysis (PCA) first, and then, by extension, develop concepts of factor analysis (FA). The coverage of these two techniques in a single chapter is based on their mathematical commonalities. It is a mistake, however, to carry the relationship too far. All too often PCA and FA are confused. "This is unfortunate, as factor analysis is both conceptually and mathematically very different from principal component[s] analysis" (Bailey and Gatrell, 1995, p. 225).

Our goals here are to develop the mathematical structure of PCA and FA, describing the similarities and differences, and to extend these techniques into an explicitly spatial setting. As part of our spatial extension, we introduce the "regression roots" as a useful avenue of handling various computational complexities that geographic space introduces into data analyses.

To begin with, a standard definition of each technique helps clarify basic similarities and differences between these two data reduction techniques. Principal components analysis is a "model in which the factors are based on the total variance. With component analysis, unities (ones) are used in the diagonal of the *correlation matrix*; this procedure computationally implies that all the variance is common or shared" (Hair et al., 1995, p. 365). Factor analysis is a "model in which the factors are based on a reduced correlation matrix. That is, [nonnegative values less than one] are inserted in the diagonal of the *correlation matrix*, and the extracted factors are based only on the *common variance*, with *specific* and *error variance* excluded" (Hair et al., 1995, p. 365).

Many of the other techniques explored in this book (e.g., regression, various analysis of variance approaches, discriminant analysis, canonical correlation) are designed to explain the variation in a set of dependent variables. In contrast, PCA and FA are designed so that the synthetic variables produced maximize the explanation of the variation in the entire data set. In other words, in PCA and FA there is no partitioning of the data set into dependent and independent variables. Rather, the entire data set is considered simultaneously with each variable relating to every other variable (see Hair et al., 1995, Chapter 7, for a lengthy discussion of this issue).

6.0. Introduction

Redundant (duplicate) information contained in data (introduced in Chapter 2) can enter through the inclusion of multiple variables (or attributes) in an analysis, where the attributes themselves are repetitious at least to some degree. Such redundant information causes multicollinearity to be present in matrix **X**. Alternatively, this redundancy can enter through the latent spatial autocorrelation frequently found in a geo-referenced variable. The multivariate problem is concerned with avoiding *double-counting* of any repeated attribute information. The techniques of principal components and factor analysis address this concern. The accompanying *spatial statistics* problem looks at avoiding *double-counting* of any duplicated locational information. This spatial concern is addressed by techniques presented in Chapter 11 that, at least conceptually, parallel principal components and factor analysis.

The problem statement associated with ***principal components analysis*** (PCA) and ***factor analysis*** (FA) may be stated as follows:

> Does a large set of variables measure different attributes or the same attribute (that is, how much redundant attribute information is contained in the data set)?

This problem statement generates three specific questions, all of which are answered by undertaking a PCA or FA either of the correlation matrix **R** (based on matrix **X** for attribute information), or of the geographic weights matrix **C** (for locational information).

> *PCA Question #1*: What are the patterns of linear relationships exhibited by a correlation matrix?

Conventionally, answering this question is equivalent to assessing the extent of multicollinearity in matrix **X** and often is addressed with a standard PCA. In terms of spatial statistics, answering this question is equivalent to treating (**I** + **C**) as a correlation matrix[1] to assess the extent of multicollinearity latent in locational information.

> *PCA Question #2*: Can a given correlation matrix be parsimoniously described?

Answering this question is equivalent to eliminating the redundant information contained in matrix **X** and often is addressed with FA, which in the geographical literature frequently is nothing more than a modified version of PCA. As such, FA alters the PCA solution by geometrically rotating usually a subset of the PCA eigenvectors. In terms of spatial statistics, the principal eigenvector of matrix **C** furnishes an index of topological centrality, which is the only parsimonious description of matrix **C** that has been widely recognized and exploited. Eigenvector rotation is not employed in this case because it would destroy this topological index interpretation.

> *PCA Question #3*: Are certain dimensions latent in a given correlation matrix?

Answering this question is equivalent to hypothesis testing and is addressed with FA, which, again, in the geographic literature usually is a modified version of PCA. Few geographic problems are sufficiently well understood to support hypotheses about the number and nature of underlying dimensions. Two exceptions include urban social areas and interpoint distance matrices.

---

[1]The identity matrix **I** is added to matrix **C** in order for the diagonal elements of the resulting pseudo correlation matrix to equal one; $r_{jj} = 1$ by definition, because any variable must be perfectly positively correlated with itself. Recall This operation increases each eigenvalue of matrix **C** by one but does not affect the eigenvectors of matrix **C** (Chapter 3).

6.1. Selected Mathematical Features of PCA

The mathematical solution to the PCA problem involves eigenvalues and eigenvectors (Chapter 3) that are selected in such a way as to measure spread along a latent direction in the attribute data space by the variance of the areal unit attribute values in that direction. Given a set of variables $\{X_j, j = 1, 2, ..., P\}$ converted to z-score mode (to remove the impact of measurement scale magnitude) and with $DET(\mathbf{Z}^T\mathbf{Z}) \neq 0$, a set of uncorrelated variables $\{F_j, j = 1, 2, ..., P\}$ always can be derived by a set of linear transformations **E**, where

$$\Lambda = \mathbf{E}^T(\mathbf{Z}^T\mathbf{Z})\mathbf{E} \ .$$

Matrix **E** is the eigenvector matrix extracted from the correlation matrix **R**, and Λ is the diagonal matrix of eigenvalues λ_j (the ordering of eigenfunctions in matrices **E** and Λ must be the same). Recall that the correlation matrix **R** is symmetric, resulting in orthogonal eigenvectors (i.e., the pairwise angles between them are 90°; property P9, Table 3.1) and the matrix **F** of uncorrelated synthetic variables $\{F_j, j = 1, 2, ..., P\}$. In other words, the original attribute data space is rotated to a new one to maximize separation in this new synthetic data space; the objective is to find a more informative frame of reference within which to view the n areal unit attributes. Matrix **F** exhibits many appealing properties, with those of most interest in multivariate analysis being

Property 1: the first synthetic variate, or component, F_1 is selected to account for the greatest amount of variance in matrix $\mathbf{Z}^T\mathbf{Z}$, the second component F_2, to account for the greatest amount of residual variance (that variance remaining after the first component is removed), and so on through to the choice of F_p;

Property 2: there are exactly the same number of Fs as there are Xs (a correct computer algorithm will allow for the existence of multiple roots, including multiple zero eigenvalues, and will always calculate mutually orthogonal eigenvectors); and,

Property 3: for a large sample size n, most of the variance in the z-score matrix **Z** will be accounted for by a relatively small number of principal components.

This discussion of common variance can be cast in terms of set theory. Recall the discussion in §4.2.2 concerning multiple regression analysis in which the squared correlation between two variables, r^2, equals the percentage of common variance for the pair of variables. Let V_j be the variance associated with variable Z_j. $V_j = 1.0$, because all X_j variables have been converted to their z-score counterparts (and the variance of a z-score = 1.0 by definition). The pictorial representation of this set theoretic description (Figure 6.1) illustrates that the task at hand is to partition the total set of variances into groups of common variation and unique components.

Another visual portrayal of this situation can be sculpted in terms of the eigenfunctions of the correlation matrix **R**. For a two-variable case, the geometric interpretation of the eigenfunctions is presented in Figure 6.2. The original graphical axes are Z_1 and Z_2. The principal eigenvector represents a line that passes through the scatter of z-score points, as in regression analysis. But now the perpendicular distances between each z-score point and this line are minimized, rather than the horizontal or vertical distances that are minimized in OLS regression. The relative length of this line is given by the principal eigenvalue. The second eigenvector must be constructed perpendicular to the first eigenvector with its relative length being given by the second eigenvalue. The second eigenvector intersects the principal eigenvector in such a way that it also minimizes the perpendicular distances between each z-score point and itself, subject to the constraint of being orthogonal to the principal eigenvector. The coefficients attached to variables Z_1 and Z_2 refer to the angles θ_1 and θ_2 formed by the intersections of the eigenvectors with the original z-score axes. These coefficients are used to create the synthetic variates F_j and are the elements contained in the individual eigenvectors.

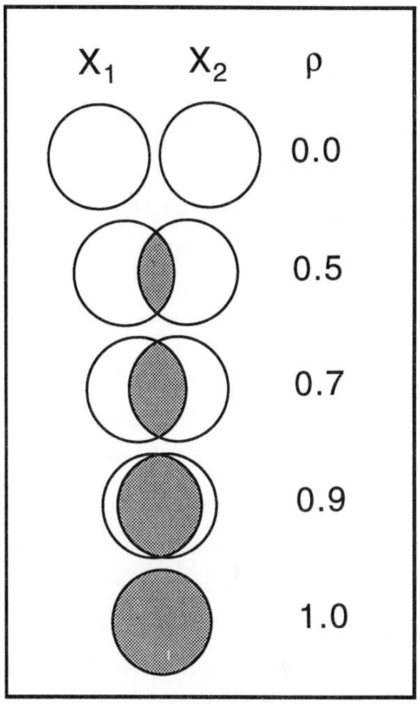

Figure 6.1. A set theoretic portrayal of common variance.

The geometry of PCA can be further exemplified by Figure 6.2. If the relationship between variables X_1 and X_2 is completely random, having an accompanying correlation coefficient of 0, then the relationship between Z_1 and Z_2 also will be random. In this case, the scatter of z-score points in the population will be such that concentric rings will describe the two-dimensional distribution of these points. The eigenvalues extracted will each equal 1, whereas the eigenvectors extracted will be the original z-score axes. If a partial relationship exists between X_1 and X_2, such that $0 < |r| < 1$, then these concentric rings become elliptical in form, and the scatter of z-score points in the population will be such that an ellipse will describe the two-dimensional distribution of these points. Finally, if the relationship between X_1 and X_2 is perfect, such that $|r| = 1$, then these concentric rings collapse to a straight line with a relative length equal to 2. This final result is consistent with regression analysis, as well as with the description of a two-dimensional scatter of z-score points when the relative length of the second eigenvector is 0.

This geometric description of data can be extended to three dimensions (Figure 6.3). Now if the pairwise relationships between variables X_1, X_2, and X_3 are completely random, with an accompanying correlation matrix of **I**, then the three-dimensional scatter of z-score points in the population will be described by concentric nested spheres. Again, the eigenvalues extracted equal 1, whereas the eigenvectors extracted equal the original z-score axes. If a partial relationship exists between X_1, X_2, and X_3, then these nested concentric spheres become ellipsoidal in form (Figure 6.3). If any pair of correlations equals 1 in absolute value, then one of the dimensions of the ellipsoid will collapse to a plane, and the resulting figures will look like those appearing in Figure 6.2. If all pairs of correlations equal 1 in absolute value, then the entire ellipsoid collapses to a straight line with relative length equal to 3. Consequently, the objective of PCA may be viewed as identifying the orthogonal axes of an ellipsoid in hyperspace.

Measures for the synthetic variates can be calculated as ***principal component scores.*** Recalling that

$$\Lambda = \mathbf{E}^T\mathbf{R}\mathbf{E} = \mathbf{E}^T(\mathbf{Z}^T\mathbf{Z})\mathbf{E} = (\mathbf{Z}\mathbf{E})^T(\mathbf{Z}\mathbf{E}) ,$$

where $\mathbf{E}^T\mathbf{E} = \mathbf{I}$, $\mathbf{E}$ is the matrix of eigenvectors, and matrix $\mathbf{ZE}$ is the projection of the z-score matrix $\mathbf{Z}$ onto the new component axes. Matrix $\mathbf{ZE} = \mathbf{F}$ defines the mutually orthogonal linear combinations of the original variables. These are the component scores as well as the synthetic variates. Similarly,

$$\mathbf{R} = \mathbf{Z}^T\mathbf{Z} = \mathbf{E}\Lambda\mathbf{E}^T = (\mathbf{E}\Lambda^{1/2})(\mathbf{E}\Lambda^{1/2})^T,$$

with $\mathbf{E}\Lambda^{1/2}$ designated as the component loadings or ***component structure matrix***. Matrix $\mathbf{E}\Lambda^{1/2}$ refers to the angles between the original z-score axes and the resultant eigenvectors, and its entries are the correlation coefficients for correlations between the synthetic variates and the z-score variables. Typically this structure matrix identifies a first principal component reflecting the average magnitude of all P original variables, whereas often it reveals that the remaining principal components basically represent contrasts between subsets of the original variables. Finally, because the sum of the eigenvalues equals the trace of matrix $\mathbf{R}$ (property P3, Table 3.1), which is P, and this trace equals the total generalized variance embraced by the matrix $\mathbf{Z}$ of variables (each Z_j variable has a variance of 1, by construction), $\frac{\lambda_j}{P}$ may be interpreted as the percentage of variance accounted for by principal component j.

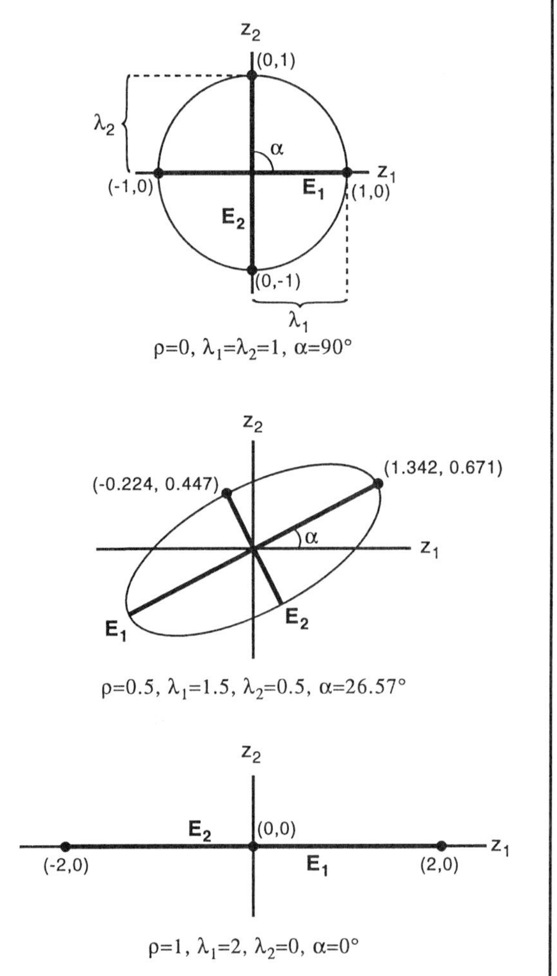

Figure 6.2 A two-dimensional geometric representation of eigenfunctions; zero correlation (r=0) [top], moderate positive correlation (r>0) [center], and perfect correlation (r=1) [bottom].

6.2. Multicollinearity

PCA originally was used to extract a single index, based upon the principal eigenvalue and its associated eigenvector, from a correlation matrix $\mathbf{R}$ derived from a data matrix $\mathbf{X}$ (hence reference to it, at times, in the singular—principal component analysis). This technique constructs a linear combination of a set of variables, much like a multiple regression equation is a linear combination of variables, to represent this set of variables in a reduced form. The first principal component, the initially hoped for single index, typically reflects the dominant trend latent in each of the P original variables, often rendering it more or less unintelligible; other principal components frequently uncover contrasts between subsets of the original variables. In PCA, these linear combinations are the synthetic variates constituting matrix $\mathbf{F}$. Rather than the ordinary least squares

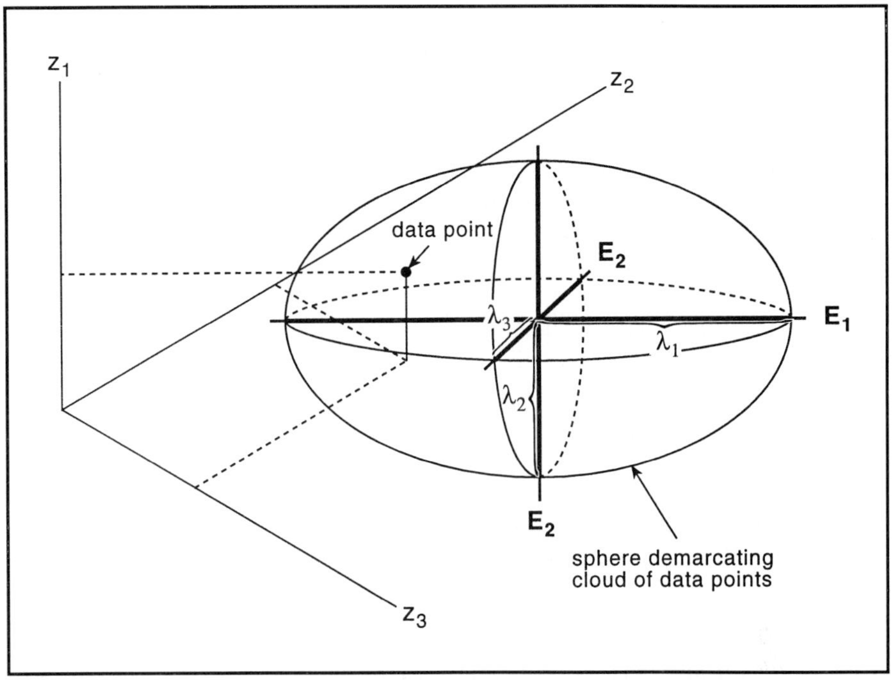

Figure 6.3. A three-dimensional geometric representation of eigenfunctions.

criterion, though, eigenvectors provide the set of coefficients that are used to construct this linear combination. In addition, whereas in regression there is a dependent variable Y used in the determination of the coefficients, in PCA only the set of independent variables X_j is used (without X_0, the intercept vector of ones; recall that the mean of a z-score is 0).

PCA can be used to answer *PCA Question #1* in two different ways. On the one hand, it can be used to reduce redudant information in a correlation matrix of n observations in matrix **X**; a topic addressed in Chapter 8.[2] On the other hand, it can be used to reduce redundant information in a correlation matrix of the P variables making up matrix **X**. Hence the correlation matrix **R** is a square P-by-P matrix. This goal is equivalent to a discrimination between groups of similar variables (and is called R-mode in the applied statistics literature). Variables are grouped together by PCA in such a way that the correlations among them within a group are maximized, whereas the correlations among variables in different groups are minimized.

EXAMPLE PR-6.1. Demonstrate that PCA removes the multicollinearity in the 5-by-5 correlation matrix of distances between municipios and the major urban areas in Puerto Rico (see Map 2.1).

Let DISTSJ denote distance from San Juan, DISTA denote distance from Arecibo, DISTM denote distance from Mayaguez, DISTP denote distance from Ponce, and DISTC denote distance from Caguas. SAS designates the synthetic variables produced by **ZE** as PRIN1, PRIN2, ..., PRIN5.

Using the SAS procedures

[2]In this case, the correlation matrix **R** is a square n-by-n matrix, where the correlation between observations I and j is calculated with the pairs of variable values. This analysis is equivalent to a discrimination between groups of similar areal units (and is called Q-mode in the applied statistics literature). Areal units are grouped together in such a way that the correlations among them within a group are maximized, whereas the correlations among areal units in different groups are minimized. Often such grouping is attained through the use of cluster analysis (Chapter 8).

```
PROC PRINCOMP OUT=TEMP1 (REPLACE=YES);
   VAR DISTSJ DISTA DISTM DISTP DISTC;
RUN;
PROC CORR;
   VAR PRIN1-PRIN5;
   WITH DISTSJ DISTA DISTM DISTP DISTC;
RUN;
```

produced SAS output that includes

Correlation Matrix

	DISTSJ	DISTA	DISTM	DISTP	DISTC
DISTSJ	1.0000	-.3559	-.8714	-.2577	0.9690
DISTA	-.3559	1.0000	0.7110	0.7577	-.4406
DISTM	-.8714	0.7110	1.0000	0.5616	-.8961
DISTP	-.2577	0.7577	0.5616	1.0000	-.2266
DISTC	0.9690	-.4406	-.8961	-.2266	1.0000

Eigenvalues

PRIN1	PRIN2	PRIN3	PRIN4	PRIN5
3.47747	1.23146	0.24277	0.03419	0.01411

Eigenvectors

	PRIN1	PRIN2	PRIN3	PRIN4	PRIN5
DISTSJ	-.465321	0.420471	0.279882	0 .377079	-.621416
DISTA	0.405898	0.500558	0.689455	- .280527	0.175056
DISTM	0.528475	-.056869	0.047387	0 .840141	0.096939
DISTP	0.335862	0.638074	-.659122	- .109758	-.183220
DISTC	-.476061	0.402831	-.098133	0 .247452	0.735004

Principal Components Structure Matrix

	PRIN1	PRIN2	PRIN3	PRIN4	PRIN5
DISTSJ	-0.86773	0.46660	0.13790	0.06972	-0.07383
DISTA	0.75692	0.55547	0.33971	- 0.05187	0.02080
DISTM	0.98550	-0.06311	0.02335	0.15535	0.01152
DISTP	0.62631	0.70808	-0.32476	- 0.02029	-0.02177
DISTC	-0.88776	0.44703	-0.04835	0.04575	0.08732

The pairwise correlations reported by SAS indicate that multicollinearity may well be a serious problem in this data set (see *RULE OF THUMB 4.4*). This result is not surprising; because San Juan and Caguas are very close neighbors, distances between each of these cities and a given municipio will be very similar (hence, $r_{SJ,C} = 0.9690$). Further, as distance increases between either of these cities and municipios lying west of them, distance between Magaguez and the same municipios decreases (hence, $r_{C,M} = -0.8961$). Meanwhile, two of the eigenvalues from matrix **R** are relatively close to 0. Multiplying each eigenvector entry by the square root of its corresponding eigenvalue yields the structure matrix, $\mathbf{E}\Lambda^{1/2}$; this also can be obtained by the foregoing SAS PROC CORR procedure. In other words, for the loading of DISTSJ on PRIN1, $(3.47747)^{1/2}\times(-.465321) = -0.86773 = r_{DISTSJ,PRIN1}$.

Limits for PCA solutions can be established by studying the two possible extreme cases. If every entry in matrix **R** is unity (i.e., there is complete and perfect multicollinearity), then one group will be produced. If matrix **R** is the identity matrix **I** (i.e., there is no multicollinearity), then P distinct groups will be produced. In practice, most correlation matrices fail to have either absolutely no multicollinearity or complete/perfect multicollinearity, and thus the number of groups, say G, produced is such that 1 < G < P.

EXAMPLE PR-6.2. Calculate the limiting cases for multicollinearity in a 5-by-5 correlation matrix, based on distances between municipios and the five major urban areas in Puerto Rico.

Consider the synthetic variates (i.e., PRIN1, PRIN2, ..., PRIN5) that are produced by the PCA referred to in Example PR-6.1. Using the SAS procedures

```
PROC PRINCOMP DATA=TEMP1;
   VAR PRIN1 PRIN1 PRIN1 PRIN1 PRIN1;
RUN;
PROC PRINCOMP DATA=TEMP1;
   VAR PRIN1-PRIN5;
RUN;
```

produced SAS output that includes

Correlation Matrix

	PRIN1	PRIN1	PRIN1	PRIN1	PRIN1
PRIN1	1.0000	1.0000	1.0000	1.0000	1.0000
PRIN1	1.0000	1.0000	1.0000	1.0000	1.0000
PRIN1	1.0000	1.0000	1.0000	1.0000	1.0000
PRIN1	1.0000	1.0000	1.0000	1.0000	1.0000
PRIN1	1.0000	1.0000	1.0000	1.0000	1.0000

Eigenvalues

PRIN1	PRIN2	PRIN3	PRIN4	PRIN5
5.00000	0.00000	0.00000	0.00000	0.00000

Correlation Matrix

	PRIN1	PRIN2	PRIN3	PRIN4	PRIN5
PRIN1	1.0000	0.0000	0.0000	0.0000	0.0000
PRIN2	0.0000	1.0000	0.0000	0.0000	0.0000
PRIN3	0.0000	0.0000	1.0000	0.0000	0.0000
PRIN4	0.0000	0.0000	0.0000	1.0000	0.0000
PRIN5	0.0000	0.0000	0.0000	0.0000	1.0000

Eigenvalues

PRIN1	PRIN2	PRIN3	PRIN4	PRIN5
1.00000	1.00000	1.00000	1.00000	1.00000

The pairwise correlations in the first part of this example, where variable PRIN1 is repeated five times, all equal 1 because every variable is perfectly correlated with itself. Every entry in the resulting correlation matrix is exactly 1. As expected, PCA identifies a single nonzero eigenvalue (which equals P = 5) for this limiting case. Meanwhile, the pairwise correlations in the second example yield the identity matrix, **I**. Here the five synthetic variables from Example PR-6.1 are guaranteed to be mutually orthogonal because they have been constructed with an earlier PCA. As expected, PCA identifies five identical eigenvalues (all of which equal 1) for this second limiting case.

6.3. Moving From PCA to Factor Analysis: Seeking Parsimony

Factor analysis can be viewed as a technique through which a few prespecified synthetic variables are constructed so that, once all of these hypothetical variables are controlled for, correlations between the residual parts of the original P variables become 0. One fundamental difference between PCA and FA, then, is that PCA is concerned with accounting for variances of the original P variables through axis rotation, whereas FA is concerned with explaining correlations among these original variables with a model consisting of the posited number of common factors. There are many facets of factor analysis. Only one will be discussed here. One form of factor analysis begins with a PCA of the data matrix **X**. Principal components analysis constructs components in a sequential way such that the first component accounts for a maximum of common variance, the second component accounts for a maximum of residual common variance (once this variance has been adjusted so that the variance accounted for by the first component is removed) subject to its being orthogonal to the first, and so on. But the resulting solution is not necessarily interpretable. Moreover, this criterion of sequential variance maximization is somewhat arbitrary, only identifying the principal patterns of variation without necessarily identifying distinctive clusters of variables. Interpretability often is enhanced and distinctive cluster of variables uncovered by rotating the principal components solution. In terms of interpretability, such simplicity is a preferred solution. Geometrically speaking, rotation refers to the procedure called orthogonal transformation (Figure 6.4). This rotation takes one into the realm of factor analysis.

Rotation replaces the sequential variance maximization criterion by some alternate objective criterion that attempts to simplify the structure revealed by the components structure matrix in some unique manner. Once rotated, this matrix is relabeled the ***factor structure matrix***. Various principles of simple structure can be invoked to achieve this rotation. Generally speaking, these principles should attempt to alter the initial loadings matrix in such a way that (1) each row has at least one of its loadings (which are correlation coefficients) close to 0, (2) there are at least as many variables with near-0 loadings as there are factors, (3) for

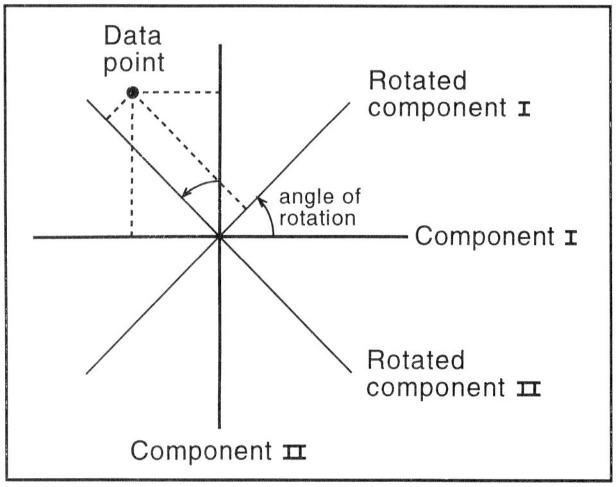

Figure 6.4. Graphical depiction of axis rotation.

every pair of columns there are several variables with high loadings in one column but not in the other, (4) a large proportion of the variables have near-0 loadings on any pair of columns, and (5) only a small number of column pairs have high non-0 loadings for the same variables. There are three rotation principles of interest, namely

(1) The set of eigenvectors should be selected so that all variances accounted for are simultaneously maximized. This rotation emphasizes column simplification in the structure matrix. It is called the VARIMAX rotation and is the most widely used rotation principle in multivariate geography applications.

(2) The set of eigenvectors should be selected so that variables load highly onto one and only one factor. This rotation emphasizes row simplification in the structure matrix and is called the QUARTIMAX rotation. In QUARTIMAX, many variables can load high on the same factor. Frequently, QUARTIMAX yields a prominent general factor that captures most variables and subsequent factors that tend to encompass subclusters of variables.

(3) A compromise between VARIMAX and QUARTIMAX that attempts the simultaneous simplification of rows and columns is called EQUIMAX.

Algebraically the rotation problem is one of finding some matrix $\mathbf{L}$, such that $\mathbf{L}^T\mathbf{L} = \mathbf{I}$ (the rotation matrix is orthogonal and its columns are normalized). This matrix then is applied to the PCA solution such that

$$\mathbf{L}^T\Lambda\mathbf{L} = \mathbf{L}^T\mathbf{E}^T\mathbf{R}\mathbf{E}\mathbf{L} = \mathbf{L}^T\mathbf{E}^T(\mathbf{Z}^T\mathbf{Z})\mathbf{E}\mathbf{L} = (\mathbf{E}\mathbf{L})^T(\mathbf{Z}^T\mathbf{Z})(\mathbf{E}\mathbf{L}) \ .$$

The necessity for matrix $\mathbf{L}$ to have normalized and orthogonal columns becomes clear by inspecting what happens to this equation if the original variables are uncorrelated, meaning $\mathbf{R} = \mathbf{I}$. As such,

$$\mathbf{L}^T\mathbf{E}^T\mathbf{I}\mathbf{E}\mathbf{L} = \mathbf{L}^T\mathbf{E}^T\mathbf{E}\mathbf{L} = \mathbf{L}^T\mathbf{L} \ ,$$

which needs to be the identity matrix. Because matrix $\mathbf{L}$ operates on the diagonal matrix of eigenvalues, Λ, it will cause a redistribution of the value P over the diagonal entries. Since PCA extracts the maximum set possible, subject to their summing to P, then matrix $\mathbf{L}$ will tend to decrease the larger eigenvalues and increase the smaller eigenvalues. These new eigenvalues have the corresponding eigenvectors $\mathbf{EL}$ associated with them. There are literally an infinite number of possible $\mathbf{L}$ matrices, but the three structure (rotation) principles described allow most of them to be dismissed.

6.4. PCA/FA and Hypothesis Testing

In a quest for parsimony (*PCA Question #2*), the factor analytic modification of PCA focuses attention on a small (often substantially less than P) number of the components. Thus, criteria are needed for selecting those components to be rotated. Assuming that matrix **X** is a random sample drawn from a multivariate normal distribution having mean $\boldsymbol{\mu}$ and variance-covariance matrix Σ and that the λ_js are nonnegative and distinct (no repeating eigenvalues), then significance tests can be applied to PCA results. The purpose of this type of test is to distinguish between those principal components summarizing important commonalities, which then can be retained for subsequent analysis, and those principal components that can be dismissed as nothing more than random noise (the notion of parsimony). This test may be conducted in several different ways. First, one could test the pair of hypotheses

$$H_0:\ \lambda_j = 1 \quad \text{versus} \quad H_A:\ \lambda_j = 0\ .$$

For large n, the estimated values $\hat{\lambda}_j$ are independent and normally distributed with mean $\mu = \lambda_j$ and variance $\sigma^2 = \frac{2\lambda_j^2}{n}$. Thus the test statistic is z, with this statistic being calculated as

$$z = \frac{(\hat{\lambda}_j - \lambda_j)}{\lambda_j \sqrt{\frac{2}{n}}}\ ,$$

which reduces considerably under H_0: $\lambda_j = 1$, to test whether the null hypothesis is true. Two drawbacks are affiliated with this test. First, what is a large n? Second, because a specific alternate hypothesis has been identified, the power of the test should be evaluated; that is, what is the probability of a Type II error in this case?

To circumvent the two weaknesses associated with this first testing situation, one can simultaneously test the vector of eigenvalues λ. This test is based upon the normality assumptions outlined earlier, and leads to the chi square statistic (the sum-of-squared normal random variables). This joint test assumes that the vector $\sqrt{n}\ (\hat{\lambda}_j - 1)$ is distributed multivariate normal with $\boldsymbol{\mu} = \mathbf{0}$ and a variance-covariance of $2\mathbf{I}$, yielding the chi square test statistic $\chi^2 = n\sqrt{2\sum_{j=1}^{k}(\lambda_j - 1)^2}$, with k degrees of freedom. As in all chi square computations, the marginal addition of each eigenvalue can be evaluated by successively subtracting each eigenvalue term from the remainder, as well as subtracting a degree of freedom each time. These subtractions are sequential, beginning with the largest eigenvalue and k degrees of freedom. Accordingly, k = P for the first test, k = P - 1 once the first eigenvalue has been removed, and so forth.

Another joint test is for the null hypothesis

$$H_0:\ \lambda_{P-k} = \lambda_{P-k+1} = \ldots = \lambda_P\ .$$

This test is based upon the notion of generalized variance, which begins as *DET*(**R**). Let Q be the ratio of the actual volume to the maximum possible volume of the data parallelepiped.

$$Q = \frac{\prod_{j=1}^{t} \lambda_j}{\left[(\sum_{j=1}^{t} \lambda_j) t^{-1}\right]^t},$$

Recall that redundant information converts a sphere to an ellipsoid and that the ellipsoid is smaller in volume than the sphere. In the case of PCA, H_0 represents a further reduction by collapsing the sphere onto a line. The line, which has no volume, indicates the presence of perfect and complete multicollinearity. Then the modified test statistic $-(n-P-\frac{1}{2})LN(Q)$ follows a chi square distribution having $\frac{(t+2)(t-1)}{2}$ degrees of freedom, where t sequentially equals P, P - 1, P - 2, ..., 2, 1. In other words, first t = P, and Q is tested for significance. If Q is not significant, then all eigenvalues are expected to be equal to one in the population. If Q is significant, then the first eigenvalue is removed, the numerator and denominator of Q recalculated for t = P - 1, the degrees of freedom set to $\frac{(P+1)(P-2)}{2}$, and this updated version of Q tested for significance. This sequential testing continues until the computed chi square statistic no longer is significant. The test first hypothesizes that all eigenvalues are equal, and if this hypothesis is rejected, then λ_1 is considered to differ from the others, and the subset of (P - 1) eigenvalues is tested for equality. A major drawback with this hypothesis testing procedure is that the eventual subset of equal eigenvalues that is found may have values substantially less than 1.

Less rigorous criteria have been devised for evaluating eigenvalues for several reasons: (1) the assumption of multivariate normality often is questioned, (2) sufficiently large n often are unobtainable with geo-referenced data series, and (3) the drawbacks of the preceding significance tests are of concern.

RULE OF THUMB 6.1. Only eigenvalues greater than 1 are important.

This criterion is based upon the idea that if matrix **R** is the identity matrix, then all of its eigenvalues equal 1. Because the trace of matrix **R** always equals P, regardless of the extent of multicollinearity, then as variables become increasingly intercorrelated, some eigenvalues must increase while others must decrease. Hence, only those eigenvalues experiencing an increase are deemed important here. This criterion often is the default found in commercial software packages. Unfortunately, as P increases, the number of eigenvalues that are greater than 1 also tends to increase, with many eigenvalues just marginally greater than 1. This liability sometimes is dealt with by classifying only those eigenvalues accounting for a specified minimum amount of variance as important.

RULE OF THUMB 6.2. Only eigenvalues accounting for at least 5% of the total variance are important.

The amount 5% frequently is recognized as a nontrivial, minimum amount of variance (this criterion also is used sometimes with regression; see §4.2.2). But, this 5% rule of thumb criterion also suffers when P increases. Achieving a preset percentage of the total variance becomes increasingly more difficult, whereas 100/P continually decreases, as P increases. A third general criterion is furnished by the scree diagram, a two-dimensional plot of the eigenvalues in decreasing order versus the percent of variance accounted for by each eigenvalue. Here one seeks a distinct elbow in the slope of the curve. Ideally, the plot will suggest two curves that intersect. This point of intersection identifies the number of important principal components.

> *RULE OF THUMB 6.3.* When plotting the percent of variance accounted (on the vertical axis) versus the respective ranks of the corresponding eigenvalues (ranked in descending order), only those first few eigenvalues aligning in roughly a linear fashion are deemed important.

> EXAMPLE PR-6.3. Execute a modified PCA for the eight agricultural production density variables, based upon $\lambda_j > 1$ and a varimax rotation.
>
> Using the SAS procedures
>
> ```
> PROC FACTOR DATA=STEP1 OUT=TEMP1(REPLACE=YES)
>    NFACT=4 ROTATE=VARIMAX;
> VAR DFRM DCD DMLK DSGR DCF DTB DBN DFAM;
> RUN;
> ```
>
> produced SAS output shown on the following page.

The criterion of $\lambda_j > 1$, used to identify important components, has been chosen because (1) no evidence has been obtained (yet) indicating whether these variables conform to a multivariate normal distribution, (2) a 5% variance accounted for threshold embraces a small eigenvalue (i.e., 0.4811), and (3) the scree test reveals an elbow after the fourth component (another one does materialize after the first component, too; see Figure 6.5). Meanwhile, the varimax rotation undoubtedly helps to clarify the attribute structure uncovered here. Although the initial fourth component (FACTOR4) remains unchanged (becoming FACTOR3 after rotation), DCD is moved from the initial first component while DSGR is removed from the initial second component. DCD and DSGR are coupled and appear as a resulting rotated fourth factor. Rotation also separates DMLK and DSGR. Consequently, the first rotated factor may be labeled rural farming, since it is positively related to farm density (DFRM), farm family density (DFAM), and coffee (DCF) and banana/plantain (DBN) production (two crops tending to be produced on the numerous small farms of the interior highlands). The second rotated factor may be labeled milk production. The third rotated factor may be labeled tobacco production and the fourth, sugar cane production (which is cultivated land extensive).

Noteworthy from a geographical point of view is that there is no guarantee that the maximum attribute separations represented by principal components are necessarily those that correspond to the arrangement of areal units constituting a geographic distribution. "[P]rincipal

Eigenvalue	3.1278	1.3521	1.1819	1.1250	0.4811	0.3448	0.2466	0.1407
Proportion	0.3910	0.1690	0.1477	0.1406	0.0601	0.0431	0.0308	0.0176
Cumulative	0.3910	0.5600	0.7077	0.8483	0.9085	0.9516	0.9824	1.0000

Initial Factor Structure Matrix

	FACTOR1	FACTOR2	FACTOR3	FACTOR4
DFRM	0.76396	0.04577	-0.03529	0.24263
DCD	0.56880	-0.21011	0.67176	0.20253
DMLK	-0.10458	0.70671	0.52970	-0.36604
DSGR	-0.30552	-0.73644	0.51922	0.07570
DCF	0.84003	-0.25957	-0.21500	-0.24307
DTB	0.20388	0.35273	-0.00514	0.87400
DBN	0.88165	-0.11976	-0.21258	-0.16467
DFAM	0.76928	0.24072	0.29633	-0.18790

Varimax Rotated Factor Structure Matrix

	FACTOR1	FACTOR2	FACTOR3	FACTOR4
DFRM	0.69381	-0.02840	0.40336	0.03116
DCD	0.46914	0.22602	0.27001	0.71825
DMLK	-0.14472	0.94559	-0.06297	-0.07672
DSGR	-0.29268	-0.24772	-0.24101	0.84017
DCF	0.90807	-0.18447	-0.13995	-0.01091
DTB	-0.00543	-0.04058	0.96237	-0.04554
DBN	0.92002	-0.10826	-0.00848	-0.07569
DFAM	0.73431	0.46103	0.10719	0.09796
Eigenvalues	3.018346	1.267345	1.254954	1.246135

components analysis will not necessarily be of use if the objective of an analysis is to determine which combinations of attributes demonstrate the most significant spatial pattern. Including the spatial coordinates [of the areal units] as extra variables into the principal components analysis is not particularly helpful in resolving this problem either, since the 'geography' is then treated on an equal basis [with] any other attribute" (Bailey and Gatrell, 1995, p. 224).

6.5. PCA/FA and Spatial Autocorrelation

Three types of analyses appear in the literature combining PCA with spatial autocorrelation. One poses the three PCA questions listed at the beginning of this

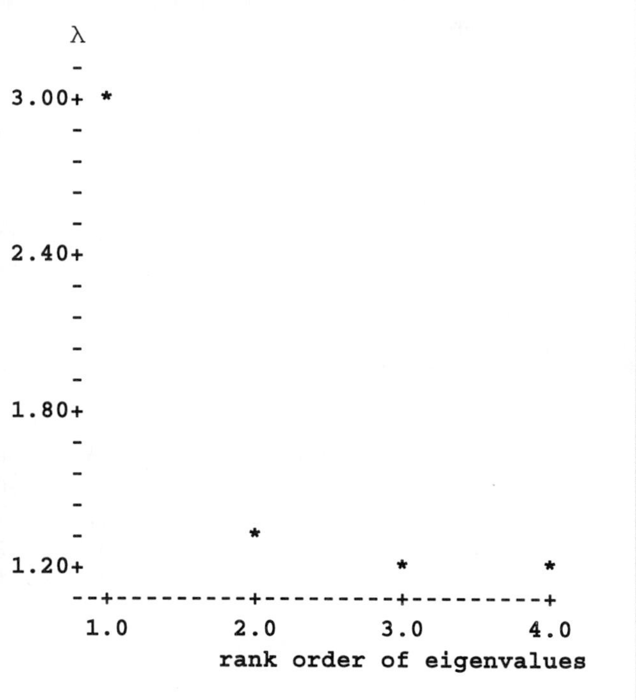

Figure 6.5. Scree plot for eigenvalues from the varimax rotated solution.

chapter in terms of a matrix of cross-MCs (recall that MC denotes the Moran Coefficient; see Wartenberg, 1985). The second addresses parsimony of the geographic weights matrix **C**. And the third variety of analysis asks these questions about a correlation matrix constructed from spatially lagged principal component scores (Switzer, 1986).

Because the expected value of a cross-MC for variables h and j is $\frac{-\rho_{hj}}{n-1}$, where ρ_{hj} denotes the traditional product moment correlation between the two variables h and j, there may well be modest to very strong similarity between PCA results for matrix **R** constructed from a set of variables, on the one hand, and for their corresponding cross-MCs, on the other hand.

EXAMPLE PR-6.4. Calculate the nature and degree of multicollinearity in cross-MCs (Example PR-2.3 revisited).

Using the SAS procedures

```
DATA COVAR(TYPE=COV);
   INFILE CROSSMC;
   INPUT MCDFRM MCDCD MCDMLK MCDSGR MCDCF MCDTB
       MCDBN MCDFAM;
   _TYPE_ = 'COV';
RUN;
PROC FACTOR DATA=COVAR METHOD=PRIN COV N=4
   ROTATE=VARIMAX;
   VAR MCDFRM MCDCD MCDMLK MCDSGR MCDCF MCDTB
       MCDBN MCDFAM;
RUN;
```

produced SAS output that includes

Eigenvalue	1.4168	0.6585	0.3181	0.1988	0.1610	0.0107	-0.0037	-0.0542
Proportion	0.5236	0.2434	0.1176	0.0735	0.0595	0.0040	-0.0014	-0.0200
Cumulative	0.5236	0.7669	0.8845	0.9579	1.0174	1.0214	1.0200	1.0000

Varimax Rotated Factor Structure Matrix

	FACTOR1	FACTOR2	FACTOR3	FACTOR4
MCDFRM	0.31099	0.92748	0.25758	-0.05147
MCDCD	0.61759	0.21178	-0.08002	-0.65248
MCDMLK	-0.10702	0.03407	-0.04145	0.92129
MCDSGR	-0.32008	-0.01151	-0.77541	-0.54793
MCDCF	0.86342	0.33625	-0.24439	-0.09983
MCDTB	-0.17909	0.24434	0.93282	-0.18383
MCDBN	0.93271	0.26552	0.20292	-0.12601
MCDFAM	0.64315	0.74055	0.04527	0.03721

In this problem the cross-MCs are treated as covariances [hence the use of COVAR(TYPE=COV)]. One of the complicating outcomes of this declaration in SAS is the presence of negative eigenvalues (further discussion of this peculiarity is beyond the scope of this book). Four factors have been retained for rotation, in keeping with Example PR-6.3. The interpretation in this case fails to be as clear-cut as that for the preceding attribute data case because DFAM, DCD, and DSGR all load markedly onto two rotated factors. DMLK and DSGR are not separated by the varimax rotation. However, the first, third, and fourth rotated factors display very similar structure to their counterparts extracted from the attribute data.

As discussed in §3.3, the purpose of a geographic weights matrix **C** is to capture the location information part of geo-referenced data. The principal eigenvalue furnishes an overall summary index of the degree of topological connectedness for a surface partitioning, whereas the associated principal eigenvector provides an areal unit-by-areal-unit index of topological centrality.

EXAMPLE PR-6.5. Calculate a parsimonious description of matrix **C** for the municipio partitioning of Puerto Rico.

Using the SAS procedures

```
DATA CORREL(TYPE=CORR);
   SET STEP2;
   _TYPE_ = 'CORR';
RUN;
PROC FACTOR DATA=CORREL
    OUTSTAT=TEMP2(REPLACE=YES)
   METHOD=PRIN N=1 EV;
   VAR C1-C73;
RUN;
PROC TRANSPOSE OUT=TEMP2(REPLACE=YES) PREFIX=VAR;
   VAR C1-C73;
RUN;
DATA TEMP2(REPLACE=YES);
   SET TEMP2;
EIGENVAL = VAR79 - 1;
EIGENVCP = VAR80;
RUN;
```

produced SAS output that includes

OBS	EIGENVAL	EIGENVCP	OBS	EIGENVAL	EIGENVCP	OBS	EIGENVAL	EIGENVCP
1	5.50193	0.26871	25	0.60500	0.06706	50	-1.50277	0.44041
2	5.17230	0.04554	26	0.37613	0.22269	51	-1.53776	0.75268
3	4.85617	0.03379	27	0.26905	0.10675	52	-1.58869	0.25633
4	4.32879	0.40924	28	0.17413	0.18641	53	-1.62383	0.12669
5	3.98437	0.50227	29	0.14188	0.23795	54	-1.73581	0.32158
6	3.95364	0.11122	30	0.10095	0.20042	55	-1.85003	0.06731
7	3.83436	0.23434	31	-0.11577	0.05464	56	-1.87425	0.02849
8	3.37842	0.08706	32	-0.12907	0.07589	57	-1.92521	0.07015
9	3.28026	0.17630	33	-0.19238	0.06331	58	-1.96425	0.11257
10	2.79956	0.70356	34	-0.29703	0.45288	59	-1.97179	0.38667
11	2.62981	0.36366	35	-0.47824	0.50485	60	-2.02906	0.11040
12	2.49041	0.06455	36	-0.49059	0.15910	61	-2.07034	0.25269
13	2.24493	0.45358	37	-0.53472	0.06445	62	-2.11464	0.35542
14	2.19156	0.13689	38	-0.60874	0.31531	63	-2.17255	0.17014
15	2.01379	0.14341	39	-0.67452	0.16501	64	-2.22608	0.27667
16	1.89030	0.49516	40	-0.81317	0.18130	65	-2.26787	0.35835
17	1.68543	0.02573	41	-0.91945	0.10068	66	-2.33547	0.15852
18	1.49305	0.49646	42	-0.94338	0.01406	67	-2.40732	0.19768
19	1.38026	0.54739	43	-1.03565	0.23918	68	-2.42148	0.41616
20	1.19199	0.63067	44	-1.10203	0.18555	69	-2.47624	0.30923
21	1.07740	0.44789	45	-1.18139	0.07787	70	-2.54057	0.18327
22	0.94561	0.54965	46	-1.22020	0.12566	71	-2.66084	0.34319
23	0.81514	0.15015	47	-1.29816	0.07706	72	-2.68042	0.17209
24	0.79106	0.00723	48	-1.38155	0.45992	73	-2.76731	0.19192
			49	-1.43706	0.06417			

The first part of the SAS code declares matrix **C** to be a correlation matrix, which conceptually is in keeping with spatial autocorrelation analysis. In a previous DATA operation the identity matrix **I** was added to matrix **C**, because all variables must be perfectly correlated with themselves (i.e., r_{jj} = 1). This modification is why the eigenvalues must have 1 subtracted from them in DATA TEMP2 (property P4, Table 3.1). The principal eigenvalue value of 5.50193 suggests that the surface partitioning of Puerto Rico topologically is far from being maximally connected, as is expected for a planar surface where n is not small. The code N=1 and EV produce the principal eigenvector, which is guaranteed to be positive. According to this index, municipio #51 (Orocovis) has the largest value of EIGENVCP and topologically is most centrally positioned in the island's surface partitioning. At the same time, municipio #24 (Fajardo) has the smallest value of EIGENVCP and topologically is most remotely positioned in this surface partitioning. These findings are consistent with a visual inspection of the boundary map of Puerto Rico (Map 1.1).

Switzer and Green (1984) argue that a conventional PCA of geo-referenced data is rather naive, because it ignores spatial properties of the data. They propose, instead, that spatially lagged principal component scores should be subjected to a PCA. Their approach seeks to separate signal from noise in geo-referenced data. They contend that components with minimum spatial autocorrelation concentrate and isolate noise. In terms of synthetic variables, synthetic variates with large eigenvalues have minimal spatial autocorrelation and mainly identify noise, whereas synthetic variates with small eigenvalues have maximal spatial autocorrelation and mainly identify geographic pattern. Switzer (1986) recommends a construction of the form $\mathbf{F}_j^* = \mathbf{F}_j - \mathbf{W}\mathbf{F}_j$ (a refinement of this specification is presented in Chapter 12). In the following example, the spatially lagged variables $\mathbf{WF}_j$ have been constructed elsewhere (Chapter 12) and then appended to the original data set.

EXAMPLE PR-6.6. Execute a PCA of spatially lagged principal component scores from Example PR-6.3.

Using the SAS procedures

```
PROC PRINCOMP DATA=STEP1 STD
    OUT=TEMP1(REPLACE=YES);
   VAR DFRM DCD DMLK DSGR DCF DTB DBN DFAM;
RUN;
  ⋮
PROC PRINCOMP DATA=STEP1 STD
    OUT=TEMP1(REPLACE=YES);
   VAR WPC1-WPC8;
RUN;
```

produced SAS output that includes

Principal Components Structure Matrix

	PRIN1	PRIN2	PRIN3	PRIN4	PRIN5	PRIN6	PRIN7	PRIN8
WPC1	0.11085	-0.36974	0.29849	0.13722	0.77574	0.10347	-0.11299	-0.34327
WPC2	0.15516	0.47217	0.47767	-0.37454	0.27914	0.23293	0.01296	0.50218
WPC3	0.27842	0.33829	0.53374	0.24357	-0.23560	-0.51452	-0.14595	-0.34972
WPC4	0.11651	0.34171	-0.43253	0.52426	0.39845	-0.30008	-0.15943	0.36537
WPC5	-0.58072	-0.05418	0.28412	0.20335	0.12085	-0.30770	0.63238	0.16902
WPC6	-0.35322	0.20632	0.21488	0.56737	-0.17312	0.61354	-0.23633	-0.04838
WPC7	0.60239	0.02823	-0.05309	0.29131	-0.07652	0.31362	0.66359	-0.06391
WPC8	0.23361	-0.60729	0.28878	0.24407	-0.23545	-0.08688	-0.19801	0.57573

Eigenvalue

1.0069	1.0060	1.0023	1.0010	0.9998	0.9968	0.9950	0.9923

Of note is that for this particular example, the spatially lagged synthetic variables are close to being orthogonal, suggesting that there is little difference in the spatial autocorrelation being concentrated on individual components. The component score mappings for PRIN1 and PRIN8 confirm this contention (Map 6.1). This finding is somewhat contradictory to that from a PCA of the cross-MCs (Example PR-6.4), for which a structural shift due to the presence of nonzero spatial autocorrelation is detected.

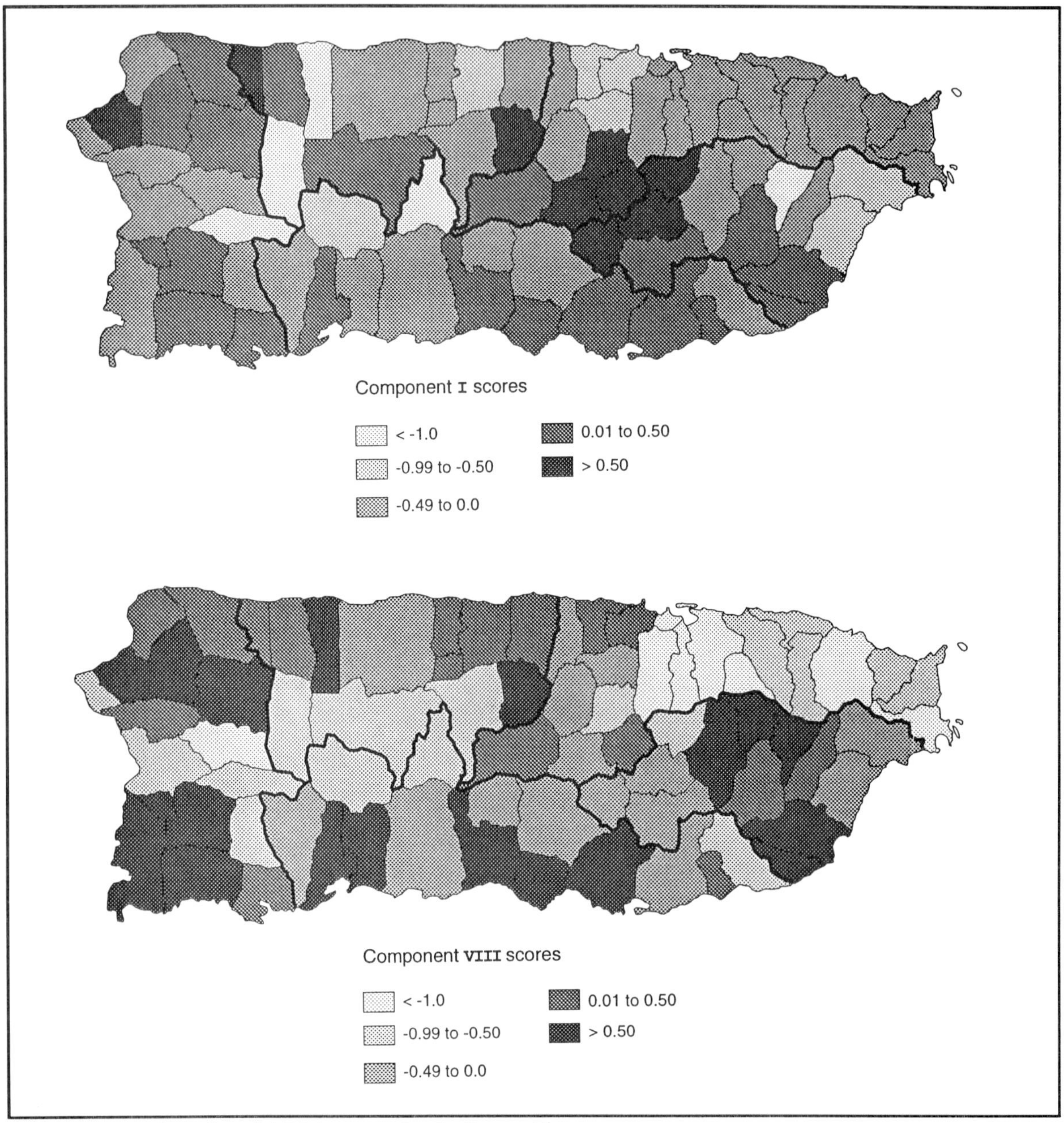

Map 6.1. Mapping of spatial PCA scores, for the first and last components.

6.6. PCA and Dimensional Hypothesis Testing

PCA can be used to answer *PCA Question #3* in those rare instances where the underlying dimensionality of geo-referenced data is theoretically known. One such case involves interpoint distance matrices. Consider an n-by-n matrix **D** whose d_{ij} elements are the squared distances, δ_{ij}^2, between the n points I and j in the geographic landscape (keep in mind that $\delta_{ii} = 0$). As Example PR-6.1 demonstrates, such a distance matrix will contain considerable, serious multicollinearity. Three known dimensions are latent in matrix **D**: one is related to topological centrality, similar to the principal eigenvector extracted from matrix **C** and two are related to the original (u, v) Cartesian coordinate system used to compute δ_{ij}.

EXAMPLE PR-6.7. Calculate the nontrivial dimensions latent in the 73-by-73 squared distance matrix for Puerto Rico.

Using the SAS procedures

```
DATA CORREL(TYPE=CORR);
   SET STEP2;
   _TYPE_ = 'CORR';
RUN;
PROC FACTOR DATA=CORREL
        OUTSTAT=TEMP2(REPLACE=YES)
         METHOD=PRIN N=73 EV;
   VAR D1-D73;
RUN;
```

produced SAS output that includes

Eigenvalues							
2.2338	1.0000	1.0000	1.0000	1.0000	1.0000	1.0000	1.0000
1.0000	1.0000	1.0000	1.0000	1.0000	1.0000	1.0000	1.0000
1.0000	1.0000	1.0000	1.0000	1.0000	1.0000	1.0000	1.0000
1.0000	1.0000	1.0000	1.0000	1.0000	1.0000	1.0000	1.0000
1.0000	1.0000	1.0000	1.0000	1.0000	1.0000	1.0000	1.0000
1.0000	1.0000	1.0000	1.0000	1.0000	1.0000	1.0000	1.0000
1.0000	1.0000	1.0000	1.0000	1.0000	1.0000	1.0000	1.0000
1.0000	1.0000	1.0000	1.0000	1.0000	1.0000	1.0000	1.0000
1.0000	1.0000	1.0000	1.0000	1.0000	1.0000	0.8896	0.8521
0.0244							

Again, because **D** is declared to be a correlation matrix in SAS, **I** had to be added to it (which increases each eigenvalue by 1). Prior to this addition, though, this matrix is divided by the maximum squared distance, because all correlation coefficients must lie in the interval [-1, 1]. Thus, the preceding eigenvalues are for $\frac{\mathbf{I}+\mathbf{D}}{MAX\ (\delta_{ij}^2)}$. Accordingly, by inspection there are only four nontrivial eigenvectors, associated with eigenvalues 2.2338 - 1 = 1.2338, 0.8896 - 1 = -0.1104, 0.8521 - 1 = -0.1479, and 0.0244 - 1 = -0.9756. The first eigenvector forms the topological configuration matrix. The correlation between its inverse and the principal eigenvector values from matrix **C** is -0.817. This relationship should be inverse, since δ_{ij}^2 in matrix **D** increases as the distance between municipios I and j increases, whereas the largest values in matrix **C** mostly are for the smallest nonzero distances. The deviation from a correlation of 1 arises because not every smallest nonzero distance is coded 1 in matrix **C**. Instead, c_{ij} is based upon whether areal units I and j are contiguous. The 71st eigenvector correlates with the north-south V coordinate obtaining a correlation coefficient of -0.987. The 73rd eigenvector correlates with the east-west U coordinate, obtaining a correlation coefficient of 1.000. In addition, an extra nuisance dimension has been uncovered here. Although it is orthogonal to the principal eigenvector, the correlation between elements in these two vectors yields a correlation coefficient of -0.987. This value arises because the means of these two vectors are not equal to zero. Therefore, as hypothesized, only three substantive dimensions can be extracted from matrix **D**.

This approach is similar to that of multidimensional scaling, which focuses on the eigenvalues of an n-by-n dissimilarity matrix. In the present problem, dissimilarity is indexed with intercentroid Euclidean distance in geographic space. The multidimensional solution to this problem will also yield a very good reproduction of the geographic distribution of centroids. Further discussion of this procedure is beyond the scope of this book.

6.7. Artificial Orthogonality: Principal Components Regression

The pairwise correlations for principal component scores constituting synthetic variates always will be 0, by construction, avoiding any double-counting of attribute information that might arise in the original P variables. The original matrix **X** can be replaced by the matrix of these scores in a multiple regression analysis where multicollinearity is a serious problem. This substitution generates the expected multicollinearity diagnostics for a mutually orthogonal situation. In this case, except for the intercept term (see *RULE OF THUMB 4.3*), the VIFs (variance inflation factors) all equal 1, the parameter estimates correlation matrix is **I**, and the eigenvalues and condition numbers all equal 1. Further, the FORWARD, BACKWARD, and STEPWISE procedures will yield exactly the same ordering of variable importance.

EXAMPLE PR-6.8. Execute the principal components regression counterpart to Examples PR-4.3 and PR-4.6.

Using the SAS procedures

```
PROC FACTOR DATA=STEP1 OUT=TEMP3 METHOD=PRIN
      NFACTORS=5 ROTATE=VARIMAX;
  VAR I0 DFRM DCD DISTSJ DISTA;
RUN;
PROC REG DATA=TEMP3;
  MODEL SQDSGR = FACTOR1-FACTOR5/COLLIN VIF CORRB;
RUN;
```

produced SAS output that includes

Parameter Estimates

Variable	DF	Parameter Estimate	Standard Error	T for H0: Parameter=0	Prob > |T|
INTERCEP	1	1.896655	0.11469532	16.536	0.0001
FACTOR1	1	0.196208	0.11548907	1.699	0.0940
FACTOR2	1	0.876451	0.11548907	7.589	0.0001
FACTOR3	1	-0.472879	0.11548907	-4.095	0.0001
FACTOR4	1	0.243739	0.11548907	2.110	0.0386
FACTOR5	1	0.706107	0.11548907	6.114	0.0001

Variable	DF	Variance Inflation
INTERCEP	1	0.00000000
FACTOR1	1	1.00000000
FACTOR2	1	1.00000000
FACTOR3	1	1.00000000
FACTOR4	1	1.00000000
FACTOR5	1	1.00000000

Correlation of Estimates

CORRB	INTERCEP	FACTOR1	FACTOR2	FACTOR3	FACTOR4	FACTOR5
INTERCEP	1.0000	-0.0000	0.0000	0.0000	-0.0000	0.0000
FACTOR1	-0.0000	1.0000	-0.0000	-0.0000	0.0000	-0.0000
FACTOR2	0.0000	-0.0000	1.0000	-0.0000	0.0000	0.0000
FACTOR3	0.0000	-0.0000	-0.0000	1.0000	0.0000	-0.0000
FACTOR4	-0.0000	0.0000	0.0000	0.0000	1.0000	0.0000
FACTOR5	0.0000	-0.0000	0.0000	-0.0000	0.0000	1.0000

Collinearity Diagnostics

Number	Eigenvalue	Condition Number
1	1.00000	1.00000
2	1.00000	1.00000
3	1.00000	1.00000
4	1.00000	1.00000
5	1.00000	1.00000
6	1.00000	1.00000

Of note here is that the $R^2 = 0.63994$ value exactly equals that for Example PR-4.3.

Not surprisingly, this complete removal of multicollinearity from matrix **X** is not without cost; otherwise, it would constitute the standard regression technique. The first noteworthy feature of PCA regression concerns the identification of important components. The eigenvalues for Example PR-6.8 are

Eigenvalue	1.9922	1.4050	0.6994	0.5475	0.3559
Proportion of Variance Accounted for	0.3984	0.2810	0.1399	0.1095	0.0712

If the criterion of $\lambda_j > 1$ is used (*RULE OF THUMB 6.1*), then the last three F_js would be dropped. But as the t-statistics in Example PR-6.8 signify, all three of these components have coefficients that are significantly different from 0 at the 5% level. Thus, the risk with attempting to achieve interpretability in a FA mode is that a small amount of very important information can be discarded; selecting only those principal components accounting for large amounts of variance may leave out components that, while not accounting for much of the overall variance in the attribute data, are useful in terms of statistical explanation. Adapting Friedman's famous New York *Times* quote, "Sometimes the [information] is in the [pattern], and sometimes the [information] is in the [noise]." The second noteworthy feature of PCA regression concerns bias. Recall that each principal component score is a linear combination of the original variables. If matrix **X** is complete to begin with, then discarding either seemingly trivial components/factors or ones with nonsignificant t-statistics will introduce complicated parameter estimation bias.

EXAMPLE PR-6.8 (continued). Execute all-possible principal components regressions.

Using the SAS procedures

```
PROC RSQUARE DATA=TEMP3 MSE ADJRSQ CP;
   MODEL SQDSGR = FACTOR1-FACTOR5;
RUN;
```

produced SAS output shown on the following page.

Comparing these results with those of Example PR-4.5 illustrates our point about bias. Of course, the last step in both examples is exactly the same. For all other variable combinations, the best C_p index values for each number of variables is higher using principal components than when using the original variables containing multicollinearity. In fact, for Example PR-6.8 this generalization holds for all four of the criteria statistics.

Another drawback of PCA regression is that because the principal components are mutually orthogonal, standard errors of PCA regression coefficients will be inversely related to the component eigenvalues. Hence, regression coefficients for the smaller principal components will have the larger theoretical standard errors.

6.8. A Multivariate Normality Diagnostic

The final PCA problem to be treated in this chapter pertains to assuming multivariate normality. The importance of this assumption for the application of significance tests to MANOVA is outlined in §5.2.3, and the importance for eigenvalues of the correlation matrix **R** is mentioned in §6.3. Moreover, this assumption is the foundation for most multivariate significance tests.

Univariate normality tests are reviewed in §1.3, with special emphasis on the Shapiro-Wilk statistic. This particular univariate test can be directly extended to an informal multivariate normality test. In this extension, a chi square (which actually is a squared normal variable) probability plot replaces the usual normal probability plot used in the univariate case. And, ranked

In	Rsq	Adj Rsq	C(p)	MSE	Variables in Model
1	0.30951	0.29978	59.488	1.7379	FACTOR2
1	0.20089	0.18963	79.700	2.0112	FACTOR5
1	0.09010	0.07728	100.3	2.2901	FACTOR3
1	0.02394	0.01019	112.6	2.4566	FACTOR4
1	0.01551	0.00165	114.2	2.4778	FACTOR1
2	0.51040	0.49641	24.106	1.2499	FACTOR2 FACTOR5
2	0.39961	0.38245	44.722	1.5327	FACTOR2 FACTOR3
2	0.33344	0.31440	57.034	1.7016	FACTOR2 FACTOR4
2	0.32502	0.30573	58.601	1.7231	FACTOR1 FACTOR2
2	0.29099	0.27073	64.934	1.8100	FACTOR3 FACTOR5
2	0.22483	0.20268	77.245	1.9789	FACTOR4 FACTOR5
2	0.21640	0.19401	78.813	2.0004	FACTOR1 FACTOR5
2	0.11403	0.08872	97.862	2.2617	FACTOR3 FACTOR4
2	0.10561	0.08006	99.429	2.2832	FACTOR1 FACTOR3
2	0.03945	0.01200	111.7	2.4521	FACTOR1 FACTOR4
3	0.60049	0.58312	9.341	1.0346	FACTOR2 FACTOR3 FACTOR5
3	0.53433	0.51409	21.652	1.2060	FACTOR2 FACTOR4 FACTOR5
3	0.52591	0.50529	23.220	1.2278	FACTOR1 FACTOR2 FACTOR5
3	0.42354	0.39848	42.268	1.4929	FACTOR2 FACTOR3 FACTOR4
3	0.41512	0.38969	43.836	1.5147	FACTOR1 FACTOR2 FACTOR3
3	0.34896	0.32065	56.147	1.6861	FACTOR1 FACTOR2 FACTOR4
3	0.31492	0.28514	62.480	1.7742	FACTOR3 FACTOR4 FACTOR5
3	0.30650	0.27635	64.048	1.7960	FACTOR1 FACTOR3 FACTOR5
3	0.24034	0.20731	76.359	1.9674	FACTOR1 FACTOR4 FACTOR5
3	0.12955	0.09170	96.975	2.2543	FACTOR1 FACTOR3 FACTOR4
4	0.62443	0.60234	6.886	0.9870	FACTOR2 FACTOR3 FACTOR4 FACTOR5
4	0.61601	0.59342	8.454	1.0091	FACTOR1 FACTOR2 FACTOR3 FACTOR5
4	0.54984	0.52336	20.766	1.1830	FACTOR1 FACTOR2 FACTOR4 FACTOR5
4	0.43905	0.40606	41.382	1.4741	FACTOR1 FACTOR2 FACTOR3 FACTOR4
4	0.33044	0.29105	61.594	1.7595	FACTOR1 FACTOR3 FACTOR4 FACTOR5
5	0.63994	0.61307	6.000	0.9603	FACTOR1 FACTOR2 FACTOR3 FACTOR4 FACTOR5

sum-of-squared principal component scores replace the single ranked variable in the univariate case. The quantitative index remains the correlation coefficient between these two rankings.

EXAMPLE PR-6.9. Execute a multivariate normality evaluation for the eight agricultural production density variables of Example PR-6.3.

Using the SAS code and procedures

```
DATA TEMP1(REPLACE=YES);
   SET TEMP1;
   DSQ = USS(OF PRIN1-PRIN8);
RUN;
PROC RANK OUT=TEMP1 (REPLACE=YES);
   RANKS DSQRANK;
   VAR DSQ;
PROC SORT DATA=TEMP1 OUT=QQ;
   BY DSQ;
DATA TEMP1 (REPLACE=YES);
   SET TEMP1;
   Z = (DSQRANK - .5)/73;
   CHISQ = CINV(Z,8);
PROC PLOT;
   PLOT DSQ*CHISQ;
RUN;
PROC CORR;
   VAR DSQ CHISQ;
RUN;
```

produced SAS output that includes

```
Pearson Correlation Coefficients / Prob > |R| under Ho: Rho=0 / N
= 73
                             DSQ              CHISQ
     DSQ                 1.00000            0.88470
     CHISQ               0.88470            1.00000
```

The test statistic is 0.8847, noticeably less than unity. Johnson and Wichern (1992, p. 160) advocate plotting DSQ versus CHISQ (as in Figure 6.6). They note that although the distances involved are neither independent nor exactly chi square distributed, this plot nevertheless should resemble a straight line when the data conform to a multivariate normal distribution. In contrast, a systematically curved plot suggests lack of multivariate normality. Therefore, there is evidence to indicate that the multivariate frequency distribution of the Puerto Rican agricultural production density variables does not conform to a multivariate normal distribution in the population. This finding suggests that a prudent course of action is to use a rule of thumb, rather than a significance criterion in Example 6.3, to identify important principal components.

Individual univariate transformations (Chapter 1) render a joint multivariate test statistics of 0.97946 in this case, indicating a considerable improvement in the joint distribution conforming to MVN (multivariate normality).

EXAMPLE PR-6.10. Execute a multivariate normality evaluation for the eight transformed agricultural production density variables of Example PR-6.9.

The following SAS code was used to transform the individual variables (see Chapter 1):

```
TDFRM = (DFRM-.7)**0.8;
TDCD = (DCD + 28.1)**1.6;
TDMLK = (DMLK)**0.45;
TDSGR = LOG(DSGR+3);
TDCF = (DCF + 0.75)**(.3);
TDTB = 1/(DTB + 1);
TDBN = 1/SQRT(DBN+0.75);
TDFAM = (DFAM + 0.25)**.2;
```

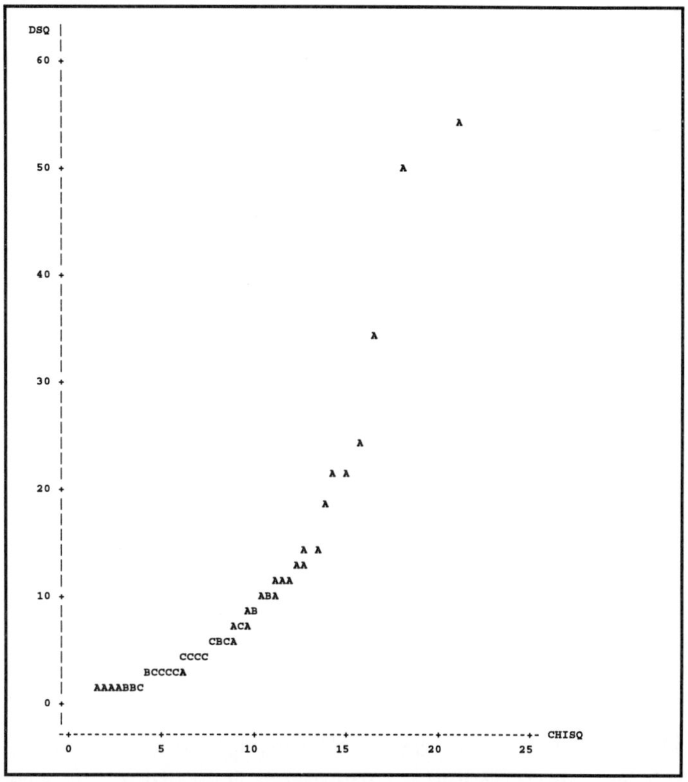

Figure 6.6. Chi square versus ordered distances plot for Puerto Rico's eight agricultural production density variables.

Using the appropriately modified version of the SAS code appearing in Example PR-6.9, with these transformed variables, produced SAS output that includes

Pearson Correlation Coefficients / Prob > |R| under Ho: Rho=0 / N = 73		
	DSQ	CHISQ
DSQ	1.00000	0.97946
CHISQ	0.97946	1.00000

The accompanying plot of DSQ versus CHISQ (Figure 6.7) is very linear, much more so than the one appearing in Figure 6.6. It also suggests the possible presence of one or two outlier areal units.

6.9. Problems

P-6.1. Given the *Old Maps* (Table 10.1 in Andrews and Herzberg, 1985) data set, identify the underlying dimension(s) for the 11 map latitudes, and then identify the underlying dimension(s) for the 11 map longitudes. How do these results compare with those for multicollinearity diagnostics computed for P-4.1?

P-6.2. Given the *Insurance Availability in Chicago* (Table 67.1 in Andrews and Herzberg, 1985) data set, identify and label the conspicuous dimension(s) underlying the 7 socioeconomic variables reported in this data set.

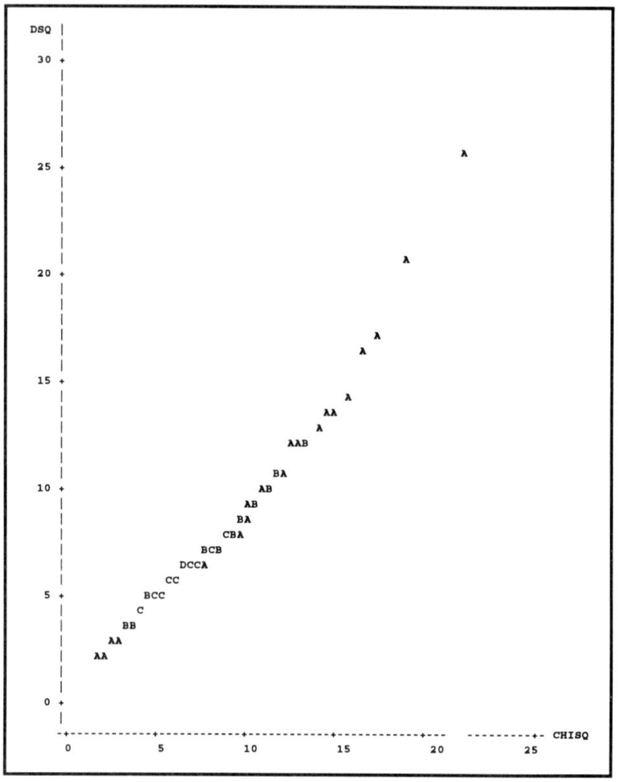

Figure 6.7. Chi square versus ordered distances plot for Puerto Rico's eight transformed agricultural production density variables.

6.10. References

Andrews, D. and H. Herzberg. 1985. *Data: A Collection of Problems from Many Fields for the Student and Research Worker*. Berlin: Springer-Verlag.

Bailey, T. and A. Gatrell. 1995. *Interactive Spatial Data Analysis*. Harlow, U.K.: Longman Scientific Publishing.

Hair Jr., J., R. Anderson, R. Tatham and W. Black. 1995. *Multivariate Data Analysis*. 4th ed. Englewood Cliffs, N.J.: Prentice Hall.

Johnson, R. and D. Wichern. 1992. *Applied Multivariate Statistical Analysis.* 3rd ed. Englewood Cliffs, N.J.: Prentice Hall.

Switzer, P. 1986. "Min/Max Autocorrelation Factors for Multivariate Spatial Imagery." In *Computer Science and Statistics*, Proceedings of the 16th Symposium on the Interface, edited by L. Billard 13–16. Amsterdam: North Holland.

Switzer, P. and A. Green. 1984. "MIN/MAX Autocorrelation Factors for Multivariate Spatial Imagery." *Technical Report No. 6*, Stanford University, Department of Statistics.

Wartenberg, D. 1985. "Multivariate Spatial Correlation: A Method for Exploratory Geographical Analysis." *Geographical Analysis* 17: 263-83.

CHAPTER 7

DISCRIMINANT FUNCTION ANALYSIS

Chapter Preview. As with multivariate analysis of variance, widespread use of discriminant function analysis (DFA) relies on the availability of adequate raw computing power. The goal of DFA is to generate a linear combination of the independent variables that is capable of discriminating among categorical dependent variable(s).

DFA is similar to regression analysis in that a linear combination of the independent variables is sought that best explains the variance in the dependent variable. In the case of DFA, the dependent variable is categorical rather than interval/ratio, as in regression. In Chapter 10 we compare the results of DFA with those generated by ordinary least squares (OLS) and nonlinear least squares (NLS). The three sets of results effectively demonstrate the strengths and weaknesses of the three approaches. In addition, in Chapter 12 a special form of regression, known as logistic regression, is introduced. If all the assumptions of both logistic regression and DFA are met, then the two techniques tend to give similar results. Logistic regression, however, is less sensitive to violations of its assumptions than DFA is to violations of its assumptions.

With an increase in the sophistication of data sets available for marketing purposes, DFA has found applications in the segmentation (targeting) of different geographic populations. For example, DFA can be used to identify the combination of characteristics that best predicts membership in different client groups, so as to permit a more focused allocation of resources (delivery of services, advertising, etc.)

7.0. Introduction

Frequently, multivariate analysis involves classification of observations. Spatial scientists repeatedly encounter or are called upon to regionalize areal data. ***Discriminant function analysis*** (DFA) can be an appropriate analytical technique for such undertakings when several distinct parent populations of areal units are *known and expected* to exist. This situation is in contrast to a situation in which prespecified multiple populations are suspected but unknown. In this case, the classification chore becomes one of uncovering latent *natural* classes by sorting areal units into potential populations. This latter multivariate problem is analyzed using cluster analysis (Chapter 8).

Two distinctive goals are affiliated with the DFA classification task. First, a given population classification itself can be evaluated. The initial step in such classification evaluation is ANOVA for the univariate case and MANOVA for the multivariate case (Chapter 5). If MANOVA indicates the presence of significant differences between multivariate regional means, then the relevant question is which variables contribute to the detected average regional differentiation. The curiosity motivating this question is similar to that prompting the ANOVA question of which individual regional means lead to rejection of its null hypothesis. Because of the inclusion of P variables here, separate ANOVAs could yield any one of the 2^P possible combinations of null hypothesis decisions; this scenario parallels the multiple comparisons notion outlined in Chapters 4 and 5. The relationship between individual ANOVAs and MANOVA is further complicated by the presence of multicollinearity among the P variables. DFA particularly addresses these two features of MANOVA.

EXAMPLE PR-7.1. Calculate DFA equivalent to the MANOVA of Example PR-5.6.

Using the SAS procedure

```
PROC DISCRIM POOL=TEST MANOVA;
   CLASS ADM;
   VAR DCD LNDBN;
RUN;
```

produced SAS output that includes

Discriminant Analysis
Multivariate Statistics and F Approximations

Statistic	Value	F	Num DF	Den DF	Pr > F
Wilks' Lambda	0.69768348	3.30329	8	134	0.0018
Pillai's Trace	0.30749169	3.08853	8	136	0.0031
Hotelling-Lawley Trace	0.42589706	3.51365	8	132	0.0010
Roy's Greatest Root	0.40770332	6.93096	4	68	0.0001

As a comparison with Example PR-5.6 quickly discloses, these DFA results are identical to its MANOVA results. The individual ANOVAs yield F-ratios of 6.90 ($Pr > F = 0.0001$) for DCD and 0.79 ($Pr > F = 0.5325$) for LNDBN. Both the MANOVA and DFA summary statistics indicate that the differentiation supplied by DCD dominates a joint analysis with these two variables. A more definitive assessment is desirable, though, and can be obtained from the eigenfunctions of DFA. The results reported for Example PR-7.1 should be quite robust, since the homogeneity of variance test (obtained with the POOL=TEST option) yields a chi squared value of 14.47426 ($Pr > \chi^2_8 = 0.2715$), and within-regions bivariate normality Q-Q plot correlations all are in excess of 0.97808 (Chapter 1).

DFA determines optimal linear combinations of variables, much like those computed with principal components (PCA) or factor (FA) analysis (Chapter 6), that reveal the dimensions along which the most conspicuous mean differences occur. Just as in PCA or FA, a structure matrix can be constructed for DFA revealing the nature and degree of relations between the original variables and these discriminating dimensions, suggesting the directions along which these most conspicuous mean differences occur. In this case, the goal of DFA is to answer the following question:

DFA Question #1: What is the most powerful linear combination of P predictor variables that can be constructed to maximally separate regional means?

Again, reminiscent of PCA or FA scores, synthetic variables are computed in DFA and are referred to as DFA scores (some of these types of synthetic variates are used in §5.2.2).

The second goal of the DFA classification task concerns optimally assigning a new areal unit to the prespecified populations. DFA furnishes the mathematical equations that allow identification of the regional population that the new areal unit most closely resembles. This goal is encountered less frequently in spatial analyses than in some other disciplines. In spatial analysis, the goal of DFA more commonly is modified either as (1) assessment of the classification probabilities for the analyzed areal units or (2) validation analysis based upon a reserved subset of the areal units. The goal of the modified DFA, then, is to answer the following question:

> *DFA Question #2*: What is the computational rule that can be used to optimally allocate new areal units to existing regional populations?

Stated in this way, DFA can be linked to the location-allocation problem (Chapter 8).

An additional, highly specialized use of DFA occurs in remote sensing where it is common practice to estimate discriminant functions with a relatively small systematic sample of pixels from an image and then to treat the remaining pixels as new observations. Historically, this convention was prompted by scarce computer resources and technology limitations and may well disappear in light of continuing computing advances. This application is not covered explicitly in this book (see Fisher, 1993, for details).

7.1. Properties of the Linear DFA Model

Selected salient features of the linear DFA model merit attention before its applications can be fully appreciated and understood. Foremost are its underlying assumptions. Second is the associated eigenfunction problem and third is statistical significance testing.

7.1.1. Notable Assumptions for the Linear DFA Model

The primary mathematical assumptions are direct extensions of those for ANOVA and parallel those of MANOVA: (1) equality of regional covariance matrices, (2) regional multivariate normality (MVN), and (3) independent errors. Of course, there are indisputable conceptual assumptions, too: (1) the set of areal units should belong to two or more mutually exclusive regions, (2) the P predictor variables are measured at the interval/ratio scale, (3) each within-region sample variance is nonzero, and (4) group sizes are not grossly different. DFA can be performed when many of these assumptions are violated. The results, however, must be interpreted with great caution or may well be uninterpretable in the context of a particular problem.

DFA is a robust technique in the face of modest mathematical assumption violations. If regional covariance matrices are unequal, the computed discriminant functions may fail to maximize the separation of multivariate regional means and may yield distorted probabilities of regional membership. This specification error consequence can be circumvented, to some extent, by employing an approximation strategy similar to the one used with the univariate difference of means test (Griffith and Amrhein, 1991, pp. 297–99). As such, rather than basing a DFA on the covariance matrix pooled across all regions, each individual covariance matrix is used, in turn, with its respective region. One major drawback of utilizing this approximation is that the analysis tends to have an increased risk of exploiting sample-specific information. Furthermore, it primarily impacts upon probability calculations.

Not surprisingly, the assumption of regional MVNs is vital for tests of significance. DFA seems to be insensitive to minor violations of this assumption, suffering some reduction in efficiency and precision. The MVN assumption also is important with regard to those probabilities of regional membership produced by the resulting DFA classifier. Even modest deviations from MVN can magnify some regional membership probabilities while necessarily suppressing others (because the classification scheme is collectively exhaustive of the sample). Consequently, DFA will tend not to minimize the number of areal unit misclassifications.

Independence of the errors almost always is violated with geo-referenced data because of the presence of nonzero spatial autocorrelation. More traditional complications arise from the presence of serious multicollinearity among the predictor variables, nonconstant variance of the errors, and outliers and influential observations. DFA as a regression problem, treated in §7.4, will help illuminate this situation.

7.1.2. The DFA Eigenfunction Problem

The eigenfunction problem of MANOVA is (see §5.2.1)

eigenvalues: $DET(\mathbf{W}^{-1}\mathbf{B} - \lambda\mathbf{I}) = 0$, and (7.1)

eigenvectors: $DET(\mathbf{W}^{-1}\mathbf{B} - \lambda_j\mathbf{I})\mathbf{E}_j = \mathbf{0}$. (7.2)

Testing for the difference of multivariate regional means is based upon the solution to equation (7.1). Not surprisingly, then, these equations provide the foundation for DFA. Whereas the eigenvectors of the correlation matrix, **R**, supply coefficients for computing PCA scores (see §6.1), the elements of eigenvector $\mathbf{E}_j$ in equation (7.2) supply coefficients for calculating the j-th discriminant function (DFj).

EXAMPLE PR-7.2. Compute the DFA coefficients using MANOVA that are in the optimal linear combinations latent in Example PR-7.1.

Using the SAS procedure

```
PROC ANOVA;
   CLASS ADM;
   MODEL DCD LNDBN = ADM;
   MANOVA H=ADM;
RUN;
```

produced SAS output that includes

```
                    Analysis of Variance Procedure
                  Multivariate Analysis of Variance
       Characteristic Roots and Vectors of: E Inverse * H, where
        H = Anova SS&CP Matrix for ADM    E = Error SS&CP Matrix

  Characteristic    Percent        Characteristic Vector  V'EV=1
       Root
                                             DCD           LNDBN
    0.4077033235      95.73          0.01193414     -0.00535267
    0.0181937374       4.27         -0.00335764      0.08058495
```

The optimal linear combination equations revealed by these MANOVA results for Example PR-7.1 give the pair of DF scoring variables

$$\text{DF1} = 0.91243\times\text{DCD} - 0.40924\times\text{LNDBN} \text{ , and}$$
$$\text{DF2} = -0.04163\times\text{DCD} + 0.99913\times\text{LNDBN} \text{ .}$$

Calculation of these coefficients may be exemplified as follows:

$$0.91243 \approx \frac{0.01193414}{\sqrt{0.01193414^{\,2} + (-0.00535267)^2}} \text{ and}$$

$$-0.40924 \approx \frac{-0.00535267}{\sqrt{0.01193414^{\,2} + (-0.00535267)^2}} \text{ .}$$

Some redundancy of information is detected in the appearance of LNDBN in both DFs. The second set of scores, however, is little more than one of the original variables (LNDBN).

As axis rotation in FA symbolizes (§6.3), an eigenfunction solution generally is not unique but is related to a certain optimization criterion. In PCA, this criterion is the variance whereas in DFA the appropriate criterion is the ratio of between- to within-regions sums-of-squares. PCA adjusts for the multicollinearity between individual variable ANOVAs, producing results more comparable with MANOVA, especially if an adjustment for simultaneous testing is made. However, there is no reason to suspect that a PCA solution will be substitutable for or equivalent to the DFA solution for the same set of P variables or that the variable transformations used to help satisfy the assumptions of PCA will also be suitable in terms of the more restrictive assumptions of DFA.

EXAMPLE PR-7.3. Execute an ANOVA and a MANOVA of PCA scores for Example PR-6.10.

Using the SAS procedures

```
PROC PRINCOMP STD OUT=PCASCORE NOPRINT;
   VAR TDFRM TDCD TDMLK TDSGR TDCF TDTB TDBN TDFAM;
RUN;
PROC ANOVA;
   CLASS ADM;
   MODEL PRIN1-PRIN8 = ADM;
   MANOVA H=ADM;
RUN;
```

produced SAS output that includes

Variable	Source	df	Mean Squared Error	F-ratio	Pr > F
PC1	Model	4	1.4884003	1.53	0.2026
	Error	68	0.9712706		
PC2	Model	4	9.858681	20.59	0.0001
	Error	68	0.478901		
PC3	Model	4	3.438287	4.01	0.0056
	Error	68	0.856571		
PC4	Model	4	1.0251337	1.03	0.3999
	Error	68	0.9985215		
PC5	Model	4	5.398568	7.28	0.0001
	Error	68	0.741261		
PC6	Model	4	1.2680324	1.29	0.2833
	Error	68	0.9842334		
PC7	Model	4	2.3982800	2.61	0.0428
	Error	68	0.9177482		
PC8	Model	4	3.326419	3.85	0.0070
	Error	68	0.863152		

Statistic	Value	F	Num DF	Den DF	Pr > F
Wilks' Lambda	0.10182098	6.07419	32	226.552	0.0001
Pillai's Trace	1.56676676	5.15123	32	256	0.0001
Hotelling-Lawley Trace	3.74797802	6.9689	32	238	0.0001
Roy's Greatest Root	2.42865988	19.4293	8	64	0.0001

If a 5% level of significance is applied to individual variables in this example, then the comparable *Bonferroni* level, given eight variables, is 5%/8 = 0.625% (see §5.2). Accordingly, regional means differ for PC2, PC3 and PC5, but not for PC7 or PC8. These three components imply that only about one-third (19.1% + 14.1% + 5.0% from Example PR-6.10) of the statistical information is useful for separatory purposes. This finding is at odds with those obtained from ANOVAs on the individual transformed variables, in which five of the eight cases are significant, with four being highly significant. In addition, the MVN test statistics, by agricultural administrative region, are

	San Juan	Arecibo	Mayaguez	Ponce	Caguas
Q-Q plot correlati on	0.97158	0.83262	0.90304	0.94683	0.92297

The Arecibo region appears to violate this MVN assumption. Finally, Bartlett's homogeneity of variance test yields a chi squared test statistic of 223.9800, which is highly significant. In other words, removal of multicollinearity via PCA does not ensure that the underlying linear DFA model assumptions are met. The need for considerable remedial action is indicated by these reported diagnostic results.

Nevertheless, a number of interesting parallels do exist between the PCA and DFA procedures. Although the solution to equation (7.2) involves an oblique rather than an orthogonal rotation, the DF scores are mutually uncorrelated (but not orthogonal).

EXAMPLE PR-7.4. Calculate a correlation matrix for DFA scores extracted from the eight original variables used in Example PR-7.3.

The normalized eigenvectors reported in the MANOVA results yield the following DF scoring equations:

```
DF1 =  -0.0055563*TDFRM + 0.0005858*TDCD  - 0.0346882*TDMLK +
        0.1679145*TDSGR + 0.0600858*TDCF  + 0.0324747*TDTB +
        0.9776857*TDBN  + 0.1001808*TDFAM;
DF2 =   0.0116054*TDFRM - 0.0004438*TDCD  + 0.0275745*TDMLK +
        0.1384758*TDSGR + 0.0180561*TDCF  - 0.2593887*TDTB  +
        0.7443987*TDBN  + 0.5984901*TDFAM;
DF3 =  -0.0222418*TDFRM + 0.0006678*TDCD  + 0.1088275*TDMLK +
        0.0462698*TDSGR + 0.0303753*TDCF  - 0.4982187*TDTB  -
        0.6850889*TDBN  - 0.5167495*TDFAM;
DF4 =   0.0046310*TDFRM + 0.0000408*TDCD  + 0.0462307*TDMLK -
        0.0751240*TDSGR + 0.0292316*TDCF  + 0.1647183*TDTB +
        0.4554935*TDBN  - 0.8699062*TDFAM;
```

Using the SAS procedure

```
PROC CORR;
   VAR DF1-DF4;
RUN;
```

Except for rounding error, these equations produce the following correlation structure (nearly the identity matrix, **I**):

	DF1	DF2	DF3	DF4
DF1	1.00000	0.00001	0.00001	-0.00003
DF2	0.00001	1.00000	0.00001	-0.00000
DF3	0.00001	0.00001	1.00000	-0.00004
DF4	-0.00003	-0.00000	-0.00004	1.00000

The maximum number of nonzero eigenvalues from equation (7.1) is *MIN*[(K-1), P], where K-1 is the number for matrix **B**, and P is the number for matrix **W**. The associated eigenvectors sequentially account for a decreasing proportion of the total variance displayed in matrix $\mathbf{W}^{-1}\mathbf{B}$, with the combined set of nonzero eigenfunctions preserving information (in total, accounting for 100% of the variation in regional means). More specifically, because the dimensions are optimal and uncorrelated, the first DF accounts for the maximum regional differentiation, the second DF accounts for the largest regional differences not accounted for by DF1, the third DF accounts for the most conspicuous regional differences not accounted for by DF1 and DF2, and so on (i.e., a sequential optimization criterion, as in that for PCA).

EXAMPLE PR-7.5. Compare MANOVA results of the DFA scores with those based upon the original data, and those based upon the principal components scores of Example PR-7.3.

Using the SAS procedures

```
PROC ANOVA;
   CLASS ADM;
MODEL TDFRM TDCD TDMLK TDSGR TDCF TDTB TDBN
         TDFAM = ADM;
   MANOVA H=ADM;
RUN;
PROC ANOVA;
   CLASS ADM;
   MODEL DF1-DF4 = ADM;
   MANOVA H=ADM;
RUN;
```

together with the code appearing in Example PR-7.3, produced SAS output given on the following page. Whether the MANOVA is conducted on the original data, DFA scores, or PCA scores, in all three cases, the number of nonzero eigenvalues is *MIN*(5-1, 8) = 4, and each discriminant dimension accounts for the same amount of variance, with the eigenvalues occurring in decreasing order. This ordering of the eigenvalues allows a ranking of the orthogonal discriminating criteria. Using PROC ANOVA with the individual DF scores, the respective univariate F-ratios for the DFs are 41.29 (PR > F = 0.0001), 12.56 (PR > F = 0.0001), 7.18 (PR > F = 0.0001), and 2.69 (PR > F = 0.0383). In all three cases, the total variance is the same (which, as with PCA, is indexed by the sum of the eigenvalues), the information preserving property of DFA and PCA. This decomposition for DFA and PCA is as follows:

Character-istic Root	Percent of Variance Accounted	Characteristic Vector V'EV=1			
original data		TDFRM	TDCD	TDMLK	TDSGR
		TDCF	TDTB	TDBN	TDFAM
2.4286598770	64.80	-0.00314569	0.00033165	-0.01963877	0.09506494
		0.03401763	0.01838561	0.55351765	0.05671747
0.7385629192	19.71	0.00850671	-0.00032531	0.02021200	0.10150218
		0.01323501	-0.19013071	0.54564059	0.43869033
0.4226238536	11.28	-0.00866465	0.00026014	0.04239542	0.01802511
		0.01183315	-0.19408881	-0.26688694	-0.20130772
0.1581313679	4.22	0.00376070	0.00003317	0.03754247	-0.06100582
		0.02373805	0.13376244	0.36989153	-0.70642266
PCA scores		PRIN1	PRIN2	PRIN3	PRIN4
		PRIN5	PRIN6	PRIN7	PRIN8
2.4286598770	64.80	0.05168067	0.17377595	0.07596141	0.00372371
		0.00184730	-0.02097560	0.03940206	0.08349486
0.7385629192	19.71	0.01783580	-0.05332282	0.04490513	-0.04566870
		0.11381128	0.02314886	0.02523004	0.05251414
0.4226238536	11.28	0.04459532	-0.05633973	0.06424322	-0.00015837
		-0.06720774	-0.02182115	0.07361368	-0.00751819
0.1581313679	4.22	-0.02964990	-0.00951481	-0.03003595	0.04860379
		-0.02191303	0.07465898	0.04536670	0.06114524

The eigenvectors of DFA result in a nearly orthogonal solution for matrix $\mathbf{W}^{-1}\mathbf{B}$ based upon the DFA scores. Finally, principal components with larger eigenvalues usually are expected to be normally distributed, with non-normality shifting to components with smaller eigenvalues. The same result might be anticipated for DFs, but this example does not corroborate this contention. Instead, the following is obtained:

principal component	*Shapiro-Wilk*	*proba-bility*	*discriminant function*	*Shapiro-Wilk*	*proba-bility*
PRIN1	0.9430	0.0047	DF1	0.9397	0.0027
PRIN2	0.9702	0.2384	DF2	0.9649	0.1231
PRIN7	0.9762	0.4488	DF3	0.9852	0.8427
PRIN8	0.9828	0.7445	DF4	0.9712	0.2695

A number of striking differences between DFA and PCA scores merit mentioning. First, PCA scores are computed by using the normalized eigenvectors as coefficients applied to z-score transforms of the original variables. In contrast, DFA scores are computed by using the normalized eigenvectors as coefficients applied to the original variables themselves. Second, as Examples PR-7.4 and PR-6.2 demonstrate, both sets of scores yield an identity matrix for the correlation matrix **R**. But, Example PR-7.5 demonstrates that DFA scores yield a nearly orthogonal $\mathbf{W}^{-1}\mathbf{B}$ matrix, whereas PCA scores do not. Third, DFA scores tend to reduce the MANOVA variance heterogeneity and PCA scores do not. For instance, in Example PR-7.3 the chi squared statistic is

223.9800, the same as for the original variables. In Example PR-7.5 the chi squared statistic is reduced to 64.6709.

7.1.3. Significance Testing in DFA

As mentioned, DFA scores are computed from a linear combination of the original predictor variables. Once each linear combination is constructed it becomes an individual transformed, or synthetic, variable in the analysis. These synthetic variables are mutually uncorrelated. Conceptually speaking, an appealing approach is to complete a separate ANOVA for testing the difference of regional means with each set of DF scores, in turn, and then to calculate the necessary *Bonferroni* levels for a joint comparison. When applied to the results of Example PR-7.5, this strategy suggests that all four of the discriminant functions are significant. But each DF summarizes a different percentage of variance among the regional means. A more formal test derives from Wilk's Λ criterion, the ratio of the generalized within-regions covariation to total covariation, which has been defined in §5.2 as $\frac{1}{\Lambda} = \prod_{j=1}^{G} (1 + \lambda_j)$, where λ_j are the G [= *MIN*(K-1, P)] eigenvalues of equation (7.1). Now let $V_r = \left[\frac{n-1-(P+K)}{2}\right] \times \ln(1+\lambda_r)$. Then the test statistic becomes

$$V = \left[\frac{n-1-(P+K)}{2}\right] \sum_{j=1}^{R} \ln(1+\lambda_j) \quad ,$$

which is approximately chi squared distributed with P(K - 1) degrees of freedom. If V is found to be significant, then at least the first DF produces a significant difference between regional means. Bartlett found that the remaining DFs can be evaluated by successively performing chi squared tests, where

test statistic	*degrees of freedom*	*null hypothesis*
V	P(K - 1)	$\lambda_1 = \lambda_2 = \lambda_3 = \ldots = \lambda_G = 0$
$V - V_1$	(P - 1)(K - 2)	$\lambda_1 \neq 0;\ \lambda_2 = \lambda_3 = \ldots = \lambda_G = 0$
$V - V_1 - V_2$	(P - 2)(K - 3)	$\lambda_1 \neq 0,\ \lambda_2 \neq 0;\ \lambda_3 = \ldots = \lambda_G = 0$
$\vdots$	$\vdots$	$\vdots$
$V - \sum_{i=1}^{G-1} V_j$	(P - G + 1)(K - 1)	$\lambda_1 \neq 0,\ \lambda_2 \neq 0,\ \ldots,\ \lambda_{G-1} \neq 0;\ \lambda_G = 0$.

This sequential testing terminates once the decision about a residual Wilk's Λ criterion is "fail to reject." This significance testing situation is reminiscent of that for PCA (Chapter 6).

EXAMPLE PR-7.6. Calculate significance test of each of the DFs reported in Example PR-7.5.

$$
\begin{aligned}
V &= [73 - 1 - \frac{(8+5)}{2}] \times [\ln(1+2.42866) + \ln(1+0.73856) + \ln(1+0.42262) + \ln(1+0.15813)] \\
&= 149.6369 > \chi^2_{32,0.95} \\
&= 46.1942 \\
V - V_1 &= 65.5 \times [\ln(1+0.73856) + \ln(1+0.42262) + \ln(1+0.15813)] \\
&= 68.9298 > \chi^2_{21,0.95} \\
&= 32.6706 \\
V - V_1 - V_2 &= 65.5 \times [\ln(1+0.42262) + \ln(1+0.15813)] \\
&= 32.7046 > \chi^2_{12,0.95} \\
&= 21.0261 \\
V - V_1 - V_2 &= 65.5 \times [\ln(1+0.42262) + \ln(1+0.15813)] \\
&= 32.7046 > \chi^2_{12,0.95} \\
&= 21.0261
\end{aligned}
$$

The fourth DF is found not to be significant here, a result that conflicts with the conceptually appealing Bonferonni-adjusted ANOVA analysis outlined earlier, illustrating the increase in latent data analysis complexity when moving from univariate to multivariate analysis. Resorting to *RULE OF THUMB 6.2*, this outcome is not surprising because this fourth DF accounts for less than 5% of the relative variation accounting for regional differences (see the output from Example PR-7.5).

Recall that the first step in DFA is to find a significant difference among multivariate regional means and that the second step is to determine which optimal linear combinations of variables produce these significant differences. Neither of these steps, however, indicates which variables contribute most to the uncovered differentiation. In keeping with the contention that each multivariate technique relates to multiple regression, the stepwise variable selection criteria used in regression analysis (§4.2.2.) is used.

EXAMPLE PR-7.7. Execute a stepwise DFA for Example PR-7.3.

Using the SAS procedure

```
PROC STEPDISC METHOD=STEPWISE;
   CLASS ADM;
   VAR TDFRM TDCD TDMLK TDSGR TDCF TDTB TDBN TDFAM;
RUN;
```

produced SAS output that includes

Stepwise Discriminant Analysis
Stepwise Selection: Summary

Step	Variable Entered	Variable Removed	Number In	Partial R**2	F Statistic	Prob > F	Wilks' Lambda	Prob < Lambda
1	TDMLK		1	0.3363	8.612	0.000	0.66374756	0.0001
2	TDCD		2	0.2737	6.313	0.0002	0.48205283	0.0001
3	TDTB		3	0.2792	6.391	0.0002	0.34746662	0.0001
4	TDSGR		4	0.1988	4.033	0.005	0.27838297	0.0001
5	TDCF		5	0.2680	5.858	0.0004	0.20377775	0.0001
6	TDBN		6	0.2691	5.799	0.0005	0.14893933	0.0001
7	TDFRM		7	0.2346	4.751	0.002	0.11399565	0.0001
8	TDFAM		8	0.1068	1.823	0.135	0.10182098	0.0001

This stepwise summary suggests that only the transformed density of farm families (TDFAM) does not contribute to the differentiation of regional means, given that each of the other original variables already has been entered into the DFA model.

7.2. Interpretation of the Dimensions of Regional Differences

Several other aids, besides significance tests of dimensions and variables, are available for interpreting DFs. One is the DFA structure matrix, which parallels the PCA and FA structure matrices and allows a researcher to infer meaning from the pattern of correlations (loadings) between the DFs and the original variables. The DFA structure matrix avoids spurious effects solely attributable to the particular units used to measure each original variable, and just as in PCA and FA, it may provide useful information, especially for interpretation.

EXAMPLE PR-7.8. Construct the DFA structure matrix for Example PR-7.7.

Using the SAS procedures

```
PROC CANDISC DATA=SPRDATA2;
   CLASS ADM;
   VAR TDFRM TDCD TDMLK TDSGR TDCF TDTB TDBN TDFAM;
RUN;
PROC CORR DATA=SPRDATA2 OUT=CORRDATA;
   VAR DF1-DF4;
   WITH TDFRM TDCD TDMLK TDSGR TDCF TDTB TDBN
        TDFAM DF1-DF4;
RUN;
PROC CORR DATA=SPRDATA2 OUTP=CORRDATA NOPRINT;
   VAR TDFRM TDCD TDMLK TDSGR TDCF TDTB TDBN TDFAM;
   WITH DF1-DF4;
```

```
RUN;
DATA CORRDATA(REPLACE=YES);
   SET CORRDATA;
IF _N_=1 THEN DELETE;
IF _N_=2 THEN DELETE;
IF _N_=3 THEN DELETE;
DROP _TYPE_;
RUN;
DATA CORRDATA(REPLACE=YES TYPE=FACTOR);
   _TYPE_='PATTERN';
   SET CORRDATA;
RUN;
PROC FACTOR DATA=CORRDATA NFACTORS=4
   ROTATE=VARIMAX PREROTATE=NONE;
RUN;
```

produced SAS output that includes

	unrotated structure matrix				*varimax rotated structure matrix*			
	DF1	DF2	DF3	DF4	DF1	DF2	DF3	DF4
TDFRM	0.03847	0.38416	0.02026	-0.07023	-0.01355	-0.36573	-0.12454	0.07033
TDCD	0.49696	-0.05948	0.54884	-0.23502	-0.08212	-0.18277	0.60606	0.44667
TDMLK	-0.55284	0.31390	0.46013	0.33150	0.80554	-0.24513	-0.09986	-0.08264
TDSGR	0.58747	0.17380	-0.17307	-0.20382	-0.53949	-0.18761	0.00138	0.34724
TDCF	0.57258	-0.01267	0.18398	0.06194	-0.18762	0.00209	0.18822	0.54319
TDTB	0.27214	-0.43145	-0.45738	0.38480	-0.27132	0.66117	-0.27040	0.18333
TDBN	-0.12936	0.17976	-0.29697	0.22126	0.02637	0.01653	-0.42630	-0.05914
TDFAM	0.12919	0.36081	0.35027	-0.37441	-0.02761	-0.56255	0.28849	0.09637

The first thirteen lines of code create the unrotated structure matrix. The next fourteen lines of code subject this initial structure matrix to a varimax rotation in an attempt to facilitate interpretation. Similar to what happens in FA, the shift here is from a sequential to a simultaneous optimization of the relative sums-of-squares (relative to the amount of variability present within the regions). What constitutes a strong correlation in this situation is controversial. In general, practice has resulted in adopting the convention of $|r| > 0.3$. Accordingly, the principal discriminant function relates to milk (TDMLK) and sugar cane productions (TDSGR), which tend not to coexist (hence their differing signs). This DF might be labeled an urban-oriented farming differentiator. The second discriminant function relates to density of farms (TDFRM), farm families (TDFAM), and tobacco production (TDTB) and might be labeled a rural farming differentiator. Of note here is that TDFAM loads on to the second, and a significant, DF, yet it appears to be unimportant according to the stepwise DFA. Density of cultivated land (TDCD) and banana/plantain (TDBN) production are highly correlated with the third dimension that might be labeled a coastal lowlands/interior highlands differentiator. The fourth DF is included, even though it is nonsignificant, to show that it is not a TDFAM dimension. In fact, the second, fourth, and fifth variables selected in the stepwise DFA are the ones that load highest on this last rotated dimension.

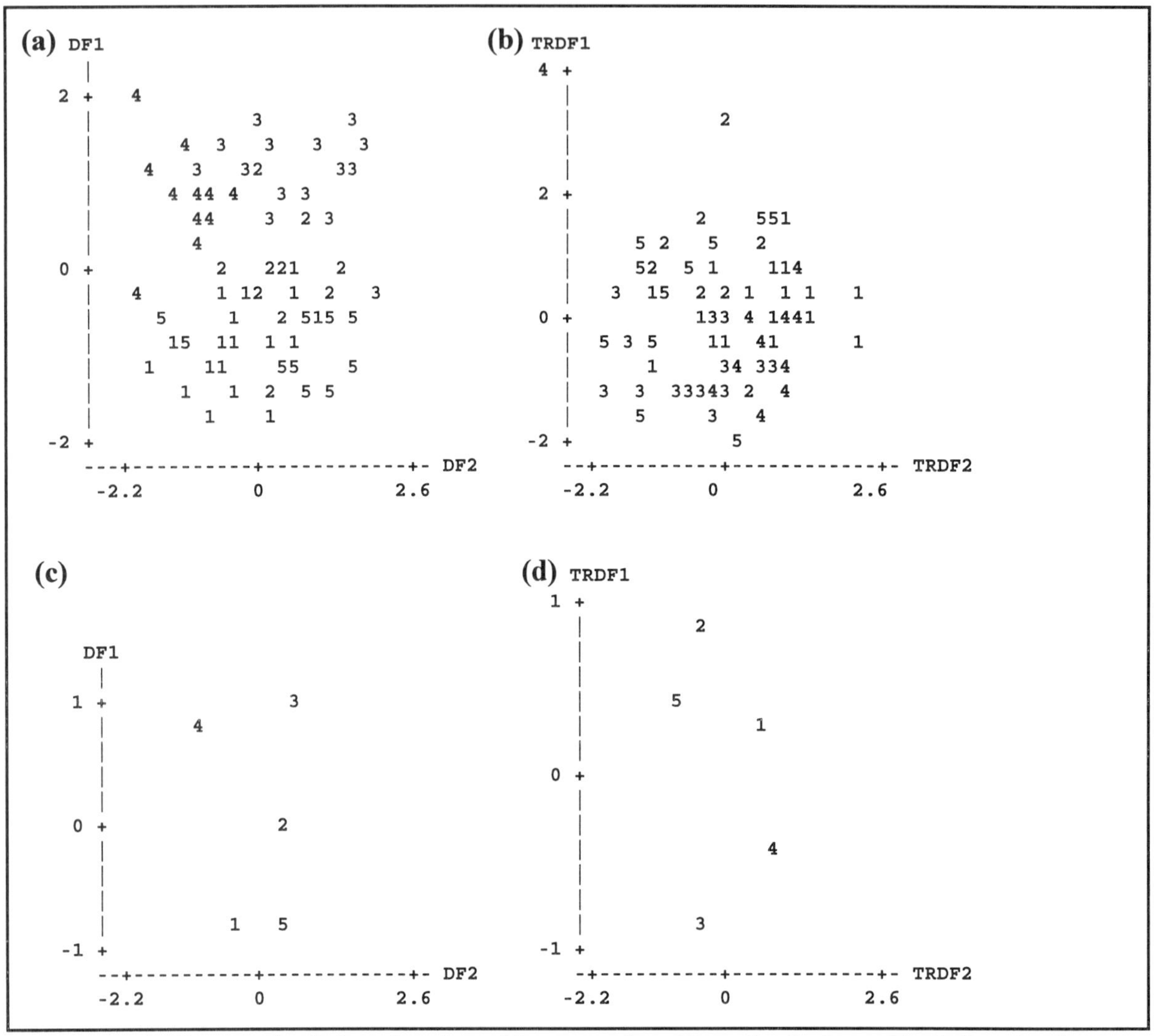

Figure 7.1. The first two discriminant functions. (a): scatter plot with areal units denoted by their regional membership, raw attribute data scores. (b): scatterplot with areal units denoted by their regional membership, transformed attribute data scores. (c): scatterplot of regional means, raw attribute data scores. (d): scatterplot of regional means, transformed attribute data scores.

A second aid to interpretation comes from plots of the DF scores. Often, only the first two dimensions are plotted, since they tend to account for almost all of the relative variation in the means. Illustrative plots for Example PR-7.8 appear in Figure 7.1. The top pair of scatterplots, Figure 7.1a–b, are for the individual areal unit DF scores. In Figure 7.1a, DF1 separates areal units in the third and fourth regions (Mayaguez and Ponce) in the top half of the plot from those in the first and fifth regions (San Juan and Caguas) in the bottom half of the plot. In the same figure, DF2 separates areal units in the first and fourth regions from those in the third and fifth regions. Jointly, these two DFs successfully differentiate among areal unit membership in these four regions but are unable to distinguish membership in the second region (Arecibo) from that in the rest. The varimax rotation alters this discrimination, shifting areal units in the second region from their central position in the discriminant function geometric space. In Figure 7.1b, TRDF1 (the first rotated DF) moves areal units in the second region to the top of the plot away from the others, whereas TRDF2 (the second rotated DF) separates areal units in the third and fourth regions in the bottom right-hand corner from the others. Jointly,. these two rotated DFs successfully differentiate among areal unit membership in these three regions, but are unable to distinguish membership in the first and third regions from the others. The bottom pair of scatterplots in Figure 7.1c–d displays the positions of the regional means. These points are the optimally separated points. Comparing Figures 7.1c–d, the varimax rotation reflects the positions of these means but repositions the second regional mean and further separates the first and fifth regional means. However, the trade-off is that the first and fourth regional means move closer together.

Drawbacks to rotation probably exist. Many multivariate statistics methodologists warn that DFA rotation is experimental and caution that it should not be done by the novice. A glimpse of why this reservation is stated is obtained from an inspection of a DFA of the rotated DF scores.

EXAMPLE PR-7.9. Execute a DFA of the varimax rotated DF scores for Example PR-7.8.

Using the SAS procedures

```
PROC STANDARD MEAN=0 STD=1 OUT=ZOUT (REPLACE=YES);
   VAR DF1-DF4;
RUN;
DATA ZOUT(REPLACE=YES);
   SET ZOUT;
TRDF1= -0.57981*DF1 + 0.08751*DF2 + 0.61207*DF3 + 0.53060*DF4;
TRDF2=  0.01641*DF1 - 0.85876*DF2 - 0.24941*DF3 + 0.44727*DF4;
TRDF3=  0.15733*DF1 - 0.47378*DF2 + 0.68110*DF3 - 0.53562*DF4;
TRDF4=  0.79925*DF1 + 0.17438*DF2 + 0.31507*DF3 + 0.48117*DF4;
RUN;
PROC ANOVA DATA=ZOUT;
   CLASS ADM;
   MODEL TRDF1-TRDF4 = ADM;
   MANOVA H=ADM;
RUN;
```

produced SAS output that includes

Character-istic Root	Percent of Variance Accounted	Characteristic Vector V'EV=1 TRDF1	TRDF2	TRDF3	TRDF4
2.4286598770	64.80	-0.12652630	0.00358158	0.03433221	0.17441255
0.7385629191	19.71	-0.01359849	0.13344429	0.07362194	-0.02709534
0.4226238536	11.28	0.08603666	-0.03505889	0.09573844	0.04428719
0.1581313676	4.22	0.06729454	0.05672627	-0.06792712	0.06102981

A comparison of this result with that reported in Example PR-7.5 indicates that the near-orthogonality of matrix $\mathbf{W}^{-1}\mathbf{B}$ achieved by the unrotated DF scores is lost. Of interest, though, is that the discriminatory power of the rotated DFAs remains unaltered, that is, the percent of variance accounted for by each discriminatory dimension is the same.

7.3. Probabilities for Regional Membership Classifications

Once the task of calculating DF scores is complete, the linear DF classifiers can be used to compute probabilities of regional membership. A simple tabulation of these classification results facilitates evaluation and interpretation of the DFs. This classification relies on Bayesian notions (see Griffith and Amrhein, 1991, pp. 162–64), with the relevant posterior probabilities being calculated as

$$\Pr(\text{an areal unit belongs to region } k \mid \mathbf{X}) = \frac{\Pr(\mathbf{X} \mid \text{region } k)}{\sum_{k=1}^{K} \Pr(\mathbf{X} \mid \text{region } k)} .$$

These Bayesian probabilities can be compared with the following classification probabilities:

Pr (correctly classifying an areal unit as belonging to a given region)
= Pr (the areal unit comes from this given region *and* is correctly classified as belonging to it), and

Pr (misclassifying an areal unit as belonging to a given region)
= Pr (the areal unit comes from a different region *and* is incorrectly classified as belonging to this region) .

The classifier allocation is made when the vector product $\mathbf{E}_r^T\mathbf{X}_k$ exceeds a given threshold value. Frequently this classification rule reduces to computing the Bayesian posterior probabilities and classifying each areal unit into that region for which it has the largest probability of membership. Of course, other thresholds can be selected, but this particular one relates to the maximum likelihood criterion.

EXAMPLE PR-7.10. Evaluate the regional classification of areal units for Example PR-7.8.

Using the SAS procedure

```
PROC DISCRIM POOL=YES CROSSLIST;
   CLASS ADM;
   VAR TDFRM TDCD TDMLK TDSGR TDCF TDTB TDBN TDFAM;
RUN;
```

produced SAS output that includes

From ADM		Number of Observations and Percent Classified into ADM:				
	1	2	3	4	5	Total
1	14	2	0	0	3	19
	73.68	10.53	0.00	0.00	15.79	100.00
2	0	9	1	0	1	11
	0.00	81.82	9.09	0.00	9.09	100.00
3	0	1	13	2	0	16
	0.00	6.25	81.25	12.50	0.00	100.00
4	1	0	0	13	0	14
	7.14	0.00	0.00	92.86	0.00	100.00
5	1	3	0	0	9	13
	7.69	23.08	0.00	0.00	69.23	100.00
Total	16	15	14	15	13	73
Percent	21.92	20.55	19.18	20.55	17.81	100.00
Priors	0.2000	0.2000	0.2000	0.2000	0.2000	

Example probability calculations upon which this tabulation is based are to be printed with the CROSSLIST option. For example, for Adjuntas, Aguada, and Barranquitas, these probabilities appear as

Obs	From ADM	Classified into ADM	Posterior Probability of Membership in ADM: 1	2	3	4	5
1	1	5 *	0.0623	0.0111	0.0010	0.0004	0.9252
2	1	1	0.8828	0.0100	0.0004	0.0028	0.1039
10	1	5 *	0.3487	0.1707	0.0422	0.0814	0.3570

These sample posterior probabilities illustrate that the classification is done according to the largest probability, regardless of how close the second largest probability is. The asterisk denotes a misclassified areal unit. These specific probabilities can be improved by allowing the individual regional within-regions covariance matrices to be used (since heterogeneous variance was detected earlier for this case). But this alternative may "mine" sample-specific information, failing to guarantee that these classifiers would perform better in the population.

Two features of this classification tabulation deserve comment. First, the prior probabilities used here are uniform ones, the best choice in the presence of ignorance about the underlying classification probability structure. Arguments can be made for these probabilities to be proportional to sample or population size, or some other variable. Second, this tabulation should be contrasted with its extreme possibilities. On the one hand, if the classifiers are random ones, then the allocation of areal units to regions is equally likely, resulting in the following expected uniform tabulation:

From ADM	Number of Observations and Percent Classified into ADM: 1	2	3	4	5	Total
1	3.80	3.80	3.80	3.80	3.80	19
2	2.20	2.20	2.20	2.20	2.20	11
3	3.20	3.20	3.20	3.20	3.20	16
4	2.80	2.80	2.80	2.80	2.80	14
5	2.60	2.60	2.60	2.60	2.60	13

On the other hand, if the classifiers are exact ones, then the allocation of areal units to regions is perfectly correct, resulting in the following diagonal-dominant tabulation:

From ADM	Number of Observations and Percent Classified into ADM: 1	2	3	4	5	Total
1	19	0	0	0	0	19
2	0	11	0	0	0	11
3	0	0	16	0	0	16
4	0	0	0	14	0	14
5	0	0	0	0	13	13

As can be seen by noting these extremities, the DFs constructed in Example PR-7.10 are quite accurate. The complete results show that the ADMs are correctly classified into their respective regions nearly 80% of the time. But this percentage still suggests a false degree of accuracy as DFs tend to overoptimize on any given sample.

7.4. DFA as a Regression Problem: Two-regions DFA

The relationship between linear DFA and linear regression is illustrated by inspecting the two-regions (K = 2) DFA case. Advantages to studying this relationship are that the more demanding problem of computing the eigenfunctions of matrix $\mathbf{W}^{-1}\mathbf{B}$ can be avoided and that a direct connection with spatial autocorrelation can be established. In principle, the simplification that materializes arises, in part, from the existence of a single nonzero eigenfunction for matrix $\mathbf{W}^{-1}\mathbf{B}$; moreover, *MIN*(2-1, P) = 1. Tatsuoka (1988, pp. 225–31) outlines the algebraic manipulations necessary to move from $DET(\mathbf{W}^{-1}\mathbf{B} - \lambda_1\mathbf{I})\mathbf{E}_1 = \mathbf{0}$ to $(\mathbf{X}^T\mathbf{X})^{-1}\mathbf{X}^T\mathbf{Y}$, where matrix **X** in the regression model is made up of the set of predictor variables used to construct the DF classifier, and vector **Y** is the binary (0-1) classification variable used to partition the DFA covariance matrices that sum to **W** and **B**. Two noteworthy properties are that (1) $\mathbf{1}^T(\mathbf{X} - \mathbf{1}\bar{\mathbf{X}}^T) = \mathbf{0}$ and implies the matrix used to construct matrices **W** and **B** is orthogonal to the vector **1** used in regression to compute an intercept term, and (2) $\mathbf{E}_1$ of $\mathbf{W}^{-1}\mathbf{B}$ is proportional to $\mathbf{b} = (\mathbf{X}^T\mathbf{X})^{-1}\mathbf{X}^T\mathbf{Y}$.

EXAMPLE PR-7.11. Calculate the relation between $\mathbf{E}_1$ of DFA based upon the binary variable LND and the no-intercept coefficients obtained by regressing LND on the same set of variables.

Using the SAS procedures

```
PROC ANOVA;
   CLASS LND;
   MODEL TLDFRM TLDCD TLDMLK TLDSGR TLDCF TLDTB
              TLDBN TLDFAM = LND;
   MANOVA H=LND;
RUN;
PROC REG;
   MODEL LND = TLDFRM TLDCD TLDMLK TLDSGR TLDCF
                  TLDTB TLDBN TLDFAM/
   VIF COLLIN CORRB SPEC;
   OUTPUT OUT=DFAREG R=LNDRES;
RUN;
```

produced SAS output that includes

```
Character-     Percent          Characteristic Vector  V'EV=1
istic Root   of Variance
              Accounted
                          TLDFRM      TLDCD       TLDMLK      TLDSGR
                          TLDCF       TLDTB       TLDBN       TLDFAM
1.5726810664  100.00    0.06371940 0.00001533 0.00155437 -0.21069175
                        0.02593182 0.04568962 1.38448976  0.47041558
-------------------------------------------------------------------
                        Parameter Estimates
                  Parameter        Standard     T for H0:
Variable   DF      Estimate           Error   Parameter=0      Prob > |T|
INTERCEP    1     -1.244116      0.50957322        -2.441          0.0174
TLDFRM      1      0.131175      0.06999871         1.874          0.0655
TLDCD       1   0.000031561      0.00005538         0.570          0.5707
TLDMLK      1      0.003200      0.04221236         0.076          0.9398
TLDSGR      1     -0.433737      0.09505806        -4.563          0.0001
TLDCF       1      0.053384      0.03860415         1.383          0.1715
TLDTB       1      0.094058      0.07284837         1.291          0.2013
TLDBN       1      2.850153      2.59833737         1.097          0.2768
TLDFAM      1      0.968412      0.37733101         2.566          0.0126
```

Ignoring the regression intercept term (this subvector is denoted as $\mathbf{b}_*$), and normalizing each of these vectors of coefficients yields $\mathbf{b}_*^T\mathbf{b}_* = 3.04602^2$ and $\mathbf{E}_1^T\mathbf{E} = 1.47963^2$ and the following normalized coefficients:

TLDFRM	TLDCD	TLDMLK	TLDSGR	TLDCF	TLDTB	TLDBN	TLDFAM
0.043064	0.000010	0.001051	-0.142395	0.017526	0.030879	0.935698	0.317927

where, for instance, 0.043064 = 0.06371940/1.47963 = 0.131175/3.04602 for TLDFRM. Moreover, these two analyses render exactly the same results!

The findings from Example PR-7.11 seem to be quite robust, appearing not to suffer from MVN and variance homogeneity assumption violations. The MVN test statistics are

LND = 0 (coastal lowlands): 0.99214 , and
LND = 1 (interior highlands): 0.98749 .

The difference of variances chi squared statistic is 55.205312 (Prob > χ^2_{36} = 0.0213). In terms of the pooled regression residuals, the S-W statistic equals 0.968324 (Prob < W = 0.1913), and the constant variance test (using the option SPEC) produces a chi squared statistic of 50.83315764 (Prob > χ^2_{44} = 0.2224).

Perhaps the two most appealing advantages accompanying the linear regression solution to the two-regions DFA problem are that the impact of multicollinearity can be assessed explicitly and the nature and degree of spatial autocorrelation latent in the geo-referenced data can be ascertained.

EXAMPLE PR-7.12 (Example PR-7.11 continued) Calculate the multicollinearity regression diagnostics.

Using the SAS regression procedure appearing in Example PR-7.11 produced SAS output that includes

Variable	Variance Inflation	Eigenvalue	Condition Number
INTERCEP	0.00000000	7.50570	1.00000
TLDFRM	2.65626929	0.64236	3.41828
TLDCD	1.47542932	0.38666	4.40584
TLDMLK	1.82312678	0.20110	6.10924
TLDSGR	1.65203439	0.15289	7.00656
TLDCF	4.42424899	0.06585	10.67650
TLDTB	1.42429596	0.02921	16.02879
TLDBN	5.26638854	0.01250	24.50062
TLDFAM	1.99928927	0.00372	44.91207

Correlation of Estimates

	INTERCEP	TLDFRM	TLDCD	TLDMLK	TLDSGR	TLDCF	TLDTB	TLDB
TLDFRM	-0.1456							
TLDCD	-0.0175	0.1055						
TLDMLK	0.0567	-0.0444	0.0544					
TLDSGR	0.2119	0.1285	-0.2595	0.1459				
TLDCF	-0.4148	-0.2726	0.0163	0.3566	-0.2493			
TLDTB	0.0008	-0.3342	-0.1545	-0.0950	0.1088	0.1617		
TLDBN	-0.7375	0.1594	0.2184	-0.1654	-0.4953	0.5905	0.0797	
TLDFAM	-0.6566	-0.2539	-0.2635	-0.3784	-0.1810	-0.0882	-0.0240	0.2033

RULES OF THUMB 4.3–4.4, which interpret these diagnostics, produce the following conclusions: (1) the variance inflation factors suggest that it is questionable whether multicollinearity is unduly influencing the OLS standard error estimates; (2) the largest eigenvalue is near 9 and occurs with at least one very small eigenvalue, plus four condition numbers in excess of 10, which together suggest that the multicollinearity present may well be introducing serious calculation error into the regression analysis; and, finally, (3) none of the pairwise parameter correlations exceed 0.9 in absolute value, implying that the regression model does not contain serious multicollinearity effects due to an ill-specified parameterization of the DFs.

Meanwhile, because the geographic distribution of LND places ones in the center of the island, encircled by zeroes (see Map 2.1), its MC should, and indeed does, indicate the presence of moderate, positive spatial autocorrelation; MC = 0.241635 (GR = 0.809721). After all, similar values are tending to cluster together, but the concentric elliptical arrangement of the two regions increases the number of adjacent 0-1 combinations along the regional boundary. For the regression residuals obtained in Example PR-7.11, MC = 0.044795. The spatial autocorrelation displayed by the individual predictor variables is

	TLDFRM	TLDCD	TLDMLK	TLDSGR	TLDCF	TLDTB	TLDBN	TLDFAM
MC	0.340063	0.304424	0.412302	0.427315	0.587406	0.343576	0.466532	0.301901
GR	0.676814	0.625006	0.534174	0.676840	0.452227	0.702605	0.638447	0.742290

The expected value of MC for the regression residuals is based upon 1 plus the sum of these eight MCs (= -0.065367). Using the asymptotic standard error for MC ($\sqrt{\frac{2}{344}} \approx 0.076249$) yields an approximate z test statistic of 1.44477, which is not significant. Therefore, the significant spatial autocorrelation detected in LND appears to dissipate with the regression model specification used in Example PR-7.11.

7.5. Problems

P-7.1. Given the *Effect of Chemicals on Earthworm Populations* (Table 52.1 in Andrews and Herzberg, 1985) data set, evaluate the plot classification (this is **Y**) from total biomass/m^2 using all three sample measures (this is **X**). Repeat this exercise predicting the subplot treatment cross-classification from total biomass/m^2.

P-7.2. Given the *Maize Fertilizer Experiments on the Islands of St. Vincent and Antigua* (Table 58.1 in Andrews and Herzberg, 1985) data set, evaluate the location classification from the number of ears and the weight of harvest.

7.6. References

Andrews, D. and H. Herzberg. 1985. *Data: A Collection of Problems from Many Fields for the Student and Research Worker*. Berlin: Springer-Verlag.

Fisher, P. 1993. "Algorithm and Implementation Uncertainty in Viewshed Analysis." *International Journal of Geographic Information Systems* 7: 331–47.

Griffith, D. and C. Amrhein. 1991. *Statistical Analysis for Geographers*. Englewood Cliffs, N.J.: Prentice Hall.

Tatsuoka, M. 1988. *Multivariate Analysis*. New York: Macmillan.

CHAPTER 8

CLUSTER ANALYSIS

Chapter Preview. Perhaps no technique has fascinated geographers more than cluster analysis (CA), with its potential for explicitly spatial analysis. An enduring quest among spatial analysts is to define objective methods for comparing spatial units to form groups or clusters of these units. Another of the techniques that rely on adequate computing power, CA did not even appear in the classic multivariate text in geography (Johnston, 1978); however, contemporary texts give CA extensive treatment (Isaaks and Srivastava, 1989; Bailey and Gatrell, 1995).

CA is something of a reversal of DFA (discriminant function analysis). In DFA, the groups are defined a priori by the categorical dependent variable. In contrast, in CA the groups or clusters are defined as the outcome of the analysis, with the assignment of observations to clusters based on some clustering function or algorithm. The clustering function, as a result, is of central importance in CA, and the definition of clustering functions has occupied a great deal of attention in geography, biology, and other disciplines.

The clustering algorithm is some function of the variables on which the clustering is to take place. The decision of which cluster an observation is assigned to is based on the "distance," in some attribute space, between observations and clusters that have already been formed. At the end of a clustering exercise, the observations within a cluster should be more similar (as measured on the clustering criterion) than observations between clusters.

Aside from the selection of the clustering criterion, the key task in CA is deciding how many clusters should be formed. To assist in this process, we promote the visual aid of a dendogram, or tree diagram, coupled with a scree plot.

Our goals in this chapter are to introduce some of the more commonly used clustering functions, and to explore how the same data set is clustered by the various functions. We examine some explicitly spatial clustering problems in the form of location-allocation analysis and some clustering regions in which spatial proximity (contiguity) is required. It is in these examples that the inherent appeal to spatial analysts is revealed.

Finally, in the last section of the chapter, we briefly introduce a highly specialized form of clustering in which small areal units are combined into fewer, larger areal units. This process is referred to as spatial aggregation and is a "hot topic" in current research efforts.

8.0. Introduction

In Chapter 7 the multivariate classification problem of uncovering latent *natural* classes, by sorting areal units into potential populations, is distinguished from the goals of DFA. In this chapter cluster analysis (CA) is used to extract groups of *similar* observations from a data set that is composed of natural classes occurring in space. These homogeneous groups are conveniently referred to as *clusters*, and frequently are based upon $p > 1$ variable scores measured for each of the observations. The CA problem is more difficult than DFA, especially because in DFA the groupings are known in advance. A brief introduction to this theme, for the univariate case ($p = 1$), is presented in Griffith and Amrhein (1991, pp. 423–38).

Several worthwhile geographical research questions can be stated. On the one hand, a researcher may be seeking some optimal grouping of geo-referenced units.

> *CA Question #1*: What are the latent *natural groups* of areal units, given group members must be coterminous (i.e., invoking a contiguity constraint)?

A second, related research question concerns location-allocation (L-A) modeling, especially *Ward's algorithm*. Ward's algorithm exploits ANOVA principles and can be coupled with centrographic measures (see Griffith and Amrhein, 1991, pp. 115–25).

> *CA Question #2*: What are the group centroids and clusters of geographical points that result from minimizing standard distances?

On the other hand, a researcher may be seeking a relatively small representative set of areal units consisting of one member from each of a larger set of natural groups of these areal units. In this context CA provides a useful data reduction.

> *CA Question #3*: What are the K strata from which an analyst can draw a relatively small representative (stratified) sample of areal units?

This sort of data reduction is similar to that achieved with principal components analysis (PCA) and not surprisingly, has motivated the use of CA to address issues of serious multicollinearity in a large set of variables, too (similar to R-mode PCA). Further, this link between CA and PCA is reflected in the use of (Q-mode) PCA for regionalization purposes (§6.2).

These two themes highlight the fundamental geographic dimension of geo-referenced data. Of course, an unexpected grouping of areal units resulting from a cluster analysis yields important insights that might lead to a valuable investigation (perhaps the generation of some hypothesis).

8.1. Principles of Cluster Analysis

Cluster analysis has evolved out of the taxonomic tradition in science and supplies dozens of numerical methods for objectively searching data to achieve this end. CA has manifested itself in the spatial sciences mostly through city classification and regionalization schemes. Often, researchers hold no preconceptions about the number of groups or group structure. Unfortunately, the selection of a specific clustering algorithm and threshold criteria for identifying the number of *true* latent groups introduce considerable subjectivity into the assessment step of cluster analysis. No single algorithm is recognized as being universally the best; and results for a given data set commonly vary from algorithm to algorithm, sometimes quite radically. The rules of group formation utilized by each algorithm respond to different idiosyncrasies and nuances of a data set. After all, there may be multiple meaningful natural groups into which a set of observations can be classified (e.g., consider a deck of cards, which can be classified into suits, face/nonface cards, or specific numbered cards). In addition, clusters can be based only on those variables included in an analysis. As with other multivariate techniques, the final clusters obtained may well be sensitive to the particular choice of variables included in an analysis. Therefore, an analyst must possess sufficient familiarity with the data, experience with the geographically distributed phenomenon under study, or an adequate conceptual framework, to distinguish between *good* and *bad* clustering results.

Two features of clustering algorithms merit review here. The first concerns the objective function employed and the optimal allocation of areal units to groups. The second concerns the establishment of a threshold objective function value for determining the number of natural groups. When distinguishing between clustering algorithms, one impartial method of evaluation is to use each algorithm to analyze a judiciously selected data set having a known group structure and then to assess how capable each algorithm is of replicating the known (i.e., true) group structure. This evaluation procedure also helps furnish useful information about suitable threshold values used to determine the number of clusters latent in data sets.

Similar to what occurs in stepwise regression, clustering can proceed in two completely opposite ways. One of these is known as the ***agglomerative hierarchical method*** (AHM), which starts with individual areal units and successively merges similar groups. By construction, at each step groups are mutually exclusive and collectively exhaustive. Generally speaking, dissimilarities within clusters are minimized, whereas dissimilarities between clusters are maximized. Not surprisingly, various measures of similarity have been proposed (Griffith and Amrhein, 1991, p. 425). Results of these algorithms can be summarized with a **dendogram**, a two-dimensional tree diagram pictorially profiling the history of the sequential grouping process. This portrayal indicates that the clusters are nested, with each cluster being a member of a larger, more inclusive cluster at a higher level in the dendogram. The branches in this tree represent clusters and come together (merge) at nodes whose positions along a quantitative vertical axis denote objective function value measures for the levels at which the fusions occur. Once a threshold value along the vertical axis is established, the dendogram is sliced, indicating the number and composition of natural clusters. The general notion often resorted to when positing a threshold value is to cut the dendogram near where a sizable change occurs in the objective function value. To minimize the introduction of biases attributable to the needs and opinions of a researcher, this procedure can be more formally implemented by making it analogous to factor analytic decision-making and using a scree plot (§6.5). As such, a marked flattening in the graph of the number of clusters (R^2 on the vertical axis for Ward's algorithm) versus the objective function value (horizontal axis) suggests that little further homogeneity is attained by subsequent mergers.

EXAMPLE PR-8.1. Execute a univariate cluster analysis using DCD for the San Juan agricultural administrative region.

Using the SAS procedure

```
PROC SORT OUT=STEP1 (REPLACE=YES);
```

```
    BY ADM;
RUN;
DATA STEP2;
    SET STEP1;
    IF ADM=1;
RUN;
PROC CLUSTER DATA=STEP2 SIMPLE METHOD=WARD NOEIGEN
        OUTTREE=DCDTREE;
    VAR DCD;
    ID NAME;
RUN;
PROC TREE DATA=DCDTREE OUT=TREEOUT;
    ID NAME;
RUN;
```

produced the SAS output used to construct the dendogram appearing in Figure 8.1. This result is comparable to the one presented for the Arecibo region in Griffith and Amrhein (1991, p. 437).

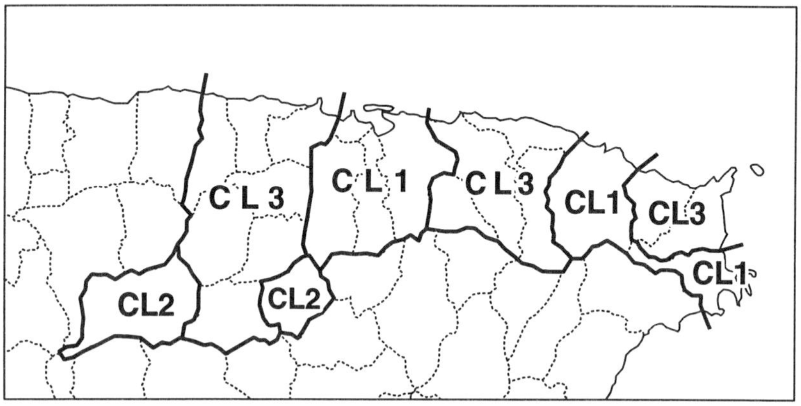

Map 8.1. Three-groups CA solution for the San Juan region based upon Ward's algorithm.

The scree plot for Example PR-8.1 (Figure 8.2) indicates that there appear to be at most four, and perhaps only three, natural groups for the 19 municipios in the San Juan region, based upon DCD. On the basis of the largest changes in R^2, three clusters can be identified (Map 8.1). One of these includes the San Juan metropolitan area, which is sensible. But, this cluster also includes Rio Grande and Ceiba, which are located toward or at the extreme eastern edge of the region. Invoking a contiguity constraint would help avoid such a non-coterminous composition. Orocovis and Comerio make up a second cluster concentrated in the interior highlands, one that also suffers from being non-coterminous. The third cluster is composed of the remaining 12 municipios in the region.

The other clustering approach is known as the ***divisive hierarchical method*** (DHM) and works in reverse of the agglomerative hierarchical method. In DHM, an initial single group is composed of all of the areal units. First this group is divided into two groups, in an attempt to maximize dissimilarity between them. Successive division into an increasing number of dissimilar subgroups continues until each areal unit forms its own group. Attention in this chapter is restricted to agglomerative methods, which dominate the applications literature and are the ones implemented in SAS.

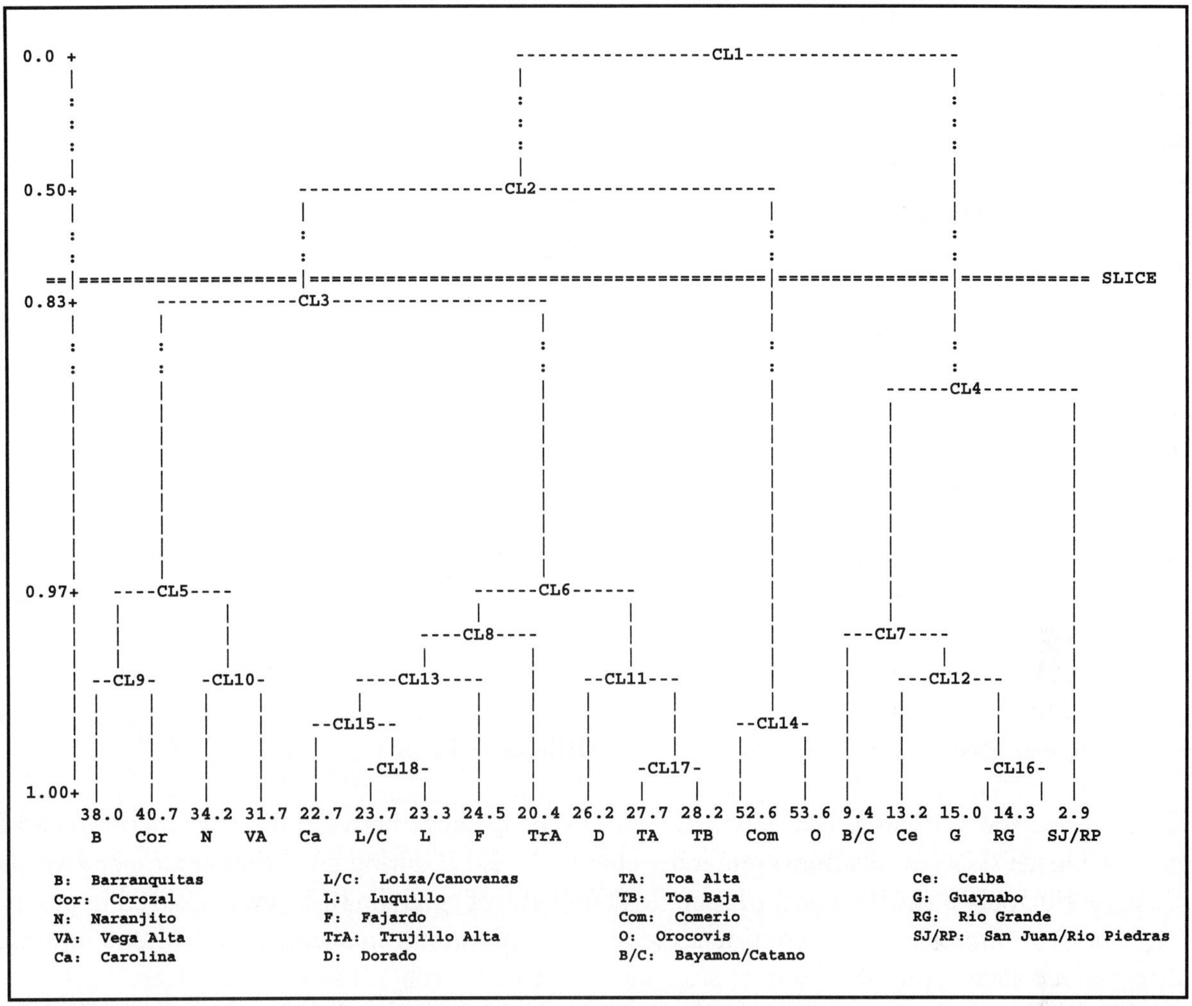

Figure 8.1. Dendogram for a univariate CA of density of cultivated land (DCD) in the San Juan AAR, using Ward's algorithm.

Computer algorithms for the AHM involve the following steps:

- Step 1: start with n clusters (the individual areal units) and a matrix of interareal unit dissimilarities;
- Step 2: search the dissimilarity matrix for the nearest pair of clusters (call this $d_{U,V}$ for clusters U and V);
- Step 3: merge clusters U and V; and,
- Step 4: repeat Steps 1-3 (n - 1) times.

These steps are guaranteed to converge, although not necessarily to the globally optimal partitioning of areal units into groups, because the total number of possible groups is a combinatorial problem.

The basic difference occurs among algorithms in Step 2, for each defines *nearest*, in terms of attribute similarities, in a distinctive and fairly simplistic way. Generally speaking, this term may be operationalized with regard to distances as

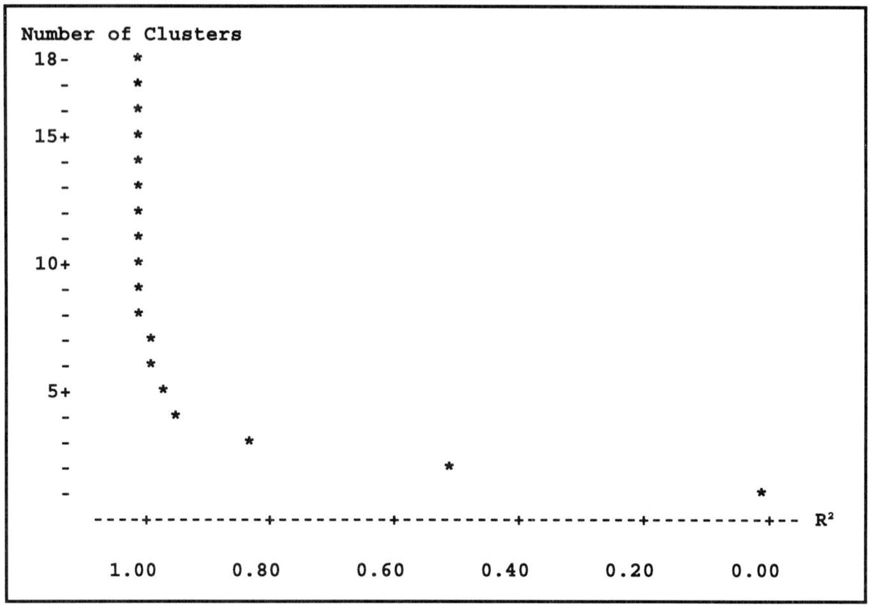

Figure 8.2. Scree plot for R^2 values presented in Figure 8.1.

$$d_{(UV)W} = \alpha_U d_{W,U} + \alpha_V d_{W,V} + \beta d_{U,V} + \tau \times ABS(d_{W,U} - d_{W,V}) \qquad (8.1)$$

(after Lance and Williams, 1967), where $d_{(UV)W}$ is the distance between the clusters (UV) and W, and the cluster (UV) results from combining clusters U and V during an earlier agglomerative step. Customarily, only a small number of specific definitions of equation (8.1) are employed in practice. The ***single linkage*** definition considers groups to be dissimilar in accordance with the minimum distance between all possible pairs of areal units, one from each of two clusters. Here $\alpha_U = \alpha_V = \tau = \frac{1}{2}$ and $\beta = 0$ in equation (8.1), yielding the objective function

$$d_{(UV)W} = MIN\{d_{UW}, d_{VW}\} \ , \qquad (8.2)$$

where the quantities d_{UW} and d_{VW} are the distances between the nearest neighbor members, in attribute space, of the clusters U and W and the clusters V and W, respectively. A *minimum spanning tree* also can be used to achieve single linkage clustering (Bailey and Gatrell, 1995). The ***complete linkage*** definition considers groups to be dissimilar in accordance with the maximum distance between all possible pairs of areal units, one from each of two clusters. In this case, equation (8.1) reduces to

$$d_{(UV)W} = MAX\{d_{UW}, d_{VW}\} \ . \qquad (8.3)$$

The ***average linkage*** definition considers groups to be dissimilar in accordance with the average distance between all possible pairs of areal units, one from each of two clusters. Here, equation (8.1) yields, for the most commonly used variant of this method, the objective function

$$d_{(UV)W} = \frac{\Sigma\,\Sigma\, d_{ik}}{(n_{UV} n_W)} \,, \qquad (8.4)$$

where d_{ik} is the distance between areal unit i in the cluster (UV) and areal unit k in the cluster W, and n_{UV} and n_W are the number of areal units in the clusters (UV) and W, respectively. The ***centroid method*** definition considers groups to be dissimilar in accordance with the squared Euclidean distance between areal units and cluster means (centroids); each areal unit is assigned to that cluster having the nearest centroid. In this case, equation (8.1) reduces to the objective function

$$d_{(UV)W} = MIN\{(\bar{\mathbf{X}}_{(UV)} - \bar{\mathbf{X}}_W)^T(\bar{\mathbf{X}}_{(UV)} - \bar{\mathbf{X}}_W)\} \,, \qquad (8.5)$$

where $\bar{\mathbf{X}}_{(UV)}$ is the mean vector of cluster (UV) and $\bar{\mathbf{X}}_W$ is the mean vector of cluster W. Finally, Ward's algorithm is based upon ANOVA regression principles (see §4.4): in Step 2 (earlier in this section) a search is made for those two groups that will add the smallest increment to the within-regions sum-of-squares. Here the objective function may be expressed as

$$d_{(UV)W} = \frac{\mathrm{MIN}\,\{\sum_{i=1}^{n_{UV}} \sum_{k=1}^{n_W} \Sigma\,(\bar{\mathbf{X}}_{(UV)} - \bar{\mathbf{X}}_W)^T(\bar{\mathbf{X}}_{(UV)} - \bar{\mathbf{X}}_W)\}}{\frac{1}{n_{(UV)}} + \frac{1}{n_W}} \qquad (8.6)$$

For the univariate case, expression (8.6) may be derived directly from the ANOVA between regions sum-of-squares term $\sum_{j=1}^{k} n_j(\bar{x}_j - \bar{x}_T)^2$ (see Griffith and Amrhein, 1991, p. 307).

Because of the lack of a consistent definition of the term nearest, expressions (8.2)–(8.6) bestow different properties on the clusters they generate. The most important ones can be indexed with measures of density (thickness of the swarm of areal units in attribute space), variance, dimension, shape (the configuration of areal units in attribute space), and separation. Furthermore, while expressions (8.2)–(8.6) presented here involve metric spaces, they can be extended to nonmetric spaces.

8.2. A Comparison of Clustering Algorithm Solutions

An evaluation of those clustering algorithms represented by equations (8.2)–(8.6) often can be very instructive, especially prior to an exhaustive analysis of some data set. A researcher should avoid obtaining results of a data analysis that principally are attributable to the algorithm employed. One important assessment dimension is the concurrence between natural clusters identified by a particular algorithm, and (true) known groups for a real-world test data set.

EXAMPLE PR-8.2. Searching for natural geographically compact groups in the Puerto Rico data, based upon DCD.

Replicating the examples appearing in Griffith and Amrhein (1991, pp. 428, 431, 433, 435, 437), except for the entire Puerto Rico data set, yields the classification results presented in the following output. Here, an entry indicates the number of municipios in an administrative region (the row label) assigned to each cluster (the column label). For example, of the nineteen municipios in the San Juan region, eleven are assigned to cluster 1, three to cluster 2, and 5 to cluster five when Ward's algorithm is the basis for clustering.

Clustering Algorithm	Agricultural Administrative Region	Cluster 1	2	3	4	5	total
Ward's	San Juan	11	3	0	0	5	19
	Mayaguez	0	7	0	0	4	11
	Arecibo	0	7	6	0	3	16
	Ponce	0	0	3	9	2	14
	Caguas	8	0	0	4	1	13
Average Linkage	San Juan	13	0	0	0	6	19
	Mayaguez	3	6	0	0	2	11
	Arecibo	0	8	0	8	0	16
	Ponce	0	0	7	6	1	14
	Caguas	0	0	4	0	9	13
Centroid	San Juan	8	10	0	0	1	19
	Mayaguez	5	0	4	0	2	11
	Arecibo	0	0	7	0	9	16
	Ponce	0	0	0	7	7	14
	Caguas	0	9	0	4	0	13
Complete Linkage	San Juan	15	4	0	0	0	19
	Mayaguez	0	5	6	0	0	11
	Arecibo	0	0	8	0	8	16
	Ponce	0	1	0	9	4	14
	Caguas	11	0	0	2	0	13
Single Linkage	San Juan	13	0	6	0	0	19
	Mayaguez	7	0	4	0	0	11
	Arecibo	2	0	6	5	3	16
	Ponce	0	5	9	0	0	14
	Caguas	0	2	11	0	0	13

After some selective column rearrangements (the numbering of clusters by an algorithm is arbitrary), diagonal concentrations in these tables are indicative of good classification results.

Fusion rules (that is, the rules for merging areal units in order to build clusters) for the Ward's, centroid, complete linkage, and single linkage algorithms lead to considerable confusion between the Arecibo and Caguas regions when the grouping variable is DCD. In other words, the clustering algorithm is unable to properly assign a municipio to its proper region in several instances. Fusion rules for the Ward's, average linkage, and complete linkage algorithms lead to confusion between the Arecibo and Mayaguez regions. Fusion rules for the average linkage and centroid algorithms lead to confusion between the Arecibo and Ponce regions. The fusion rule for the average linkage algorithm leads to confusion between the San Juan and Caguas regions, while that for the single linkage algorithm leads to confusion between the Ponce and Caguas regions. But all of these misallocations are between adjacent regions. Of more bothersome concern is that the fusion rule for the single linkage algorithm leads to confusion between the San Juan and Mayaguez regions, which are non-adjacent.

All five algorithms construct a cluster around the San Juan region (northeast portion of the island), a cluster around the Mayaguez region (western end of the island), a cluster around the Ponce region (south-central portion of the island), and a cluster around the Caguas region (eastern end of the island). All but the Centroid algorithm construct a cluster built around the Arecibo region (north-central portion of the island). Unfortunately, these known regional magnetic poles do not always render the same clusters; certain municipios are assigned to different clusters by different algorithms. Finally, the average linkage algorithm is successful in recovering the San Juan region, while Ward's and the complete linkage algorithms are successful in recovering the Ponce region (Map 8.2). Scree plots for the output from each of these algorithms appear in Figures 8.3a–b, with their visualizations presented in Map 8.2. The Ward's, average linkage and complete linkage algorithms all suggest the presence of five natural groups. The fusion index values generated by the Ward's algorithm align into a relatively smooth reverse-J decline. None of the remaining four algorithms generate a scree plot curve as smooth or as suggestive, although the complete linkage algorithm comes close. The centroid algorithm signifies the presence of either four or seven natural groups, whereas the single linkage algorithm implies that there are two natural groups.

Clusters created by each of these rules of merger can be assessed by the various properties bestowed upon them (properties that are well known to spatial analysts). Variance refers to the degree of dispersion of areal units from their cluster centroids, ranging from *tight* for relatively small dispersion situations to *loose* for relatively high dispersion situations. For comparative purposes, variance can be indexed for each cluster individually with standard distance (Griffith and Amrhein, 1991, p. 123), and overall with Bartlett's homogeneity of variance statistics (Griffith and Amrhein, 1991, p. 292). Shape refers to the form of the regional boundary taken on by each cluster. The quantitative geography literature contains various shape indices (Haggett, Cliff, and Frey, 1977, pp. 309–13), including the ***elongation ratio***. This index is a fraction in which the length of the major axis for a region, say λ_{max}, is in the numerator and the length of the minor axis, say λ_{min}, is in the denominator. Conceptually, these two variables are exactly those eigenvalues obtained with a bivariate PCA. Hence, the values of λ_{max} and λ_{min} are obtained from a PCA of the Cartesian coordinates of areal units contained in each cluster. Of course, if geographic clusters form bizarre shapes, then some of the other battery of shape indices should also be computed. Separation refers to the degree of nonoverlap by clusters or noninterdispersal of areal units allocated to different clusters and can be indexed by the percentage of successful classification obtained with a DFA applied

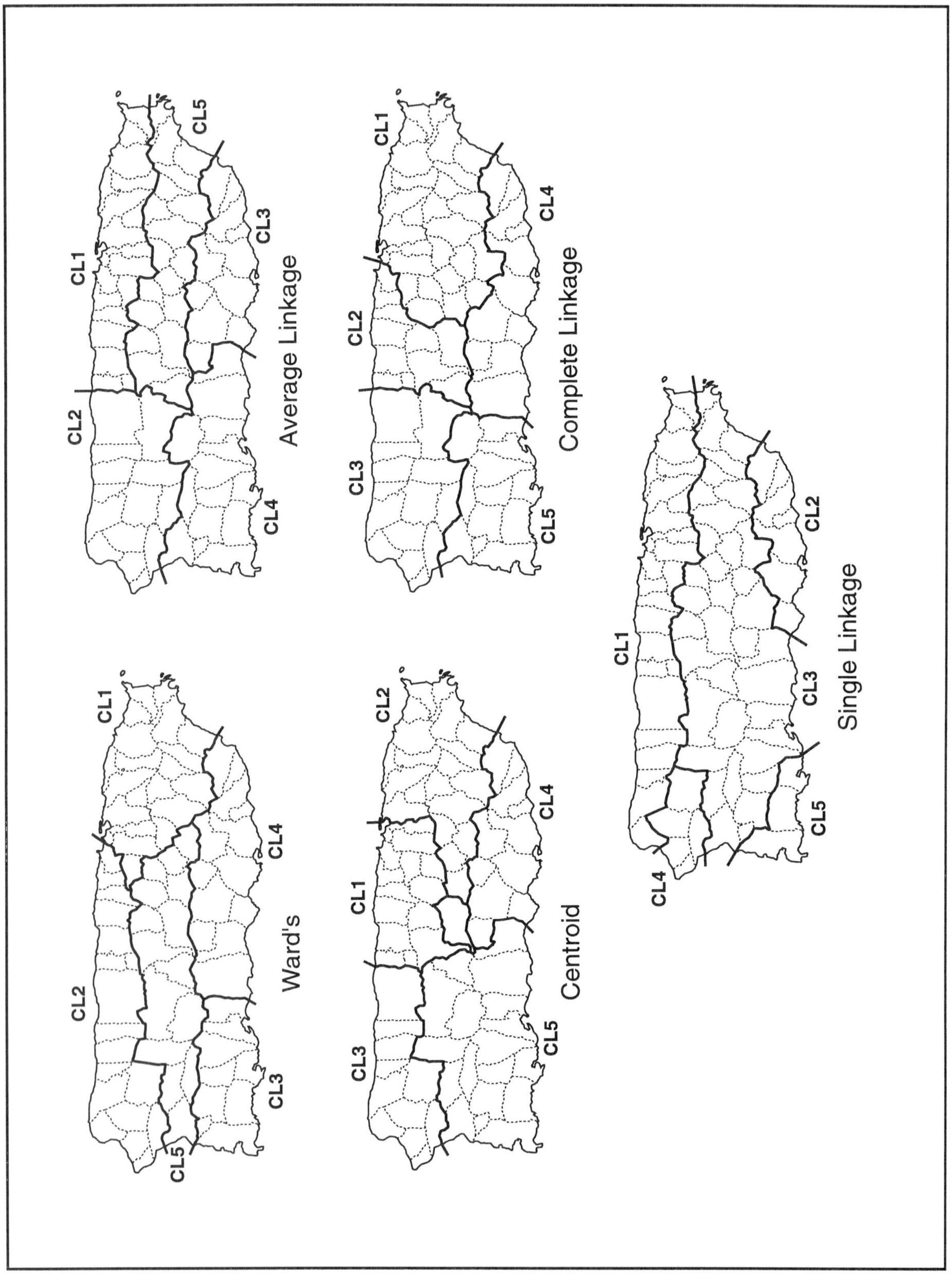

Map 8.2. The five-groups cluster solutions produced by the different algorithms using DCD as the grouping variable.

to the recovered clusters results. Finally, density refers to the thickness of clustered areal unit swarms in attribute space and may be indexed with a nearest- neighbor statistic (Griffith and Amrhein, 1991, pp. 126–30).

> EXAMPLE PR-8.2. (continued). Summarize and statistically describe the five cluster algorithm results for natural geographically compact groups in the Puerto Rico data, based upon DCD.

The overall variance measure equals 0 when all regional bivariate covariance matrices are exactly the same. The shape index equals 1 when a region is compact and circular and goes to infinity when a region is maximally elongated. Perfect separation would result in 0 mis-classifications. Finally, if the areal unit centroids had been generated at random, yielding a nearest-neighbor statistic of 1, then the standard error of the statistic would be $\sqrt{\frac{4-\pi}{n\pi}}$ (Griffith and Amrhein, 1991, p. 273). Since the island-wide nearest neighbor statistic for Puerto Rico is 1.5434, indicating a moderate tendency toward a regular spacing of areal unit centroids, the resulting random point pattern standard error value of 0.06118 is probably too large and hence provides the liberal Bonferroni-based 95% confidence interval of (1.3858, 1.7010) for evaluating the regional nearest–neighbor statistics. A "good" clustering should produce clusters of equal size (in terms of the number of smaller units combined to form a larger unit), densely located, compact (circular) in shape, and internally homogeneous.

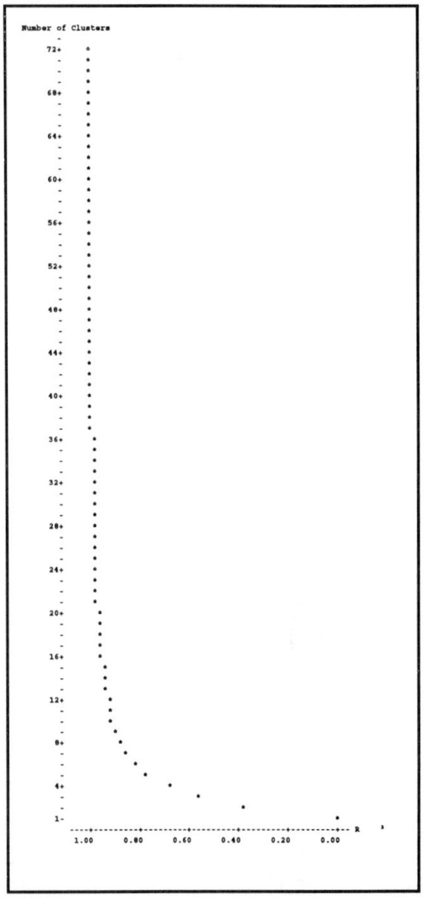

Figure 8.3. (a) Scree plot for Ward's algorithm results appearing in Example PR-8.2.

Acknowledged advantages of clusters generated by Ward's algorithm is that they tend to be relatively equal in size, to have relatively small within cluster variances, and to be relatively dense. Evaluation of the deviation from a uniform distribution of areal units across the created clusters yields

algorithm	Ward's	Average Linkage	Centroid	Complete Linkage	Single Linkage
Kolmogorov-Smirnov	0.0683	0.0898	0.1272	0.0840	0.3654

This simple examination indicates that the Ward's algorithm generates clusters that are more uniform in size than those created by any of the remaining four algorithms. Ward's algorithm ranks next-to-worst overall in terms of its standard distance type variances, whereas the accompanying regional bivariate covariance matrixes fail to be relatively equal across the clusters. Although some variability exists in the shape indices across clusters produced by this algorithm, none have a square

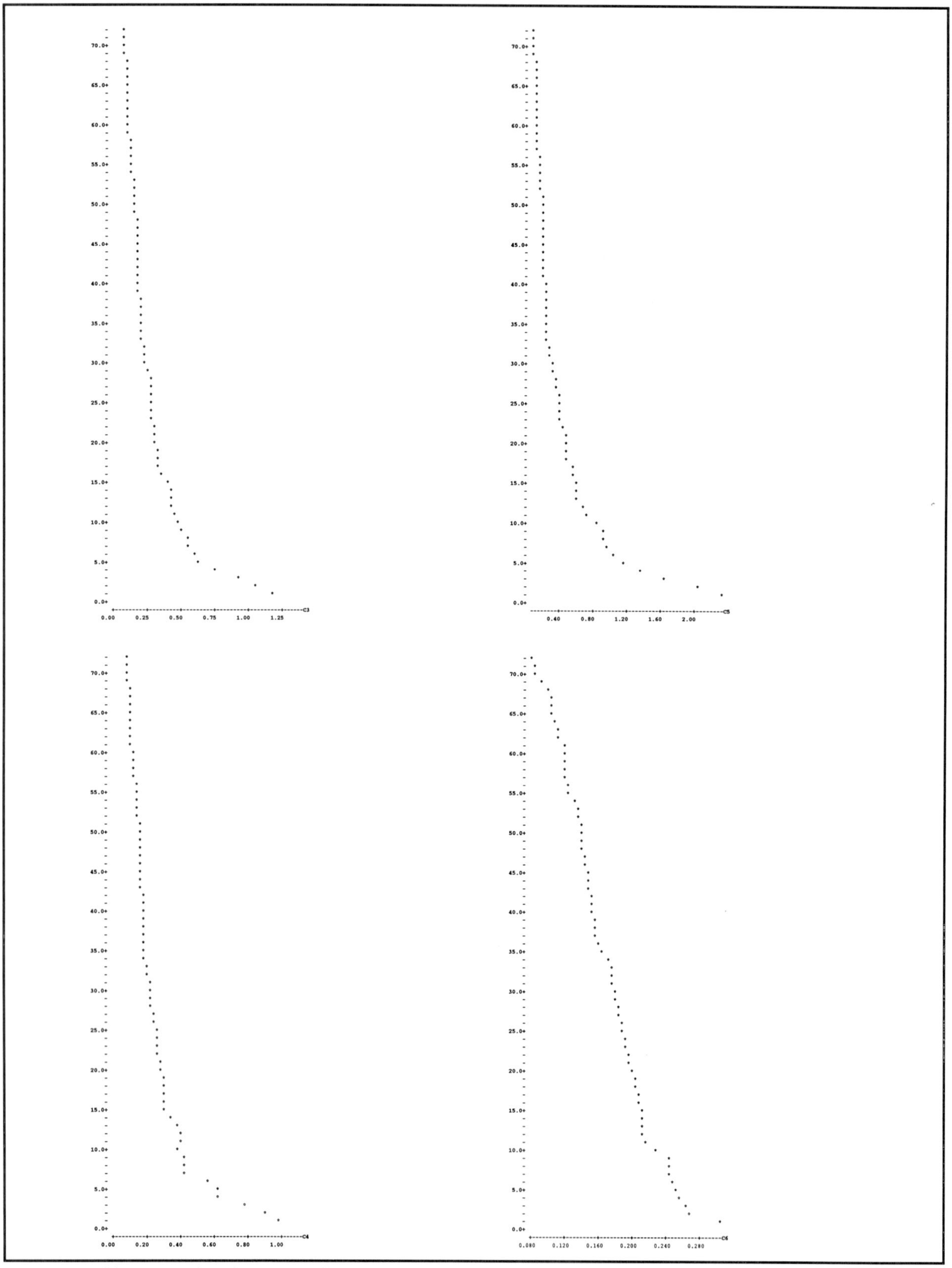

Figure 8.3. (b). Scree plots for average linkage (upper left), complete linkage (upper right), centroid (lower left), and single linkage (lower right) algorithm results appearing in Example PR-8.2.

Clustering Algorithm	Property	Cluster 1	2	3	4	5	overall
Ward's	variance	1.0799	3.6401	0.8027	1.7480	2.8852	25.88
	shape	1.74	3.75	1.53	1.61	1.85	1.17
	separation	0.1053	0.0000	0.0000	0.0769	0.0667	0.0498
	density	1.4200	1.5510	1.6697	1.5750	1.7704	1.5434
Average Linkage	variance	2.7707	1.1022	1.3071	1.4913	2.3774	21.95
	shape	6.12	1.06	1.97	1.50	1.82	1.17
	separation	0.0000	0.0000	0.0000	0.0714	0.0000	0.0143
	density	1.3665	1.6153	1.6977	1.6654	1.7106	1.5434
Centroid	variance	0.6660	1.3716	0.9622	1.3071	1.9360	11.74
	shape	1.17	1.57	2.41	1.97	1.28	1.17
	separation	0.0000	0.1053	0.0000	0.0909	0.1579	0.0708
	density	1.5421	1.5164	1.5890	1.6977	1.7159	1.5434
Complete Linkage	variance	1.6214	0.7762	1.1022	1.9903	0.8969	15.85
	shape	1.24	2.74	1.06	1.85	1.34	1.17
	separation	0.1923	0.0000	0.1429	0.0909	0.0000	0.0852
	density	1.4949	1.6667	1.6153	1.5102	1.7042	1.5434
Single Linkage	variance	6.4018	1.1099	5.3230	0.3125	0.3040	66.47
	shape	6.76	7.10	2.10	1.07	4999.00	1.17
	separation	0.1364	0.0000	0.3889	0.0000	0.0000	0.1051
	density	1.4104	1.5792	1.6103	1.7605	2.0562	1.5434

root exceeding the threshold value of 5–10 suggested by *RULE OF THUMB 4.2*. Ward's algorithm is recognized as outperforming most other clustering algorithms in terms of separation. But neither perfect separation among the five clusters nor the best separation in comparison with the four remaining algorithms is attained here (on average a 5% areal unit misclassification error rate prevails). While four of the five created clusters are reasonably dense, the fifth cluster is sparser than one would expect to obtain simply by coterminous surface partitioning. A noteworthy disadvantage not explicitly demonstrated here is that the created clusters tend to display an ordered profile.

One acknowledged advantage of the average linkage algorithm is that it is a compromise between the complete and single linkage algorithms. It also tends to maintain the nature of the original attribute space. In terms of the properties being assessed here, the average linkage algorithm clearly created the most distinctive clusters (having an overall separation error rate of only 1.43%). Unfortunately, it has the largest density index range of the five clustering algorithms, with its fifth cluster being sparser and its first cluster being denser than one would expect to obtain simply by coterminous surface partitioning. The first cluster created by this algorithm tends to be more elongated than the other four, and somewhat loosely demarcated (a sizable variance). There is little equality of cluster covariance matrices, as well.

Solutions rendered by the centroid algorithm tend to be robust in the presence of outliers but inferior to their corresponding Ward's algorithm solutions. Interestingly, in this case the centroid algorithm solution ranks best in terms of compactness (shape) and in terms of tight variances. In addition, it exhibits the best equality of variance across clusters. The density index for its fifth cluster is marginally sparser than one would expect to obtain simply by coterminous surface

surface partitioning. But its overall separation is noticeably less distinct than that achieved with Ward's algorithm.

A well-known advantage of the complete linkage algorithm is that it creates relatively compact clusters. In this case, it renders density indices whose variation is in keeping with what one would expect to obtain strictly from changing coterminous surface partitioning. Individual cluster variances are reasonably low (ranking second best), and none of the shape indices suggest marked deviation from compactness. A well-known disadvantage of complete linkage solutions is that they tend to have poor concordance with the true clusters. In this case, this feature is reflected in the sizable percentage of off-diagonal misclassifications. This algorithm also displays a poor separation capability (the overall misclassification error rate is nearly 9%).

Finally, the single linkage algorithm creates clusters that are invariant to data transformations. But, it also has a pronounced tendency to create clusters that are elongated in form, a property reflected in three of its shape index values here (one is >> 10^2). The solution generated here by this algorithm has by far the worst variance properties and performs the worst in terms of separation (its overall error rate is 10.5%, ranking it below all of the other algorithms).

> *RULE OF THUMB 8.1.* If there is no single overriding desirable property for resulting clusters to exhibit, Ward's algorithm should be selected because it tends to produce the most appealing overall results. If compact clusters are of primary concern, then complete linkage should be used. If outliers present a serious concern, then the centroid algorithm should be used. In most cases, the Single Linkage algorithm should be avoided.

Of course, there are many situations in which this heuristic should be taken with a grain of salt.

8.3. Location-allocation (L-A) and Cluster Analysis

When areal units are geo-referenced with the Cartesian coordinate pairs (u_i, v_i), they lead to the centrographic concepts of ***spatial mean***, ***spatial median*** and ***standard distance***, which are surveyed in Griffith and Amrhein (1991, pp. 116–25). These statistics are defined as follows:

$$\text{spatial mean: } (\bar{u}, \bar{v}) = \left(\frac{\sum_{i=1}^{n} f_i u_i}{\sum_{i=1}^{n} f_i}, \frac{\sum_{i=1}^{n} f_i v_i}{\sum_{i=1}^{n} f_i} \right)$$

$$\text{spatial median: } (\tilde{u}, \tilde{v}) = \text{MIN}: \sum_{i=1}^{n} f_i [(u_i - \tilde{u})^2 + (v_i - \tilde{v})^2]^{\frac{1}{2}}; \text{ and}$$

$$\text{standard distance: } SD = \left\{ \frac{\sum_{i=1}^{n} f_i (u_i - \bar{u})^2}{\sum_{i=1}^{n} f_i} + \frac{\sum_{i=1}^{n} (v_i - \bar{v})^2}{\sum_{i=1}^{n} f_i} \right\}^{\frac{1}{2}},$$

in which the notation $(\tilde{u}, \tilde{v})$ is used to distinguish the spatial median from the spatial mean.

Meanwhile, the specific question addressed in location-allocation (L-A) analysis may be stated as follows:

> *L-A Question #1*: What are the optimal weighted distance-minimizing locations of K centroids and the allocation of the n areal units to these centroids?

This question is very similar to CA Question #2. Moreover, in a two-dimensional physical space, K clusters are to be determined with the value of K not always known. The solution to this problem is a set of K spatial medians $\{(\tilde{u}_j, \tilde{v}_j); j= 1, 2, ..., K\}$. In contrast, an equivalent CA problem would search for a set of K spatial means $\{(\bar{u}_j, \bar{v}_j); j= 1, 2, ..., K\}$. The well-known relation between a mean and its affiliated median suggests that, in practice, these two solution sets rarely are exactly the same (Griffith and Amrhein, 1991, p. 82) but often may be very close to one another.

The L-A problem has the Cartesian coordinate pairs (u_i, v_i) as its data input, one for each areal unit, which links it to a bivariate CA. It also has a set of weights for each location, f_i, which is equivalent to frequencies in a CA. L-A can also have different objective function measures. For example, the maximum distance constraint is affiliated with the Complete Linkage algorithm. One noteworthy relation between these two problem areas stems from Ward's algorithm, which if cast in a L-A context would be similar to searching for a set of clusters whose spatial means have a minimum standard distance associated with them. To be precise, Ward's algorithm maximizes

$$\Sigma\,\Sigma \left\{ \frac{n_{(UV)} \times n_W[(\bar{u}_{(UV)} - \bar{u}_W)^2 + (\bar{v}_{(UV)} - \bar{v}_W)^2]}{(n_{(UV)} + n_W)} \right\}, \tag{8.7}$$

whereas the L-A algorithm used here (the ALTERN algorithm by Rushton, 1975) maximizes

$$\Sigma\,\Sigma \left\{ \frac{n_{(UV)} \times n_W[(\tilde{u}_{(UV)} - \tilde{u}_W)^2 + (\tilde{v}_{(UV)} - \tilde{v}_W)^2]}{n} \right\}. \tag{8.8}$$

One drawback of the CA algorithm is that it can generate different solutions simply by modifying the manner in which the intercluster distance matrix is searched. In other words, a global optimum is not guaranteed; sequential grouping of areal units does not necessarily optimize the criterion over all possible groupings of areal units into K clusters. A final cluster solution is conditional on the initial conditions input into the clustering algorithm and the history of interim amalgamations. Meanwhile, one drawback of ALTERN is that it is an heuristic and hence, by definition, is not guaranteed to produce a globally optimal solution.

Some cluster analysis algorithm implementations begin by executing a PCA to orthogonalize the original variables. The need for this can be seen by inspecting expression (8.7) and (8.8). Because both expressions involve sums, double counting of information (a concern in the absence of orthogonality) must be avoided. Dramatically different cluster results can be obtained with and without this preprocessing PCA. In part this difference arises because PCA is conducted on the covariance, rather than the correlation, matrix. The original covariance matrix often is preferred because standardization involved in converting a covariance matrix to a correlation matrix (i.e., all variables are transformed to z-scores) has the effect of minimizing cluster differences. In other words, using PCA scores effectively attaches equal weights to each of the components and hence treats each component as though it accounts for the same amount of variance. This equal weighting neutralizes the differentiating ability of the total attribute variance. The resulting instability of solutions suggests that a researcher should avoid an initial PCA as an *automatic* first step in CA.

> EXAMPLE PR-8.2. (continued). Evaluate the municipio centroids used to search for natural geographically compact groups in the Puerto Rico data, based upon DCD.

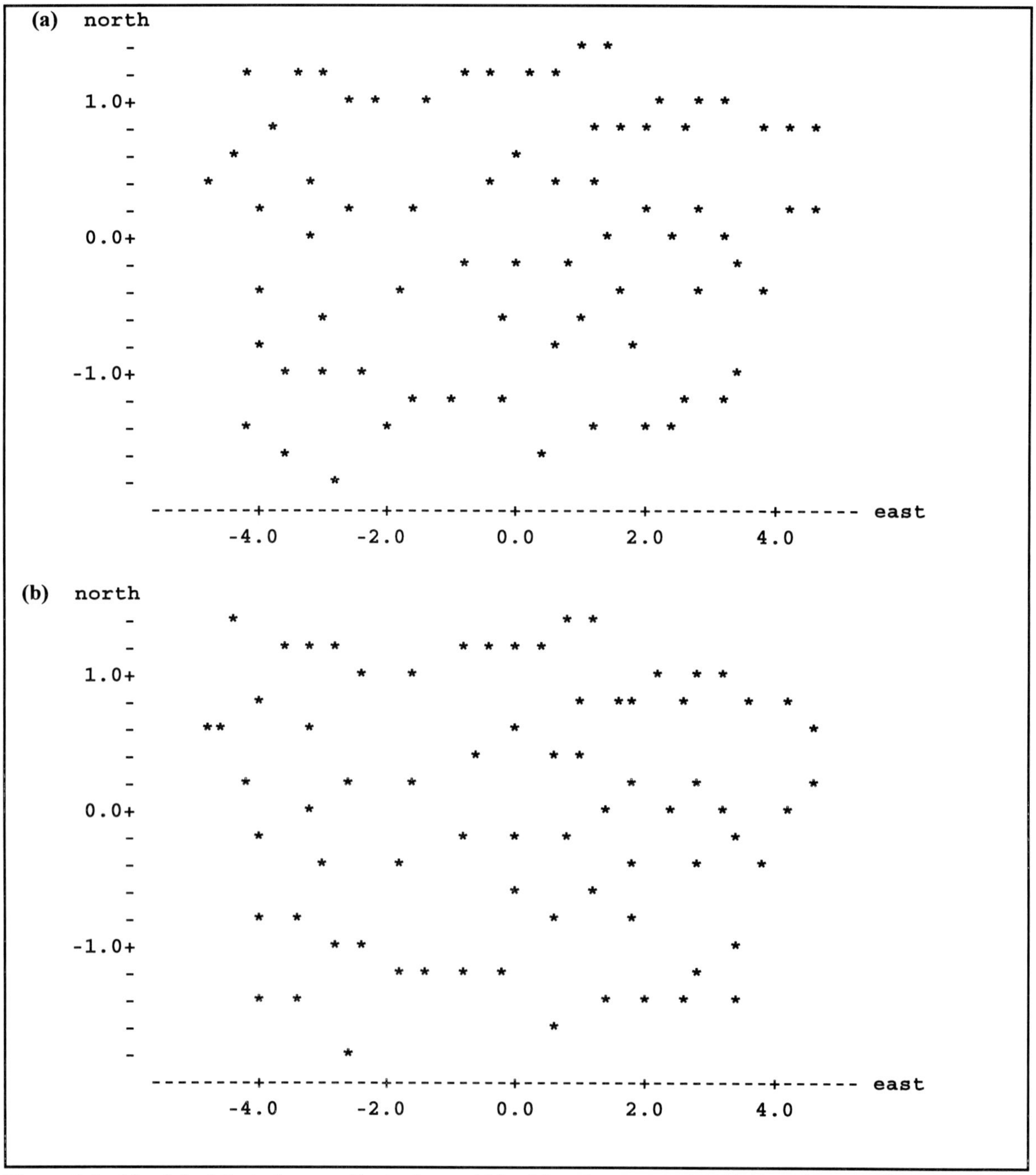

Figure 8.4. (a) original Cartesian coordinates (centroids); (b) varimax rotated PCA Cartesian coordinate scores.

Since the island of Puerto Rico is partitioned into five agricultural administrative regions, let K = 5. The Cartesian coordinates are subjected to a PCA followed by a varimax rotation. The resulting PCA scores reproduce a geographic distribution of areal unit centroids that is indistinguishable from the original ones (Figure 8.4b), in part because $r_{u,v} = 0.07875$, and so the PCA results are retained in this analysis. This is not the case for the unrotated PCA scores (see Figure 8.4c).

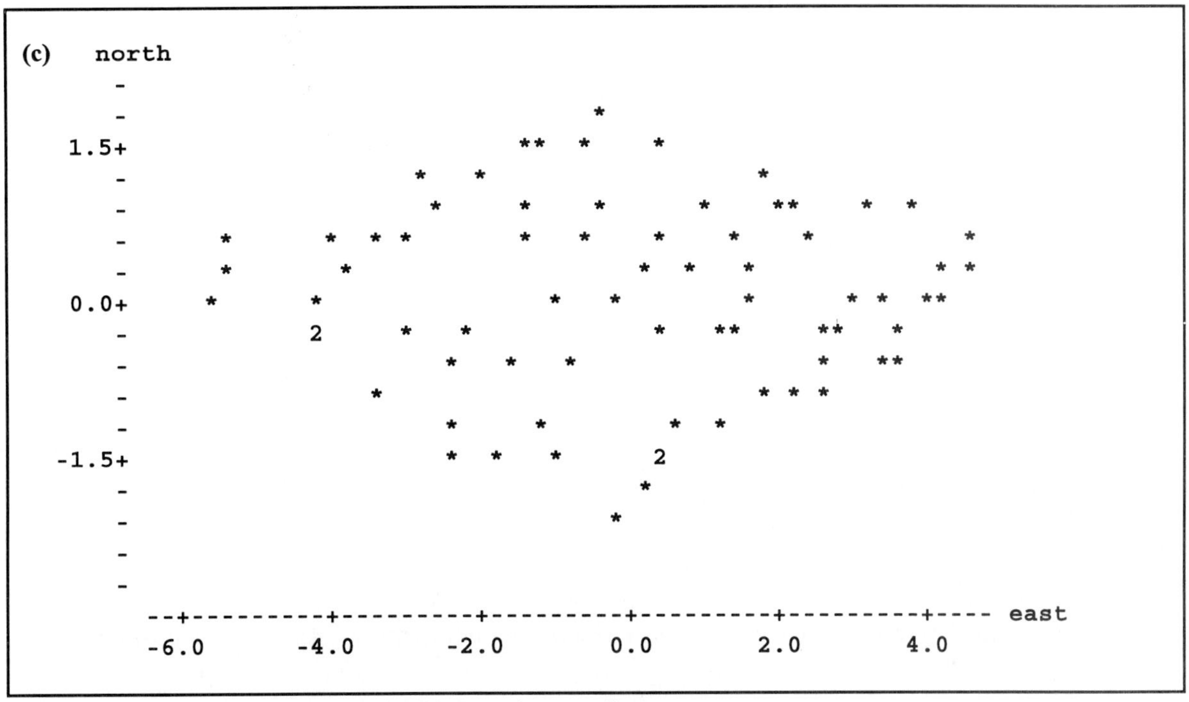

Figure 8.4 (continued). (c) unrotated PCA Cartesian coordinate scores.

Using the SAS procedure

```
PROC FACTOR METHOD=PRINCIPAL N=2 ROTATE=VARIMAX OUT=STEP1
        (REPLACE=YES);
   VAR UCOORD VCOORD;
RUN;
PROC CLUSTER DATA=STEP1 SIMPLE METHOD=WARD NOEIGEN
OUTTREE=DCDTREE;
   VAR FACTOR1 FACTOR2;
   FREQ FDCD;
   ID NUM1;
RUN;
```

produced SAS output that may be summarized as follows[1]:

[1]The AAR kernel municipios have been italicized.

Cluster 1	Cluster 2	Cluster 3	Cluster 4	Cluster 5
Aguas Buenas (5)	Aguada (3)	Cabo Rojo	Aibonito (5)	Adjuntas (4)
Bayamon/Catano	Aguadilla (3)	Guanica	Arroyo	Barranquitas (1)
Caguas (5)	Anasco (3)	Guayanilla (4)	Cayey (5)	Ciales (2)
Carolina	*Arecibo*	Hormigueros	Coamo	Cidra
Ceiba	Barceloneta/ Florida	Lajas	Guayama	Comerio (1)
Fajardo	Camuy	Penuelas (4)	Juana Diaz	Corozal (1)
Guaynabo	Dorado (1)	Sabana Grande	Maunabo (5)	Jayuya (4)
Gurabo (5)	Hatillo	San German	Patillas	Lares (2)
Humacao (5)	Isabela (3)	Yauco (4)	*Ponce*	Las Marias (3)
Juncos (5)	Manati		Salinas	Maricao (3)
Las Piedras (5)	Moca (3)		Santa Isabel	Mayaguez (3)
Loiza/Canovanas	Quebradillas		Villalba	Morovis (2)
Luquillo	Rincon (3)		Yabucoa (5)	Naranjito (1)
Naguabo (5)	San Sebastian (3)			Orocovis (1)
Rio Grande	Toa Baja (1)			Utuado (2)
San Juan/ Rio Piedras	Vega Alta (1)			
San Lorenzo (5)	Vega Baja			
Toa Alta				
Trujillo Alto				

These results, especially in terms of the denoted misclassifications, are informative:

(1) the San Juan, Arecibo, and Ponce regions are reasonable clusters (respectively, 11 of 19, 7 of 11, and 9 of 14 municipios are correctly classified);

(2) the Mayaguez region is not clearly distinguished from the Arecibo region (7 of the 16 municipios in the Mayaguez region are misallocated to the Arecibo region, whereas only 6 are correctly allocated to the Mayaguez region), and

(3) the Caguas region is not clearly distinguished from the San Juan region (8 of the 13 municipios in the Caguas region are misallocated to the San Juan region, whereas only 1 is correctly allocated to the Caguas region).

Note that the variate FDCD equals 10×DCD, since SAS requires that frequencies be integers and truncates values where they are not. The scree plot displayed in Figure 8.3a suggests that 5 is a reasonable value for the number of groups of municipios. The single fundamentally different cluster from those demarcated by the five agricultural administrative regions incorporates an east-west corridor of interior highland municipios (Map 8.3a).

The value for R^2 reported for this analysis using Ward's clustering algorithm may be computed by (1) constructing indicator variables, one for each cluster at a given step in the dendogram, (2) regressing both the u-coordinate and the v-coordinate on this set of indicator variables, and (3) computing

$$\frac{\text{sum-of-squares due to regression for the u-coordinate} + \text{sum-of-squares due to regression for the v-coordinate}}{\text{total of adjusted sum-of-squares for the u-coordinate} + \text{total adjusted sum-of-squares for the v-coordinate}}$$

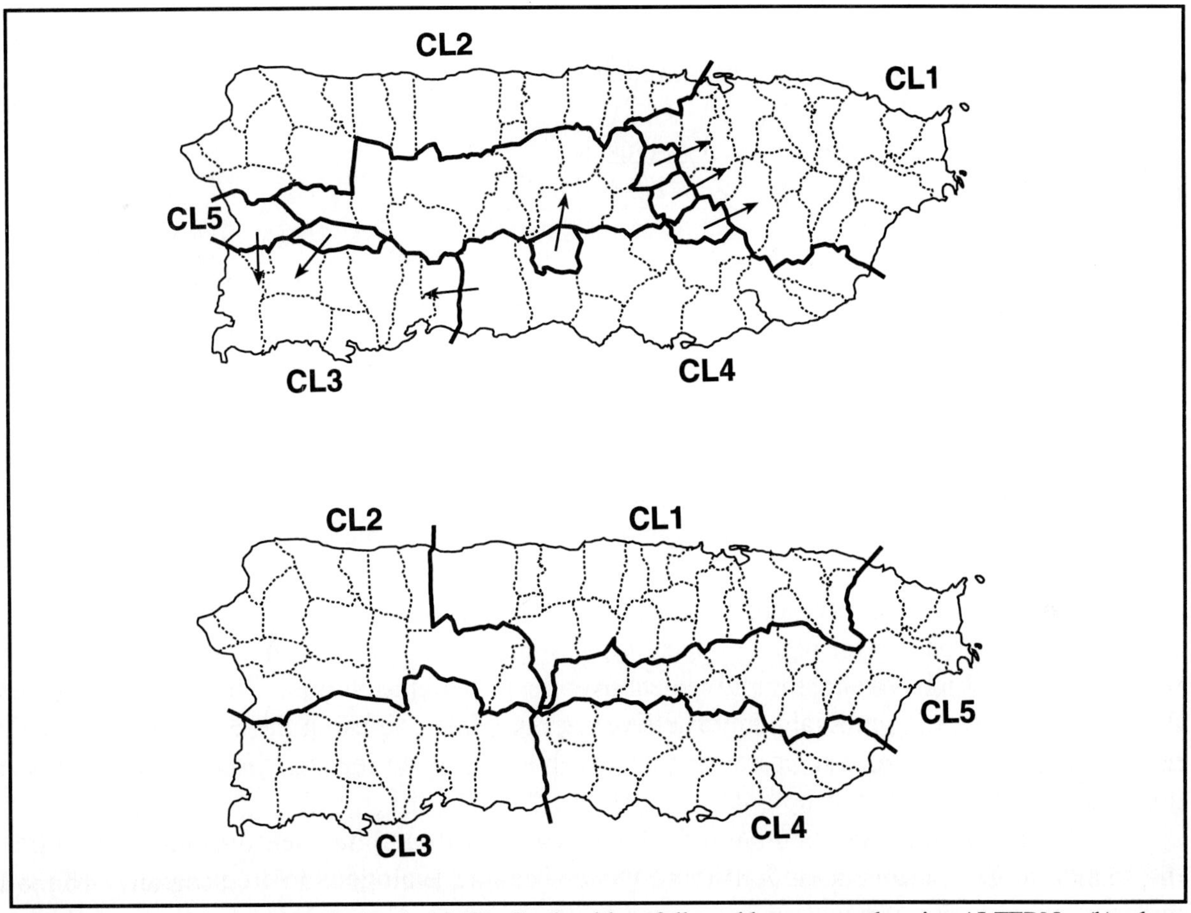

Map 8.3. (a) cluster solution obtained with Ward's algorithm, followed by an upgrade using ALTERN . (b) cluster solution obtained with ALTERN alone.

For example, constructing five indicator variables for each of the final five clusters and then arbitrarily deleting one of these variables from the regression renders $R^2 = \frac{22360.483 + 16179.686}{2.6090.735 + 23646.084} = 0.774882$.[2] Again, the need to avoid double-counting by working with

[2]These results can be obtained using the SAS procedures

```
PROC CLUSTER DATA=STEP1 SIMPLE METHOD=WARD NOEIGEN OUTTREE=DCDTREE;
   VAR FACTOR1 FACTOR2;
   FREQ FDCD;
   ID NUM1;
RUN;
PROC TREE DATA=DCDTREE NOPRINT OUT=TREEOUT NCLUSTERS=5;
   ID NUM1;
RUN;
PROC SORT;
   BY NUM1;
RUN;
DATA STEP1 (REPLACE=YES);
   SET STEP1;
   SET TREEOUT;
IF CLUSTER=1 THEN ICL1=1; ELSE ICL1=0;
IF CLUSTER=2 THEN ICL2=1; ELSE ICL2=0;
```

orthogonalized data becomes apparent.

Subjecting these results to ALTERN yields an aggregate weighted distance value of 15194 and the following reallocation of seven municipios:

municipio	old cluster	new cluster
Cidra	5	1
Comerio	5	1
Maricao	5	3
Mayaguez	5	3
Naranjito	5	1
Ponce	4	3
Villalba	4	5

While this adjustment improves the compactness of the five clusters, it still fails to recover the underlying agricultural administrative regional structure. It also highlights the similarity between means and medians.

Letting the ALTERN program automatically generate five clusters from the initial unclustered data produces the regions appearing in Map 8.3b. Now the aggregate weighted distance value is 14829, which is a slight improvement over the preceding solution. These clusters preserve the original Caguas agricultural administrative region, while confusing the San Juan and Arecibo regions, and the Mayaguez region with the Arecibo and Ponce regions. Recall that there is no guarantee that this solution is optimal.

To summarize, CAs is similar to L-A. As such, centrographic measures become central to the spatial analysis problem, and attribute values become analogous to frequencies. The well-understood weakness of unguaranteed global optimality in L-A accents the presence of this very same feature in CA. Although CA solutions may be informative and perhaps supply initial starting clusters for a L-A problem, considerably more work needs to be completed on the theme of articulating relations between these two problem sets.

8.4. Natural Groups of Coterminous Areal Units

Parallels between conventional cluster analysis of geo-referenced observations and location-allocation analysis again turn attention to the need for coterminous sets of areal units and the imposition of contiguity constraints. In other words, some geographic applications may require constraint of the clustering in such a way that geographical proximity also is optimized; attribute clusters need to form contiguous geographic regions (e.g., political districts). Although inclusion of Cartesian co-ordinates in with attributes will encourage such results, they are not guaranteed. Complications hindering the creation of such clusters (as seen in Example PR-8.1), coupled with

```
IF CLUSTER=3 THEN ICL3=1; ELSE ICL3=0;
IF CLUSTER=4 THEN ICL4=1; ELSE ICL4=0;
RUN;
PROC GLM; MODEL FACTOR1 FACTOR2 = ICL1-ICL4; FREQ FDCD; RUN;
```

the use of point pattern analysis statistics to index cluster density, return spatial scientists to familiar problem sets. Griffith and Amrhein (1991, p. 131) illustrate common point patterns that resist meaningful geographic description with traditional analysis procedures. For example, both point pattern and cluster analysis encounter difficulties in handling concentric rings of areal units. Also, two groups of areal units whose regional locations form a T-shape tend not to be well separated by CA algorithms, whereas such a situation in point pattern analysis can result in a meaningless nearest-neighbor statistic (one that exceeds 2.15). However, while point pattern techniques are applicable to random, nonclumping geographic distributions of points, in cluster analysis geographically distinct clusters of areal units are preferred. Therefore, much of the knowledge acquired by spatial scientists concerning point pattern analysis could help illuminate CA solutions.

The persistent need for coterminous areal unit cluster solutions, especially when Cartesian coordinates are not the primary variables upon which clustering is based, has motivated researchers to extend other notions from spatial analysis to CA. One noteworthy effort is represented by "The *R* Package," designed by Legendre and Vaudor (1991). This software package supplies an implementation of clustering with a two-dimensional spatial contiguity constraint. One algorithm employed by these researchers, called proportional-link linkage, allows a spectrum of solutions ranging from single to complete linkage (hence enabling intermediate possibilities beyond just average linkage). Traditional CA is achieved by letting comparisons made at each hierarchical step involve all possible pairs of areal units/groups. Coterminous CA is achieved by restricting comparisons made at each hierarchical step to only adjacent pairs of areal units/groups. The outcome is a trade-off between more homogeneous clusters, on the one hand, and coterminous clusters, on the other hand. The additional data set that is needed to accomplish this end is the standard connectivity matrix found in spatial statistics. As an aside, a centroid-type algorithm (K-means) also is available in "The *R* Package."

Recapitulating Example PR-8.1 for illustrative purposes, a scree plot of the Ward's algorithm solution suggests that either three or four clusters are latent in the San Juan region, based upon the single DCD variable. The shallowness of the slope of this plot beginning with three clusters supports the selection of this number. But the three-cluster univariate solution is composed of noncoterminous groups. Reanalyzing this data with the complete linkage algorithm renders exactly the same solution. In contrast, utilizing Legendre and Vaudor's complete linkage algorithm with a contiguity constraint alters the groupings (Map 8.4):

(1) one cluster is built around Orocovis and now also includes Barranquitas, Corozal, Naranjito, and Toa Alta;

(2) a second cluster is built around Ceiba and now also includes Fajardo and Luquillo; and,

(3) Comerio and Rio Grande now become part of the San Juan metropolitan area region.

In other words, a western, a central, and an eastern group have emerged, but without explicit reference to areal unit geo-referencing coordinates. This solution also markedly differs from what might be considered an intuitive adjustment to the solution presented in Example PR-8.1, namely

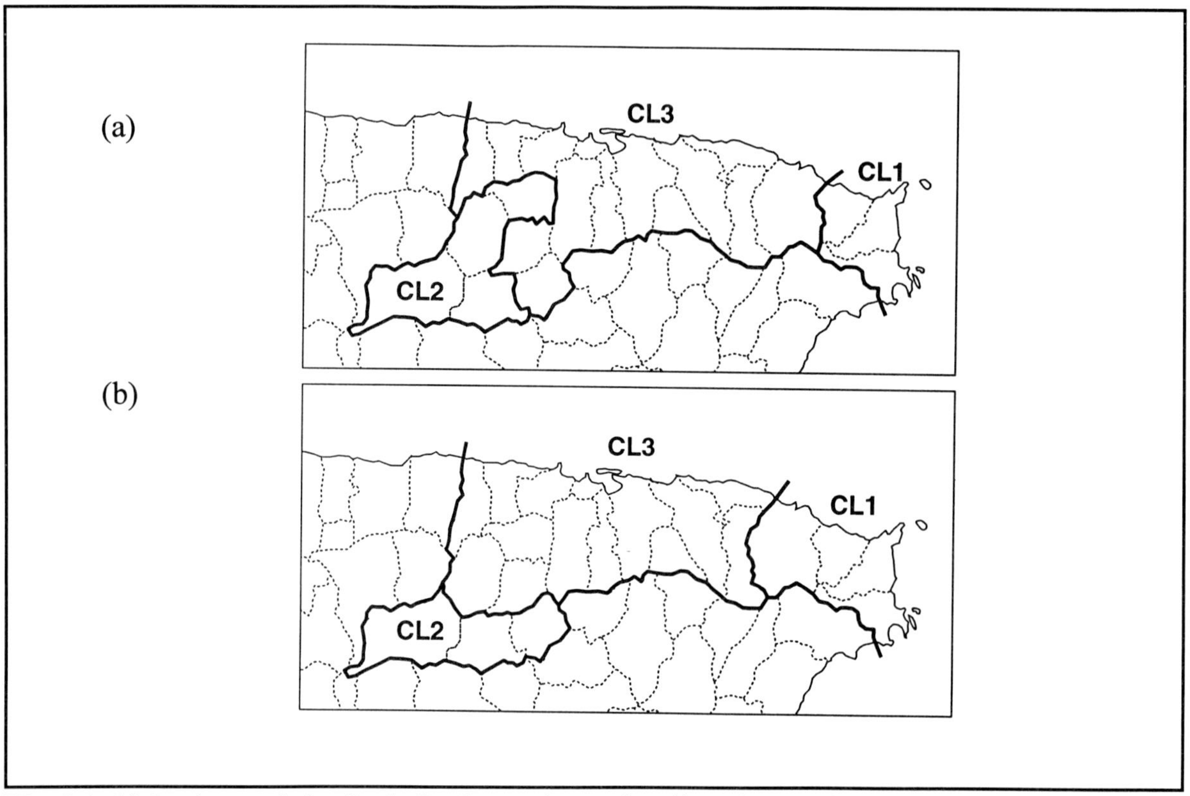

Map 8.4. (a) the cluster solution for the San Juan region produced by "The *R* Package"; (b) the intuitively adjusted cluster solution for the San Juan region from Example PR-8.1.

(1) add Barranquitas to the cluster including Orocovis and Comerio, and

(2) add Fajardo and Luquillo to the cluster including of Ceiba and Rio Grande.

Therefore, the pioneering work of Legendre and Vaudor furnishes invaluable, spatially adjusted CA algorithms for dealing with geo-referenced data other than Cartesian coordinates themselves.

8.5. Using DFA to Improve CA Solutions

CA solutions are sequentially rather than globally optimal implying that they can be improved upon. Because the pairwise differences between regional means are being maximized in Ward's algorithm, its objective relates to the search for a linear combination that maximizes these differences. One instinctive diagnostic procedure is to conduct a DFA on the recovered taxonomy. This DFA would be conducted iteratively, updating the classification scheme after each iteration, until the DFA results converged upon 100% correct classification. By doing so for Example PR-8.2, the following reallocation of seven municipios occurs:

iteration	municipio	old cluster	new cluster
1	San Lorenzo	1	4
	Cidra	5	1
	Comerio	5	1
	Naranjito	5	1
	Ponce	4	3
2	Barranquitas	5	1
	Humacao	1	5

Four of these reallocations are the same ones made by the ALTERN L-A algorithm. Moreover, their reassignment furnishes evidence that the sequential solution obtained with Ward's CA algorithm is suboptimal.

The posterior probabilities of membership in a cluster, computed with the DFA results, indicate that Ward's algorithm sometimes selects the secondmost likely cluster, even in situations where probabilities are markedly different. For example, these posterior probabilities for San Lorenzo are 0.3188 for membership in Cluster 1 and 0.6467 for membership in Cluster 4. Other situations, such as that for Comerio, have posterior probabilities with a smaller difference: 0.4676 for membership in Cluster 5 and 0.5247 for membership in Cluster 1.

Iteratively improving the CA solution with DFA will also alter the classification objective function criterion. Here R^2 increases to $\frac{22288.614 + 16754.890}{26090.735 + 23646.084} = 0.785003$. In general, improving the accuracy of the classification always will move R^2 toward 1. Meanwhile, using this modified classification scheme as the initial allocation in ALTERN yields an aggregate weighted distance of 15176, marginally better than the original classification generated by Ward's algorithm but not as good as the default solution that is produced automatically by ALTERN. Increased stability of the solution is not guaranteed by DFA fine-tuning, either. For instance, the second DFA iteration results in Barranquitas being reallocated from Cluster 5 to Cluster 1. ALTERN then reallocates this municipio to Cluster 3 at the next iteration.

To summarize, DFA offers a means of improving a CA classification scheme. It still does not, however, ensure that the global optimum is achieved. This feature of cluster analysis emphasizes the difficulty researchers encounter when trying to answer *CA Questions #1* and *#2*, and they may well find a local optimum that is only slightly better.

8.6. Aggregation Effects

It has been suspected for a long time that aggregating data contained in small areal units into fewer, larger areal units introduces error into statistical analyses (Openshaw, 1984; Amrhein, 1995). Until remedial prescriptions become available, spatial researchers at least need to qualify results of geo-referenced data analyses by acknowledging this phenomenon.

EXAMPLE PR-8.3. Demonstrate the effect of spatial aggregation using the Puerto Rico data set.

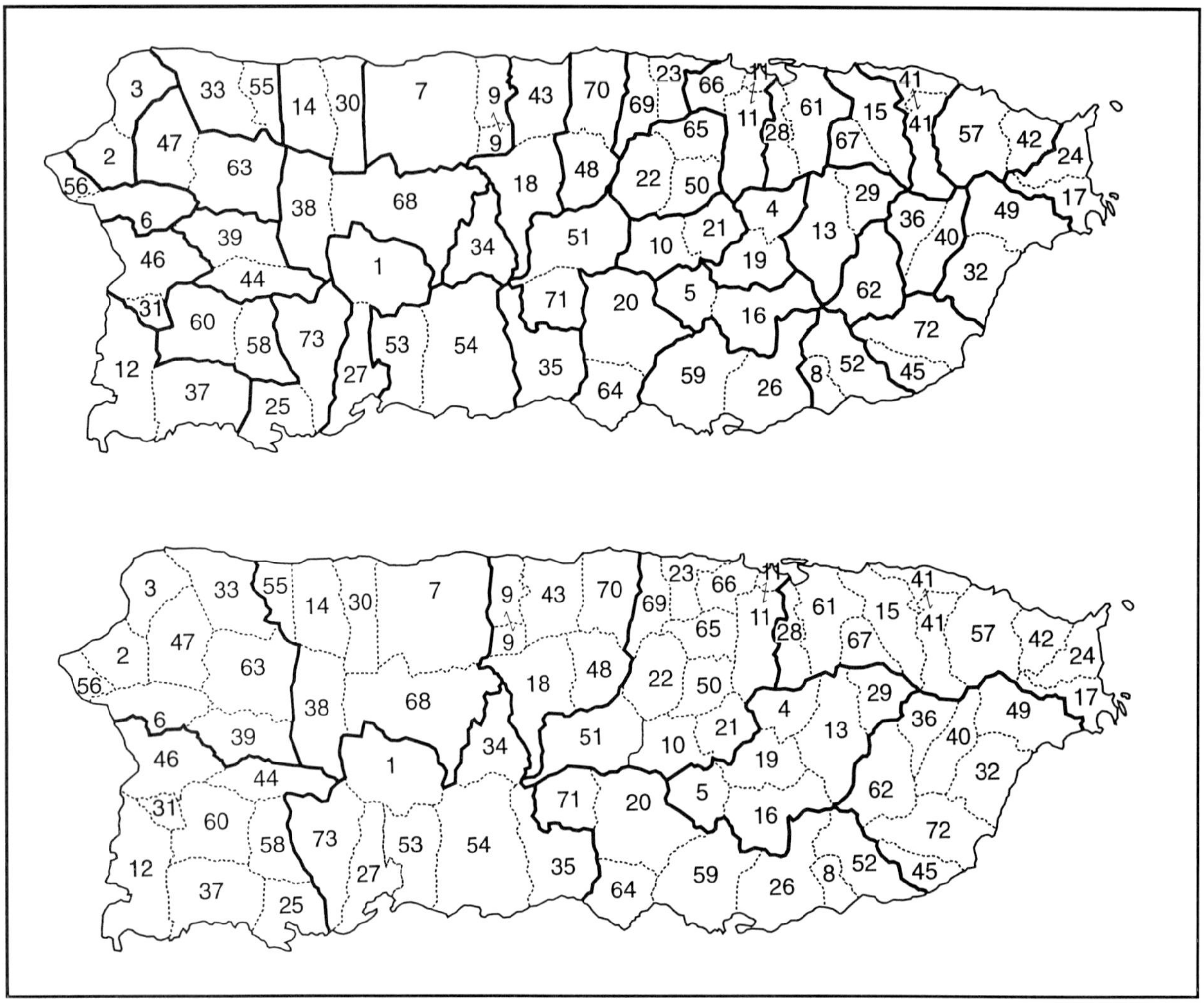

Map 8.5. (a) the aggregation of 73 municipios into 37 regions; (b) the aggregation of 73 municipios into 10 regions.

Map 1.1 shows the original 73 municipios in Puerto Rico, which have been numbered 1–73 in Map 8.5a–b and Map 2.1 shows their centroid locations. Aggregations created manually of the original 73 municipios into 37 (combining pairs of municipios), 10, and 5 zones produced the following results for the variables DCD and DMLK. Map 8.5a displays the aggregations of 73 municipios into 37 regions demarcated by heavy lines, while Map 8.5b shows the aggregation into 10 regions demarcated by heavy lines. Finally, the 5-region aggregation scheme is the existing five agricultural administrative regions (see the display of their outlines in Map 2.1). The following table based upon these different aggregations summarizes changes in various descriptive statistics (Table 8.1).

Table 8.1. Results of aggregation experiments displayed in Map 8.5.

		Number of zones			
Statistic	***Variable***	73	37	10	5
Mean	***DCD***	36.1740	36.1095	36.8290	33.2260
	DMLK	27.0863	26.0676	29.0170	28.9800
Standard deviation	***DCD***	11.9587	10.5711	7.8922	8.6477
	DMLK	49.8303	39.9695	26.8365	25.6926
Minimum	***DCD***	2.9000	8.9500	17.7800	22.7400
	DMLK	0.0000	0.0000	3.6700	3.8300
Maximum	***DCD***	62.9000	55.4000	47.4000	44.2900
	DMLK	368.6000	218.4000	87.8300	64.8100
Pearson's r	***DCD with DMLK***	0.05702	-0.00797	-0.04168	-0.79807
Prob > /r/ under H_0 Rho=0		(0.6318)	(0.9623)	(0.9090)	(0.1506)

Of particular concern in these results is the behavior of the correlation coefficient, changing from a value of roughly 0 to a very large negative value as the number of areal units decreases from 73 to 5. This result is consistent with Amrhein's findings (1995) and reinforces the earlier discussion about the inferential risks involved with small sample analysis. The changes in the values of the means and standard deviations are similar to those described for census data from Lancashire County in the U.K. (see Amrhein and Reynolds, 1996). Also noteworthy is the dramatic change in the maximum value of DMLK, which decreases from 368.6 in the 73-region case, to 64.81 in the 5-region case. Clearly the aggregation process, by using averaged smaller region values as attribute values for fewer, larger regions, has introduced measurement error into the geo-referenced data.

> *RULE OF THUMB 8.2*: Since geographic scale transformations of spatial data impact upon results, when possible data should be analyzed at the geographic scale for which they were originally collected, and the aggregation error assumptions should be carefully considered.

This rule argues that before any data collection project gets underway, the analyst needs to determine the most appropriate spatial scale for the analysis.

8.7. Problems

P-8.1. Given the *Effect of Chemicals on Earthworm Populations* (Table 52.1 in Andrews and Herzberg, 1985) data set, use total biomass/m^2 measured in *sample 1* to compare underlying natural groupings of the 40 field plots uncovered by each of the cluster algorithms. Then, compare these results with those obtained by including data from all three samples [note: each sample could constitute a separate quantitative variable].

P-8.2. Given *The Yields of Wheat on the Broadbalk Fields at Rothamsted* (Table 5.1 in Andrews and Herzberg, 1985) data set, identify underlying natural groupings of the 18 field plots based upon the 74 years of data reported for each plot.

8.8. References

Amrhein, C. 1995. "Searching for the Elusive Aggregation Effects: Evidence from Statistical Simulations." *Environment and Planning A* 27: 105–19.

Amrhein, C. and H. Reynolds. 1996. "Using Spatial Statistics to Assess Aggregation Effects." *Geographical Systems* 2: 83–101.

Andrews, D. and H. Herzberg. 1985. *Data: A Collection of Problems from Many Fields for the Student and Research Worker*. Berlin: Springer-Verlag.

Bailey, T. and A. Gatrell. 1995. *Interactive Spatial Data Analysis*. Harlow, U.K.: Longman Scientific Publishing.

Griffith, D. and C. Amrhein. 1991. *Statistical Analysis for Geographers*. Englewood Cliffs, N.J.: Prentice Hall.

Haggett, P., A. Cliff, and A. Frey. 1977. *Locational Analysis in Human Geography,* 2nd e.d. London: Edward Arnold.

Isaaks, E. and R. Srivastava. 1989. *An Introduction to Applied Geostatistics*. New York: Oxford University Press.

Johnston, R. 1978. *Multivariate Statistical Analysis in Geography*. London: Longman Scientific Publishing.

Lance, G. and W. Williams. 1967. "A General Theory of Classificatory Sorting Strategies." *Computer Journal* 9: 373–80.

Legendre, P. and A. Vaudor. 1991. "The *R* Package: Multidimensional Analysis, Spatial Analysis." Départment de sciences biologiques, Université de Montréal. Unpublished manuscript.

Openshaw, S. 1984. *The Modifiable Areal Unit Problem.* CATMOG 38. Norwich: GeoAbstracts.

Rushton, G. 1975. *Optimal Location of Facilities.* Hanover, N.H.: COMPUTe, Dartmouth College.

CHAPTER 9

CANONICAL CORRELATION ANALYSIS

Chapter Preview. While we stated in the preview to Chapter 4 that regression analysis is viewed as the "workhorse" of statistical analysis (and hence we cast each technique in terms of its "regression roots"), this view is not universally held. Regression analysis uses many independent variables to analyze one dependent variable. Both dependent and independent variables are generally interval/ratio (although we will see in Chapters 10–12 that there are varieties of regression models that relax the assumption relating to the scale of measurement). A more general linear model, and perhaps one more general even than regression, would permit more than one dependent variable as well as more than one independent variable, with either the dependent or independent variables measured at the nominal or interval/ratio scales. This more general model is canonical correlation Analysis (CCA), the focus of this chapter.

In a sense, because the three remaining chapters explore variations of the multivariate regression model introduced in Chapter 4, CCA is the last, totally new technique to be explored in this text. As such, it is perhaps appropriate that CCA "captures" the essence of all that has gone before, including PCA and FA (essentially, CCA treats each set of variables, dependent and independent, as a PCA and combines the results). Why then, if CCA is truly the most general linear multivariate technique, do we need all of the others?

CCA "places the fewest restrictions on the types of data on which it operates. Because the other techniques impose more rigid restrictions, [and hence can be shown to be special cases of CCA,] it is generally believed that the information obtained from them is of higher quality and may be presented in a more interpretable manner. For this reason, many researchers view canonical correlation as a last-ditch effort, to be used when all other higher-level techniques have been exhausted" (Hair et al., 1995, p. 328). In addition, the mathematical notation for CCA can be quite formidable, and the computational requirements large, causing many analysts to be unwilling to pay the overhead costs in time and computing power required to perform it.

If we restrict our attention to the case in which both independent (or predictor) and dependent (or criterion) variables are interval/ratio, then the goal of CCA is to reduce both sets of variables through the construction of linear combinations so that the two sets of synthetic variables are as highly correlated as possible (one pair of synthetic variates is extracted, with one of the pair coming from the dependent variable and the other one coming from the independent variable).

In the context of explicitly spatial analysis, one goal is to determine linear combinations of attributes maximally correlated with linear combinations of spatial coordinates. "The problem is that a standard canonical correlation analysis is successful in achieving a reduction in the attribute space to the two 'most spatially correlated' combinations of attributes, but suffers from the problem that the spatial directions to which these relate are an undesirable distortion of space [that] is difficult to interpret" (Bailey and Gatrell, 1995, p. 239).

9.0. Introduction

Traditional regression analysis addresses the problem of predicting a single variable, in an n-by-1 vector **Y**, from an assortment of P predictor variables (plus a vector of ones to capture the intercept term), in an n-by-(P+1) matrix **X**. One theme running through this book is that virtually every statistical technique can be reformulated as a regression problem. This contention becomes more obvious when the traditional regression technique is replaced by the general linear model (GLM), which addresses the problem of simultaneously predicting some battery of Q variables, in an n-by-Q matrix **Y**, as is done with MANOVA (§5.2) from those predictor variables in matrix **X**. Whereas multicollinearity is a concern in regression for the variables composing matrix **X**, the extension of vector **Y** to a matrix that includes Q variables raises issues about multicollinearity among this set of variables. ***Canonical correlation analysis*** (CCA) is a technique that handles this additional source of multicollinearity in the case where $P > 1$ and $Q > 1$. Accordingly, it focuses on correlations between linear combinations of variables in the two data sets. As such, it sometimes is referred to as a double-barreled PCA (geometrically, it is a PCA that then is modified by a sequence of judiciously selected axis rotations). Some researchers contend that CCA, rather than regression, is the most general multivariate statistical technique.

CCA determines matched pairs of optimal linear combinations of variables, one from each of the matrices **X** and **Y**, much like those computed with DFA (discriminant function analysis) (see Chapter 7) and PCA (principal components analysis) or FA (factor analysis) (Chapter 6), but they are maximally related. In other words, CCA uncovers the synthetic variates from matrix **X** that are most highly correlated, linearly, to their companion synthetic variates from matrix **Y**. Just as in DFA, PCA, or FA, a structure matrix can be constructed for CCA revealing the nature and degree of relations between the original variables and those common dimensions running across both data sets. Again, reminiscent of DFA, PCA, or FA, these synthetic variates constitute canonical correlation variate scores. The notion of canonical correlation refers to a measure of the strength of association between corresponding pairs of the synthetic variates. When $Q = 1$, this measure reduces to the multiple correlation coefficient of regression. The maximization feature is necessary because, as repeatedly mentioned), globally unique eigenfunction solutions are nonexistent (as is attested to by the use of various rotation criteria). Thus, this optimization aspect of the technique enables the largest correlations to be computed. Not surprisingly, the first canonical dimension often produces a canonical correlation close to one.

Genuine substantive CCA application studies are hard to find in the spatial analysis literature, primarily for three major reasons. Historically speaking, both the necessary computing resources (required for the numerically intensive calculations involved in solving a double eigenfunction problem) and software were not available. In addition, the mathematical symbolism and jargon utilized for CCA in textbooks (see, for instance, Johnson and Wichern, 1992, pp. 459–74) is quite formidable, discouraging many practitioners from studying its conceptual material or being able to interpret its results. But interpretation is no more difficult than for multiple regression, PCA, MANOVA, or DFA. After all, the same set of assumptions are employed in CCA, which seems to be just as robust against their violation as are these other multivariate techniques. Finally, there is a limited awareness of CCA among spatial scientists. We hope this chapter helps remedy unfamiliarity with this method, facilitating dissemination of its use.

Two dissimilar research questions are affiliated with CCA. The first may be phrased as:

CCA Question #1: What type of underlying dimensions span two data sets?

This question is linked to the perspective that CCA is a double-barreled PCA. The following two distinctive classes of variables constitute data sets for which *CCA Question #1* may be posed

(1) two sets of variables measured across the same set of areal units at a given point in time, such that the standard geographic matrix (Griffith and Amrhein, 1991, p. 14) of cross-sectional data is divided vertically into two groups—the row labels are identical but the column labels are different; and,

(2) a pair of variables measured across the same set of areal units at various points in time, such that two space-time series are involved—the row and column labels are identical, except for temporal position.

The second *CCA* question stems from extending the multiple regression version of the two-regions DFA problem (§7.4) to the multiple-regions case. Here matrix **Y** is made up of the set of (K - 1) indicator variables designating regional membership (recall that an arbitrarily selected regional indicator variable may be omitted to avoid the presence of perfect multicollinearity), while matrix **X** is the set of predictor variables. As with the two-regions case, the linear combination coefficients obtained are proportional to those obtained with DFA.

CCA Question #2: What underlying dimensions run across both regional membership and the set of predictor variables?

The following third distinctive class of variables composes data sets for which *CCA Question #2* may be posed

(3) one set of variables measured across regional collections of areal units at a given point in time, such that the standard geographic matrix of a single cross-section of data is divided horizontally into K groups—the column labels are identical but the row labels are different.

Data of type (1) are illustrated by the analysis reported in Griffith (1982), where two sets of two-dimensional shape measures made on a single set of city boundaries have been analyzed for common dimensions. CCA results for data of type (2) are reported in most of the examples in this chapter. Of course, some mixture of these two data set types is possible, yielding two sets of variables measured across the same areal units at two given points in time—two parallel geographic matrices having identical row and column labels for each of the two time periods in question. Finally, data of type (3) relate to DFA; results of this sort are obtainable with the SAS procedure CANDISC.

As an aside, CCA can be extended to three or more data sets, as is illustrated (Griffith, 1988).

9.1. Properties of the Linear CCA Model

Selected salient features of the linear CCA model merit attention before its applications can be fully appreciated and understood. As mentioned in §9.0 CCA involves no new underlying assumptions. Further, its statistical significance testing is similar to that already presented for DFA. But the explicit eigenfunction problem for CCA differs substantially from those previously encountered in this text.

9.1.1. The CCA Eigenfunction Problem

The eigenfunction problem of CCA is constructed by partitioning the correlation matrix **R** for the combined set of variables into a submatrix $\mathbf{R}_{XX}$ for the correlations among the designated variables in matrix **X**, a submatrix $\mathbf{R}_{YY}$ for the correlations among the designated variables in matrix **Y**, and a submatrix $\mathbf{R}_{XY}$ for the correlations among the variables of matrices **X** and **Y**. Recall that because matrix **R** is symmetric (the correlation between X_p and Y_q equals the correlation between Y_q and X_p), $\mathbf{R}_{YX} = \mathbf{R}^T_{XY}$. Consequently, the CCA eigenfunction problem may be stated as

$$\text{eigenvalues} : DET\,(\mathbf{R}_{XX}^{-1}\mathbf{R}_{XY}\mathbf{R}_{YY}^{-1}\mathbf{R}_{YX} - \lambda^2\mathbf{I}) = 0 \text{ , and} \tag{9.1}$$

$$\text{eigenvectors} : \left[\begin{pmatrix} \mathbf{0} & \mathbf{R}_{XY} \\ \mathbf{R}_{YX} & \mathbf{0} \end{pmatrix} - \lambda \begin{pmatrix} \mathbf{R}_{XX} & \mathbf{0} \\ \mathbf{0} & \mathbf{R}_{YY} \end{pmatrix}\right] \begin{bmatrix} \mathbf{U} \\ \mathbf{V} \end{bmatrix} = \mathbf{0} \text{ .} \tag{9.2}$$

Equation (9.1) is a reduced form of the determinant of its counterpart in equation (9.2). Its eigenvalue, λ^2, is the squared canonical correlation between its corresponding pair of ***canonical variates*** (CVs), computed as $\mathbf{Z}_X\mathbf{U}$ and $\mathbf{Z}_Y\mathbf{V}$ (note that the eigenvector coefficients are applied to z-score transforms of the original variables, just like in PCA). This reduced form of the eigenvalue problem highlights the reason (near-)complete or (near-)perfect multicollinearity in either matrix $\mathbf{R}_{XX}$ or $\mathbf{R}_{YY}$ is problematic, namely, that their inverses become plagued by calculation error. Furthermore, as correlations within either of these matrices increase, the ability to disentangle individual variable contributions to the CVs decreases.

The mathematical basis for equations (9.1) and (9.2) comes from the maximization problem associated with CCA. Because this technique seeks to summarize $\mathbf{R}_{XY}$ in terms of pairs of linearly combined Xs and Ys carefully chosen to achieve a maximum correlation, its optimization problem may be written as

$$MAX_{\mathbf{U}_j,\mathbf{V}_j} : \frac{\mathbf{U}_j^T\mathbf{R}_{XY}\mathbf{V}_j}{[\mathbf{U}_j^T\mathbf{R}_{XX}\mathbf{U}_j]^{\frac{1}{2}}[\mathbf{V}_j^T\mathbf{R}_{YY}\mathbf{V}_j]^{\frac{1}{2}}} \text{ .} \tag{9.3}$$

Recall that if this were a PCA, to ensure uniqueness these eigenvectors would be selected in such a way that $\mathbf{u}_j^T\mathbf{u}_j = 1$ and $\mathbf{v}_j^T\mathbf{v}_j = 1$. For convenience, in CCA uniqueness is attained by selecting the vectors so that $\mathbf{u}_j^T\mathbf{R}_{XX}\mathbf{u}_j = 1$ and $\mathbf{v}_j^T\mathbf{R}_{YY}\mathbf{v}_j = 1$, reducing the denominator of equation (9.3) to 1. The maximum number of non-0 canonical correlations is *MIN*(P,Q), because equation (9.3) involves matched pairs.

EXAMPLE PR-9.1. Construct the correlation matrix partitions for a CCA.

Griffith (1995) has conducted a CCA of geo-referenced data that are supplemental to the Puerto Rican data set (see Appendix D). The Commonwealth of Puerto Rico (CPR) collects selected agricultural production data on an annual basis, which augment data collected by the U.S. Department of Agricultural (USDA) during its periodic censuses. These two data sets are collected for the same areal units and attributes. But each of the two agencies involved uses slightly different farm, production, and growing season definitions, and the USDA must suppress certain figures for legal reasons concerning confidentiality. Suppressed USDA values have been interpolated using CPR data, following Griffith (1992). Milk production data for five consecutive USDA census years are selected, together with their corresponding CPR data.

Using the SAS procedure

```
PROC CANCORR;
        WITH CPR1-CPR5
         VAR USDA1-USDA5;
RUN;
```

produced the SAS output on the next page.

The 10-by-10 correlation matrix associated with the ten variables analyzed here is partitioned as shown in the SAS output on the next page. The first of these matrices is $\mathbf{R}_{YY}$, the second is $\mathbf{R}_{XX}$, and the third is $\mathbf{R}_{YX}$. Invoking *RULE OF THUMB 4.4* (which designates correlations greater in absolute value than 0.9 as indicating the presence of serious multicollinearity), consecutive years are found to introduce serious multicollinearity into each of the matrices $\mathbf{R}_{XX}$ and $\mathbf{R}_{YY}$. For the most part, matrix $\mathbf{R}_{YX}$ is consistent with expectations about taking repeated measurements.

When matrix $\mathbf{Y}$ is composed of indicator variables, the j-th solution for equation (7.1) equals $\frac{\lambda_j^2}{(1 - \lambda_j^2)}$. This solution is related to the solution for equation (9.1) in that λ_j^2 is the solution to equation (9.1), while $\frac{\lambda_j^2}{(1 - \lambda_j^2)}$ is the solution to equation (7.1).

Correlations Among the 'VAR' Variables

	USDA1	USDA2	USDA3	USDA4	USDA5
USDA1	1.0000	0.9451	0.8640	0.7654	0.7154
USDA2	0.9451	1.0000	0.9322	0.8585	0.8103
USDA3	0.8640	0.9322	1.0000	0.9529	0.9255
USDA4	0.7654	0.8585	0.9529	1.0000	0.9670
USDA5	0.7154	0.8103	0.9255	0.9670	1.0000

Correlations Among the 'WITH' Variables

	CPR1	CPR2	CPR3	CPR4	CPR5
CPR1	1.0000	0.9481	0.8650	0.7918	0.7194
CPR2	0.9481	1.0000	0.9539	0.8947	0.8320
CPR3	0.8650	0.9539	1.0000	0.9681	0.9181
CPR4	0.7918	0.8947	0.9681	1.0000	0.9662
CPR5	0.7194	0.8320	0.9181	0.9662	1.0000

Correlations Between the 'VAR' Variables and the 'WITH' Variables

	CPR1	CPR2	CPR3	CPR4	CPR5
USDA1	0.9874	0.9394	0.8570	0.7785	0.7111
USDA2	0.9548	0.9849	0.9328	0.8726	0.8072
USDA3	0.8729	0.9502	0.9876	0.9693	0.9206
USDA4	0.7773	0.8772	0.9533	0.9863	0.9559
USDA5	0.7226	0.8393	0.9308	0.9731	0.9784

9.1.2. Mathematical Features of Canonical Dimensions

If $\mathbf{R}_{XY} = \mathbf{0}$, then the solution to equation (9.1) is 0, and **U** more directly relates to the matrix of PCA eigenvectors for matrix $\mathbf{R}_{XX}$, while **V** more directly relates to the matrix of PCA eigenvectors for matrix $\mathbf{R}_{YY}$. This result reinforces the notion that CCA is analogous to a double-barreled PCA. Extending this analogy with PCA, the *MIN*(P,Q) eigenvectors $\mathbf{U}_j$ should be, and are, mutually orthogonal. The same condition holds for the *MIN*(P,Q) eigenvectors $\mathbf{V}_j$. In addition, the unmatched pairs of eigenvectors $\mathbf{U}_j$ and $\mathbf{V}_k$ ($j \neq k$) are orthogonal. In other words, if the synthetic variables contained in $\mathbf{Z}_X\mathbf{U}$ are correlated with those variables in $\mathbf{Z}_Y\mathbf{V}$, the resulting correlation matrix is diagonal with the individual canonical correlations as the diagonal elements. As with both PCA and DFA, the canonical dimensions are extracted in such a way that the familiar descending sequencing occurs: the first pair has the largest correlation, the second pair has the second largest correlation subject to being uncorrelated with the initial pair, the third pair has the third largest correlation subject to being uncorrelated with the first two pairs, and so on.

EXAMPLE PR-9.2. Calculate the correlation matrix for the CVs of Example PR-9.1, show orthogonality, and compute the canonical correlation coefficients:

Because both the USDA and CPR government agencies are measuring the same agricultural production, conspicuous common dimensions should underlie the two data sets described in Example PR-9.1. Five [= *MIN*(5,5)] CVs have been constructed for these data using the SAS procedure CANCORR (a more detailed discussion of the output from this procedure follows). SAS labels them V1, V2, ..., V5 for the USDA data set, and W1, W2, ..., W5 for the CPR data set. The SAS procedure CORR can be used to generate output that includes the following correlation matrices:

	V1	V2	V3	V4	V5
V1	1.00000				
V2	0.00000	1.00000			
V3	0.00000	0.00000	1.00000		
V4	0.00000	0.00000	0.00000	1.00000	
V5	0.00000	0.00000	0.00000	0.00000	1.00000

	W1	W2	W3	W4	W5
W1	1.00000				
W2	0.00000	1.00000			
W3	0.00000	0.00000	1.00000		
W4	0.00000	0.00000	0.00000	1.00000	
W5	0.00000	0.00000	0.00000	0.00000	1.00000

	V1	V2	V3	V4	V5
W1	0.99740				
W2	0.00000	0.96955			
W3	0.00000	0.00000	0.62570		
W4	0.00000	0.00000	0.00000	0.53143	
W5	0.00000	0.00000	0.00000	0.00000	0.37973

These three correlation matrices confirm the mutual orthogonality within each of the two sets of CVs as well as between the unmatched CV pairs. The third correlation matrix confirms the diagonality property of correlations between the two sets of CVs and reveals the canonical correlations (the diagonal entries) of the matched pairs of variates. As anticipated, the first canonical correlation is very close to 1. As suspected for this particular application, several of the canonical correlations are very high.

EXAMPLE PR-9.3. Compare the correlation matrix for matched and unmatched CVs (reported in Example PR-9.2) with the correlation matrices obtained from two PCAs.

Using the SAS procedure PRINCOMP on the individual data sets consisting of USDA1–USDA5 and CPR1–CPR5, followed by using the SAS procedure CORR with their resulting principal component scores, emulates a literal double-barreled PCA. In other words,

Using the SAS procedure

```
PROC PRINCOMP OUT=PC1 PREFIX=CPRPC;
   VAR CPR1-CPR5;
RUN;
PROC PRINCOMP DATA=STEP3 OUT=PC2 PREFIX=USDAPC;
   VAR USDA1-USDA5;
RUN;
DATA STEP4;
   SET PC1;
   SET PC2;
RUN;
PROC CORR;
   VAR USDAPC1-USDAPC5;
   WITH CPRPC1-CPRPC5;
RUN;
```

produced this SAS output:

	USDAPC1	USDAPC2	USDAPC3	USDAPC4	USDAPC5
CPRPC1	0.99682	-0.00843	-0.01093	0.00047	-0.02537
CPRPC2	0.01135	0.95932	0.00097	-0.00864	0.05661
CPRPC3	0.00271	0.07526	0.56902	0.14198	0.10530
CPRPC4	0.02227	-0.08130	0.13580	-0.45136	0.29729
CPRPC5	-0.00645	0.04787	0.17942	-0.19565	-0.28913

Clearly this correlation matrix is not diagonal. Furthermore, the diagonal entries are less than their CCA counterparts, with the last two correlations being negative. Canonical correlations are never negative, because they are related to the multiple correlation coefficient of regression.

Restating this CCA problem in terms of traditional regression with Q = 1, the following eigenfunction problem is equivalent to equation (9.1):

$$DET\,(\mathbf{R}_{YY}^{-1}\mathbf{R}_{YX}\mathbf{R}_{XX}^{-1}\mathbf{R}_{XY} - \lambda^2\mathbf{I}) = 0\ , \qquad (9.4)$$

because labeling of the two sets of variables is arbitrary. Because **Y** is an n-by-1 vector in this case, $\mathbf{R}_{YY}^{-1} = 1$. Equation (9.4) reduces to

$$DET\,(\mathbf{R}_{YX}\mathbf{R}_{XX}^{-1}\mathbf{R}_{XY} - \lambda^2\mathbf{I}) = 0\ .$$

But $\mathbf{R}_{YX}\mathbf{R}_{XX}^{-1}\mathbf{R}_{XY}$ may be rewritten as

$$\left[\frac{Z_Y^T Z_X}{(n-1)}\right]\left[\frac{Z_X^T Z_X}{(n-1)}\right]^{-1}\left[\frac{Z_X^T Z_Y}{(n-1)}\right]$$

$$= \frac{Z_Y^T Z_X (Z_X^T Z_X)^{-1} Z_X^T Z_Y}{(n-1)}$$

$$= \frac{Z_Y^T Z_X (Z_X^T Z_X)^{-1} Z_X^T Z_Y}{Z_Y^T Z_Y}$$

$$= \frac{b_Z^T Z_X^T Z_Y}{Z_Y^T Z_Y} \, .$$

and is the squared multiple correlation coefficient, R^2 (Chapter 4).[1] This outcome reinforces the notion that CCA is more general than regression analysis, a point illustrated further by Example PR-9.4.

EXAMPLE PR-9.4. Predict transformed density of sugar cane production (SQDSGR) from other variables (Example PR-4.3 revisited).

Using the SAS procedure

```
PROC CANCORR ALL;
   VAR SQDSGR;
   WITH DFRM DCD DISTSJ DISTA I0;
RUN;
```

produced SAS output that includes the following three segments:

---

[1]Taking this argument one step further, letting P = 1 results in $\mathbf{R}_{XX}^{-1} = 1$, $\mathbf{R}_{YX} = \mathbf{R}_{XY} = r_{XY}$, and $\mathbf{I} = 1$. Now the eigenfunction problem becomes $DET(r_{XY} r_{XY} - \lambda^2) = 0$, demonstrating that the single eigenvalue here equals the squared product moment correlation coefficient.

```
                    Canonical Correlation Analysis

                             Adjusted          Approx          Squared
               Canonical     Canonical         Standard        Canonical
               Correlation   Correlation       Error           Correlation
    1          0.799964      0.789462          0.042433        0.639942

 ---------------------------------------------------------------------------

          Likelihood
             Ratio        Approx F       Num DF       Den DF      Pr > F
    1     0.36005751      23.8163             5           67      0.0001

 ---------------------------------------------------------------------------

       Raw Canonical Coefficients for the 'WITH' Variables
                                              W1
                         DFRM          -0.070730268
                         DCD            0.0151264455
                         IO             1.1871137412
                         DISTA          0.1552034349
                         DISTSJ         0.2661750372
```

The squared canonical correlation, seen in the first part of the output, is identical to the multiple regression coefficient reported in Example PR-4.3. The F-ratio, seen in the second part of the output, is identical to the regression ANOVA F-ratio of Example PR-4.3. Finally, the set of coefficients in the single eigenvector $\mathbf{U}_1$ are proportional to those reported in Example PR-4.3, excluding the intercept term. Normalizing both sets of coefficients so that $\mathbf{b}_*^T\mathbf{b}_* = 1$ (recall from §7.4 that subvector $\mathbf{b}_*$ is the vector of regression coefficients $\mathbf{b}$, with b_0 the coefficient for the intercept term, removed) and $\mathbf{U}_1^T\mathbf{U} = 1$, yields

```
     DFRM         DCD         IO         DISTA      DISTSJ
-0.057571   0.012312   0.966249   0.126328   0.216653
```

More generally, a given squared canonical correlation *is* the multiple correlation obtained by regressing $\mathbf{U}_j$ extracted from matrix $\mathbf{X}$ on the matrix of variables $\mathbf{Y}$ or $\mathbf{V}_j$ extracted from matrix $\mathbf{Y}$ on the matrix of variables $\mathbf{X}$.

EXAMPLE PR-9.5. Execute regressions of $\mathbf{U}_1$ on matrix $\mathbf{Y}$ and of $\mathbf{V}_1$ on matrix $\mathbf{X}$ from the CCA performed for Example PR-9.2.

Canonical correlation can be directly related to regression.

Using the SAS procedure

PROC CANCORR;

```
        WITH CPR1-CPR5
         VAR USDA1-USDA5;
RUN;
PROC REG;
   MODEL V1 =USDA1-USDA5;
RUN;
PROC REG;
   MODEL W1 = CPR1-CPR5;
RUN;
```

produced SAS output that includes the following two segments:

```
Dependent Variable: V1

Root MSE        0.07465     R-square        0.9948
Dep Mean        0.00000     Adj R-sq        0.9944

                 Parameter Estimates
                Parameter      Standard     T for H0:
Variable  DF     Estimate         Error   Parameter=0      Prob > |T|
INTERCEP   1    -2.986424    0.02985771      -100.022          0.0001
CPR1       1     0.238897    0.03437119         6.951          0.0001
CPR2       1     0.151391    0.05226331         2.897          0.0051
CPR3       1     0.241080    0.05588444         4.314          0.0001
CPR4       1     0.280814    0.05267283         5.331          0.0001
CPR5       1     0.117394    0.03262483         3.598          0.0006
---------------------------------------------------------------------

Dependent Variable: W1

Root MSE        0.07465     R-square        0.9948
Dep Mean        0.00000     Adj R-sq        0.9944

                 Parameter Estimates
                Parameter      Standard     T for H0:
Variable  DF     Estimate         Error   Parameter=0      Prob > |T|
INTERCEP   1    -3.072004    0.03152133       -97.458          0.0001
USDA1      1     0.166400    0.03246554         5.125          0.0001
USDA2      1     0.208307    0.03862208         5.393          0.0001
USDA3      1     0.340962    0.04468844         7.630          0.0001
USDA4      1     0.128245    0.04040466         3.174          0.0023
USDA5      1     0.201599    0.03411876         5.909          0.0001
```

These results illustrate the relationship between regressing the first CV extracted from matrix **X** on the variables constituting matrix **Y** and vice versa, further reinforcing the notion that CCA is more general than regression analysis. The squared canonical correlation for this situation (reported in Example PR-9.2) is 0.9974^2 and equals the multiple correlation coefficient in each of these two regressions ($R^2 = 0.9948$).

9.1.3. Significance Testing in CCA

As mentioned, DFA can be viewed as a special case of CCA. It follows then that CCA significance tests parallel those outlined in Chapter 7 and derive from Wilk's Λ criterion. Letting $V_j = -\left[n-1-\frac{(P+Q)}{2}\right] \times \ln(1+\lambda_j^2)$, the test statistic becomes

$$V = -\left[n-1-\frac{(P+Q)}{2}\right] \sum_{j=1}^{G} \ln(1+\lambda_j^2) , \quad (9.5)$$

where G = *MIN*(P,Q) and is approximately chi squared distributed with P×Q degrees of freedom. If V is significant, then at least the first canonical correlation coefficient, ρ_{CC1}, is significant. The remaining canonical correlation coefficients are evaluated by successively performing chi squared tests as follows:

test statistic	*degrees of freedom*	*null hypothesis*
V	P×Q	$\rho_{CC1} = \rho_{CC2} = \rho_{CC3} = \ldots = \rho_{CCG} = 0$
$V - V_1$	(P - 1)(Q - 1)	$\rho_{CC1} \neq 0;\ \rho_{CC2} = \rho_{CC3} = \ldots = \rho_{CCG} = 0$
$V - V_1 - V_2$	(P - 2)(Q - 2)	$\rho_{CC1} \neq 0;\ \rho_{CC2} \neq 0;\ \rho_{CC3} = \ldots = \rho_{CCG} = 0$
⋮	⋮	⋮
$V - \sum_{j=1}^{R-1} V_j$	(P - R)(Q - R)	$\rho_{CC1} \neq 0;\ \rho_{CC2} \neq 0;\ \rho_{CC3} \neq 0;\ \ldots;\ \rho_{CCG} = 0$

As before, this sequential testing terminates once the decision about a residual Wilk's Λ criterion is "fail to reject."

EXAMPLE PR-9.6. Conduct significance tests for the canonical correlations reported in Example PR-9.2.

Variables contained in the USDA milk production data used in Example PR-9.2 are denoted by USDA1, USDA2, ..., USDA5. Their CPR counterparts are denoted by CPR1, CPR2, ..., CPR5. So that the statistical distributions conform better to the underlying CCA model assumptions, these production figures are transformed as

$$LN\left[\frac{USDAj_i}{4000 \times area_i + 6}\right], \text{ and } LN\left[\frac{CPR_j}{4000 \times area_i + 6}\right],$$

with each acronym for these transformations beginning with "T."

Using the SAS procedure

```
PROC CANCORR ALL OUT=CANVAR;
   VAR TUSDA1-TUSDA5;
   WITH TCPR1-TCPR5;
RUN;
```

produced SAS output that includes the CVs used in Example PR-9.2 (written to the SAS file CANVAR), as well as

Canonical Correlation Analysis

	Canonical Correlation	Adjusted Canonical Correlation	Approx Standard Error	Squared Canonical Correlation
1	0.997404	0.997144	0.000611	0.994814
2	0.969548	0.967383	0.007068	0.940024
3	0.625698	0.575880	0.071713	0.391498
4	0.531426	0.531346	0.084568	0.282413
5	0.379733	.	0.100857	0.144197

Test of H0: The canonical correlations in the current row and all that follow are zero

	Likelihood Ratio	Approx F	Num DF	Den DF	Pr > F
1	0.00011623	98.5467	25	235.5365	0.0001
2	0.02241245	30.2431	16	196.1609	0.0001
3	0.37368867	8.7702	9	158.3435	0.0001
4	0.61411265	9.1105	4	132	0.0001
5	0.85580268	11.2891	1	67	0.0013

These canonical correlation coefficients are identical to those reported in Example PR-9.2. The accompanying squared canonical correlation coefficients are the individual solutions to equation (9.1). The significance tests indicate that each canonical correlation coefficient is significantly different from 0.

9.2. Interpretation of the Canonical Dimensions Spanning Geo-referenced Data Sets

The first step in a CCA is to determine whether individual canonical correlation coefficients are large. The second step is to determine whether those coefficients found to be large are significant. The maximization problem represented by expression (9.3), however, is capable of rendering artificial dimensions with no physical meaning. To address this problem, a third tool for evaluating the meaningfulness of canonical dimensions is required. This tool is the structure matrix, which has already been used to interpret results from PCA, FA, and DFA.

A canonical structure matrix is constructed by correlating the synthetic variables $\mathbf{Z}_X\mathbf{U}_j$ with the variables contained in **X** and the synthetic variables $\mathbf{Z}_Y\mathbf{V}_j$ with the variables contained in **Y**. This structure matrix plays a crucial role in determining the importance and aiding in the interpretation of canonical dimensions. In contrast with PCA, the eigenvectors of equation (9.2) are selected to maximize correlations, rather than to account for the covariances of either $\mathbf{R}_{XX}$ or $\mathbf{R}_{YY}$. As a result, optimal linear combinations yielding extremely high canonical correlations may account for very little actual variation in the two sets of original variables; optimality does not guarantee that the linear combinations are informative. In other words, although a CCA solution is mathematically elegant and viable, there is no guarantee it is understandable or meaningful. The structure matrix continues to allow a researcher to infer meaning from the pattern of correlations (loadings) between the CVs representing canonical dimensions and the original variables. Interpretation of the two constituents of this structure matrix, one for the CVs extracted from matrix **X** and one for the CVs extracted from matrix **Y**, follow the same rules used for interpreting PCA, FA, and DFA structure matrices. A researcher inspects the relative magnitudes and the signs of the loadings, as well as the percentage of common variance for the two data sets.

EXAMPLE PR-9.7. Construct the CCA structure matrix (Example PR-9.6 continued).

These results from SAS (output on next page) help to clarify the previous results. As noted in Example PR-9.2, the USDA and CPR data sets essentially are repeated measures of the same geo-referenced phenomenon so their important canonical correlations should be close to 1. Indeed, two of the canonical correlations reported in Example PR-9.2 are very close to 1, but the results reported in Example PR-9.6 disclose that all five canonical correlation coefficients are significant. Inspection of the common variance summarized in the CCA structure matrix reveals that, in keeping with the spirit of *RULE OF THUMB 6.2*, only the first two canonical dimensions are meaningful. The first dimension (CV-I) summarizes roughly 90% of the variance displayed in both data sets, while the second dimension (CV-II) respectively summarizes 9% and 7.2% of the variance displayed in both data sets. Each of the remaining three canonical dimensions summarizes substantially less than 5% of the variance. The loadings on these first two dimensions reflect commonly found time series trends. Because of its prominence attained through displaying consistently very strong positive correlations with the original variables, CV-I may be labeled the measurement replication dimension that spans these two data sets. It focuses on similarities of density of milk production with regard to the temporal midpoint. CV-II displays weak to moderate correlations with the original variables and accents dissimilarities between the temporal endpoint geographic distributions of density of milk production.

Canonical Structure

Correlations Between the 'VAR' Variables and Their Canonical Variables

	V1	V2	V3	V4	V5
USDA1	0.8956	-0.4290	0.1165	-0.0003	-0.0182
USDA2	0.9554	-0.2484	-0.1351	0.0357	0.0767
USDA3	0.9905	0.0444	-0.0480	-0.0845	-0.0865
USDA4	0.9602	0.2326	0.0638	-0.0629	0.1257
USDA5	0.9361	0.3103	0.0860	0.1412	-0.0054

Correlations Between the 'WITH' Variables and Their Canonical Variables

	W1	W2	W3	W4	W5
CPR1	0.9043	-0.4165	0.0896	-0.0171	0.0203
CPR2	0.9690	-0.2018	-0.1357	0.0356	0.0266
CPR3	0.9892	0.0519	-0.0782	-0.0586	-0.0965
CPR4	0.9726	0.2207	0.0256	-0.0358	0.0579
CPR5	0.9295	0.3040	0.0989	0.1836	-0.0146

Standardized Variance of the 'VAR' Variables Explained by

	Their Own Canonical Variables			The Opposite Canonical Variables	
	Proportion	Cumulative Proportion	Canonical R-Squared	Proportion	Cumulative Proportion
1	0.8989	0.8989	0.9948	0.8942	0.8942
2	0.0796	0.9785	0.9400	0.0748	0.9691
3	0.0091	0.9876	0.3915	0.0036	0.9727
4	0.0065	0.9941	0.2824	0.0018	0.9745
5	0.0059	1.0000	0.1442	0.0009	0.9753

Standardized Variance of the 'WITH' Variables Explained by

	Their Own Canonical Variables			The Opposite Canonical Variables	
	Proportion	Cumulative Proportion	Canonical R-Squared	Proportion	Cumulative Proportion
1	0.9090	0.9090	0.9948	0.9043	0.9043
2	0.0716	0.9806	0.9400	0.0673	0.9716
3	0.0086	0.9892	0.3915	0.0034	0.9750
4	0.0080	0.9972	0.2824	0.0023	0.9772
5	0.0028	1.0000	0.1442	0.0004	0.9776

Axis rotation of the structure matrix is problematic, just as it is in DFA. Although the eigenfunction solution for equation (9.1) is not unique, it is constrained so that each eigenvector of equation (9.2) consists of two subvectors that are maximally correlated as well as mutually orthogonal with their unmatched subvectors. Rotation destroys these two properties and, for this reason alone, should be avoided. The fact that DFA can be cast as a CCA problem also raises further questions about whether DFA structure matrices should be rotated.

The regression formulation of CCA (as seen in Example PR-9.5) allows an assessment of ***redundancy***, the percentage of variance for each variable contained in matrix **Y** that is accounted for by the CVs of matrix **X** and vice versa. Similar to a PCA regression (§6.6), these individual multiple regressions involve an orthogonalized set of predictor variables, facilitating the evaluation of CV contributions to each statistical explanation. As in bivariate regression, where regressing variable X on Y does not yield the same coefficient as regressing variable Y on X, the resulting multiple correlation coefficients are asymmetrical with respect to the two data sets.

EXAMPLE PR-9.8. Assess redundancy via multiple regression (Example PR-9.6 continued).

The SAS output produced includes

```
Squared Multiple Correlations Between the 'VAR' Variables and
  the First 'M' Canonical Variables of the 'WITH' Variables
M            1       2       3       4       5
USDA1   0.7979  0.9709  0.9762  0.9762  0.9763
USDA2   0.9081  0.9662  0.9733  0.9737  0.9745
USDA3   0.9760  0.9779  0.9788  0.9808  0.9819
USDA4   0.9173  0.9682  0.9698  0.9709  0.9731
USDA5   0.8718  0.9623  0.9652  0.9709  0.9709

Squared Multiple Correlations Between the 'WITH' Variables and
  the First 'M' Canonical Variables of the 'VAR' Variables
M            1       2       3       4       5
CPR1    0.8136  0.9766  0.9798  0.9799  0.9799
CPR2    0.9340  0.9723  0.9795  0.9799  0.9800
CPR3    0.9734  0.9759  0.9783  0.9793  0.9806
CPR4    0.9411  0.9869  0.9871  0.9875  0.9880
CPR5    0.8594  0.9463  0.9501  0.9596  0.9597
```

Using a CV selection criterion requiring the regressor CV to account for at least 5% of the variance in a given original variable, these results indicate that USDA1, USDA2, USDA4, and USDA5 are accurately predicted from the first two CVs of the CPR data, whereas USDA3 is accurately predicted from the first CV of the CPR data. This outcome is consistent with the interpretation of CV-I as focusing on the midpoint time period geo-referenced data variable (namely USDA3). This same criterion indicates that CPR1 and CPR5 are accurately predicted from the first two CVs of the USDA data, whereas CPR3 is accurately predicted from simply the first CV of the USDA data. Again, the midpoint time period geo-referenced data variable (namely CPR3) focus emerges. In addition, the endpoint time periods contrast also emerges with much poorer descriptions of CPR2 and CPR4.

As with PCA, and in contrast with DFA, a second aid to interpretation comes from scatterplots of the pairs of CV scores. These scatterplots help reveal atypical areal units. Often only those dimensions with at least moderately large, significant canonical correlations are plotted. Such plots for Example PR-9.6 appear in Figure 9.1. One areal unit appears to be somewhat of an outlier, although not glaringly so, in terms of each of these canonical dimensions.

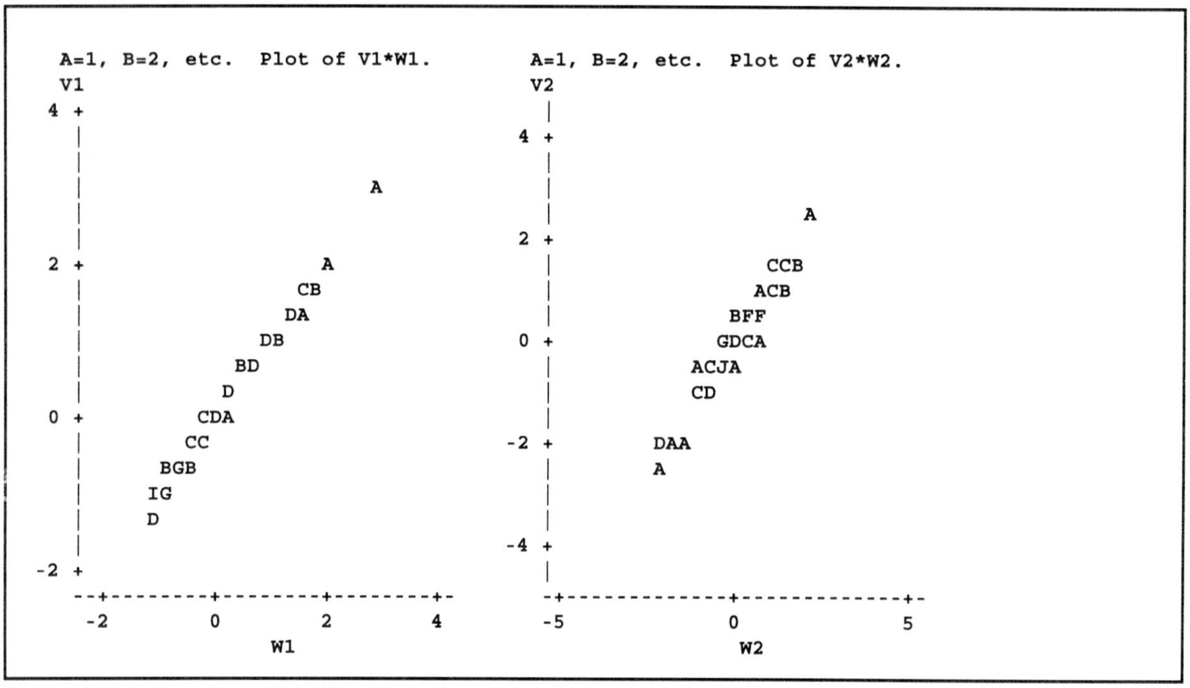

Figure 9.1. Scatterplots of the first two pairs of canonical variates for Example PR-9.6.

9.2.1. Dimensions Spanning Geo-referenced Coordinates and Areal Unit Attributes

The special problem here is to determine the two linear combinations of areal unit attributes that are most correlated with any two linear combinations of geo-referenced coordinates . The matrix **Y** is composed of the pair of Cartesian coordinates for areal unit centroids. The matrix **X** is composed of the battery of areal unit attributes. The output is a synthesis of attribute patterns into a pair of orthogonal geographic trends. The associated spatial directions that are uncovered may well be difficult to interpret. Nevertheless, this pair of canonical variates should relate directly to underlying components of spatial autocorrelation.

EXAMPLE PR-9.9. Calculate the spatially correlated components of sugar cane production in Puerto Rico.

Six years of sugar cane production, denoted SC1-SC6, are transformed to maximize their respective S-W (Shapiro-Wilk) statistics, so that the statistical distributions conform better to the underlying CCA model assumptions. These production figures are transformed as

$$\left[\frac{SCj_i}{area_i}+c_j\right]^{d_j} ,$$

with each acronym for these transformations beginning with "T."

Next, these municipio attribute values are related to the Cartesian coordinates, denoted by U and V, of the physical centroids of their corresponding municipios.

Using the SAS procedure

```
PROC CANCORR DATA=STEP3 OUT=CCOUT;
        WITH U V;
        VAR TSC1-TSC6;
RUN;
```

produce the SAS output on the coupled maps in Map 9.1.

The pairs of contour maps (Map 9.1) reflect geographic trends across the island of Puerto Rico. The first (a) is a southwest-northeast trend in Cartesian coordinate space, coupled with a north-south wavelike synthetic sugar cane surface; the second (b) is a southeast-northwest trend in Cartesian coordinate space, coupled with a ridgelike synthetic sugar cane surface with exactly the same directional orientation. The respective canonical correlations are 0.58287 and 0.45070. Both canonical variate pairs are significant. The canonical structure matrices suggest roughly a 45° axis rotation for the original Cartesian coordinates, and a principal components-type first canonical dimension that reflects the average magnitude of all six sugar cane production geographic distributions. The second canonical dimension reveals an increase in correlations through time. Both of the Cartesian coordinate canonical variates are close to what is expected in terms of maxim or near-maximum positive spatially autocorrelated surfaces. The first canonical dimension has MC = 0.76556 (GR = 0.10395) and relates most strongly (r = 0.677) to a theoretical surface having MC = 0.85621. The second canonical dimension has MC = 0.80275 (GR = 0.08418) and relates most strongly (r = 0.582) to the theoretical surface having the maximum possible positive spatial autocorrelation index of MC = 1.10362. These theoretical spatial autocorrelation surfaces are described in §9.2.2.

9.2.2 Dimensions Spanning Geo-referenced Coordinates and Eigenvectors of the Connectivity Matrix **C**

The role of the geographic connectivity matrix, **C**, has been discussed at length in Chapter 6. Tiefelsdorf and Boots (1995) show that the set of MC values for all possible *orthogonal* as well as *uncorrelated* map patterns can be obtained from the following transformed version of this matrix:

$$\left(\mathbf{I} - \frac{\mathbf{1}\mathbf{1}^{\mathbf{T}}}{n}\right)\mathbf{C}\left(\mathbf{I} - \frac{\mathbf{1}\mathbf{1}^{\mathbf{T}}}{n}\right), \quad (9.6)$$

an extension of an approach used by de Jong, Sprenger, and van Veen (1984) to establish the extreme values of MC and GR. Griffith (1996) introduces visualization features of spatial

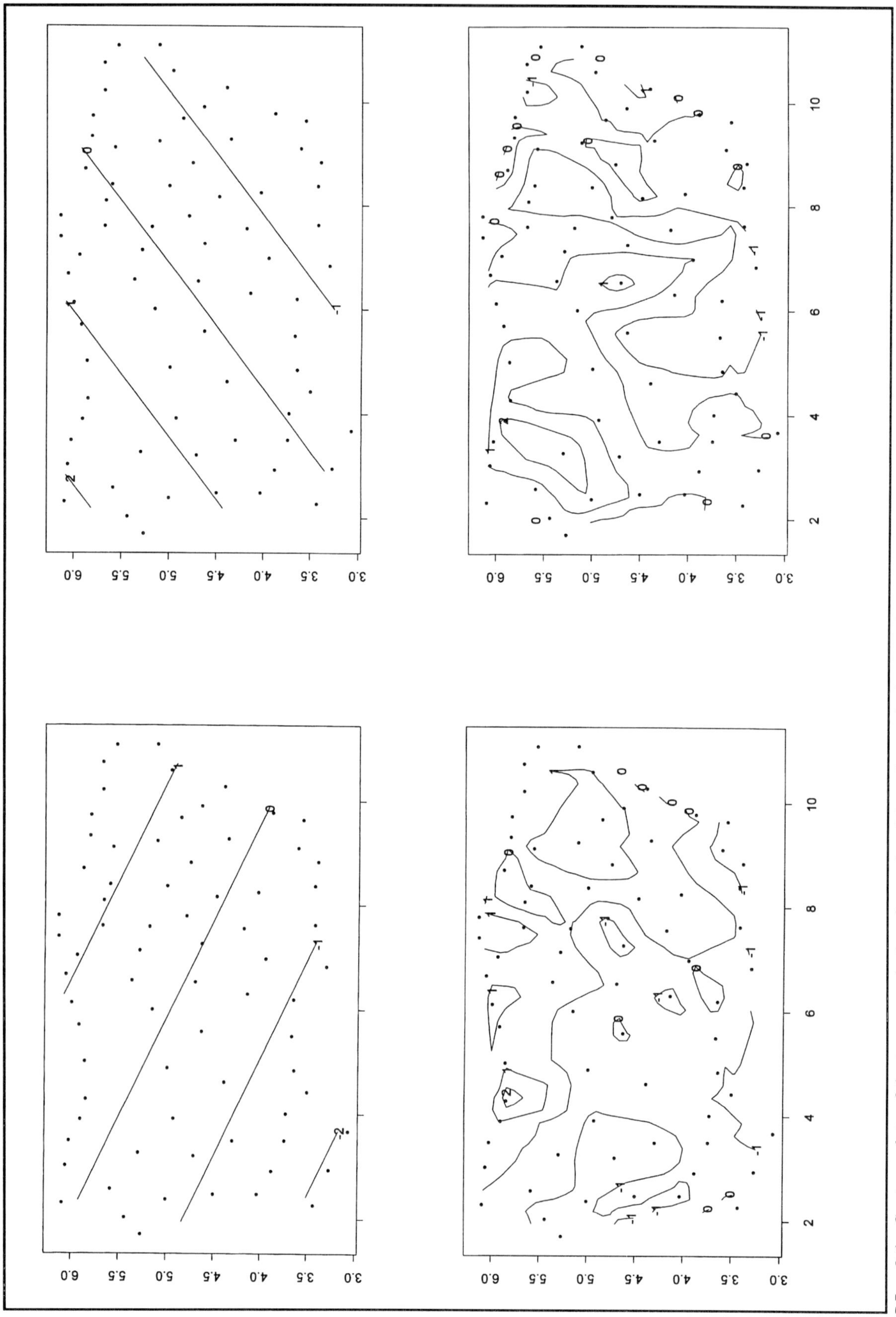

Map 9.1. Contour plots of the two canonical variate pairs. Map of canonical scores from the **Y** matrix (left) and the Cartesian coordinates (right). First canonical pair (top); second canonical pair (bottom).

autocorrelation through these eigenfunctions. Hence the special problem here addresses whether these orthogonal geographic distribution components of spatial autocorrelation extracted from expression (9.6) also span the Cartesian coordinate space.

Of interest here is that the eigenvalues of matrices **C** and expression (9.6) are essentially the same, except for λ_1 of matrix **C**, which corresponds to a 0 eigenvalue of expression (9.6). Recall that λ_1 of matrix **C** (interpreted in §6.5) is an index of planar topological connectivity. The remaining eigenvectors relate to spatial autocorrelation components of the surface partitioning, a feature revealed by the relationship

$$MC = \frac{n\lambda_j}{\mathbf{1}^T\mathbf{C}\mathbf{1}}, \quad j = 1,2,\ldots,n-1 ,$$

where λ_j are the eigenvalues of expression (9.6) and approximately the eigenvalues of matrix **C** (once λ_1 has been replace with 0). A comparison of the nonnegative elements in these two sets of eigenvalues (Table 9.1) reveals that each pair of eigenvalues (e.g., 5.17230 versus 5.20062, 0.10095 versus 0.10517) is aproximately the same value. Using the approximate standard error for MC allows the identification of eigenvectors capturing prominent components of spatial autocorrelation, because these would be ones whose eigenvalues would exceed $\dfrac{\mathbf{1}^T\mathbf{C}\mathbf{1}\left(1.96\sqrt{\dfrac{2}{\mathbf{1}^T\mathbf{C}\mathbf{1}} - \dfrac{1}{n-1}}\right)}{n}$ in absolute value.

5.50193	*****
5.17230	5.20062
4.85617	4.93782
4.32879	4.32894
3.98436	4.03475
3.95364	3.96108
3.83436	3.87261
3.37842	3.37850
3.28026	3.28331
2.79956	2.80482
2.62981	2.68169
2.49041	2.53197
2.24493	2.37040
2.19156	2.22007
2.01379	2.01441
1.89030	1.89483
1.68543	1.68648
1.49305	1.52466
1.38026	1.39184
1.19199	1.21589
1.07740	1.10923
0.94561	0.95477
0.81514	0.82548
0.79106	0.79198
0.60500	0.62168
0.37613	0.37647
0.26905	0.27129
0.17413	0.19874
0.14188	0.15048
0.10095	0.10517
*****	0.00000

Table 9.1. Positive eigenvalues of matrix **C** [left] and equation (9.6) [right].

EXAMPLE PR-9.10. Calculate the spatial autocorrelation components of the municipio surface partitioning that span the municipio centroid Cartesian coordinates for Puerto Rico.

The prominent eigenvectors are identified as those 23 having eigenvalues greater than 0.6397. These eigenvectors relate to levels of positive spatial autocorrelation and have been subjected to a canonical correlation analysis.

Using the SAS procedure

```
PROC CANCORR DATA=STEP3;
        WITH U V;
        VAR E1-E23;
RUN;
```

produced this SAS output:

```
                     Canonical Structure

Correlations Between the 'WITH' Variables and Their Canonical Variables

                             W1            W2
                U          0.9485        0.3169
                V          0.3906       -0.9206


 Correlations Between the 'VAR' Variables and Their Canonical Variables

                             V1            V2
               E1         -0.7140       -0.2796
               E2         -0.1936       -0.1806
               E3          0.2038        0.2874
               E4         -0.4383        0.6452
               E5          0.0500        0.3946
               E6          0.1008       -0.1254
               E7          0.1048        0.2970
               E8         -0.1053        0.0555
               E9         -0.1345        0.1135
               E10        -0.1433        0.1180
               E11        -0.0325        0.0649
               E12         0.2339       -0.1117
               E13         0.0884       -0.0336
               E14         0.1443        0.1883
               E15         0.0161       -0.0546
               E16        -0.1434       -0.0664
               E17         0.0106       -0.0492
               E18         0.0562        0.0079
               E19        -0.0907        0.0210
               E20         0.0653        0.0027
               E21        -0.0671        0.1421
               E22         0.1401       -0.0492
               E23        -0.0246       -0.1209
```

This output reveals that the east-west Cartesian coordinate axis correlates strongly with the eigenvector representing maximum positive spatial autocorrelation, whereas the north-south Cartesian coordinate axis correlates strongly with the fourth eigenvector, which represents very strong positive spatial autocorrelation ($MC = 73 \times 4.03473/344 = 0.85621$). These are exactly the types of dimensions to which the Cartesian coordinate canonical variates should strongly related.

9.3. DFA as a CCA Problem: Multiple-regions DFA

The relationship between linear DFA and CCA is examined by inspecting the multiple-regions ($K > 2$) DFA case.

EXAMPLE PR-9.11. Redo Example PR-7.8 as a canonical correlation problem.

Using the SAS procedure

```
PROC CANDISC;
   CLASS ADM;
   VAR TDFRM TDCD TDMLK TDSGR TDCF TDTB TDBN TDFAM;
RUN;
```

produced SAS output shown on the next page.

The eigenvalues reported in the second section of this output (i.e., 2.4287, 0.7386, 0.4226, and 0.1581) table are exactly those reported in Example PR-7.5. As mentioned earlier in this chapter, these eigenvalues are related to the canonical correlation coefficients:

$$2.4287 \approx \frac{0.841630^{\,2}}{1 - 0.841630^{\,2}}, \quad 0.7386 \approx \frac{0.651776^{\,2}}{1 - 0.651776},$$

$$0.4226 \approx \frac{0.545044^{\,2}}{1 - 0.545044^{\,2}}, \text{ and } \; -0.1581 \approx \frac{0.369513}{1 - 0.369513^{\,2}}.$$

As reported in Example PR-7.6, only the first three of these dimensions are significant. In all three cases the canonical correlations are at least moderate to strong. However, the largest is noticeably far from 1. The total canonical structure is identical to the DFA structure matrix appearing in Example PR-7.8 (except for rounding error) and can be interpreted accordingly. In summary, PR-7.8 and PR-9.9 produce exactly the same results!

9.4. The GLM and CCA

Two meaningful dimensions have been uncovered for the density of milk production data studied in this chapter. Example PR-9.5 demonstrates that the canonical correlations for these dimensions are equivalent to regressing the CVs extracted from one matrix on the variables of the other matrix. A key question about the milk production application is whether the geo-referenced USDA data can be predicted from the CPR data. The natural next question, of course, is about the nature and degree of spatial autocorrelation latent in these data. Treating CCA as a regression

```
                    Canonical Discriminant Analysis
                           Adjusted            Approx          Squared
            Canonical      Canonical          Standard        Canonical
           Correlation    Correlation           Error        Correlation
1            0.841630       0.815817          0.034372         0.708341
2            0.651776       0.590974          0.067787         0.424812
3            0.545044       0.510994          0.082841         0.297074
4            0.369513       0.337325          0.101760         0.136540

                           Eigenvalues of INV(E)*H
                             = CanRsq/(1-CanRsq)
           Eigenvalue    Difference      Proportion      Cumulative
1            2.4287        1.6901          0.6480          0.6480
2            0.7386        0.3159          0.1971          0.8450
3            0.4226        0.2645          0.1128          0.9578
4            0.1581          .             0.0422          1.0000

                Test of H0: The canonical correlations in the
                  current row and all that follow are zero
           Likelihood
              Ratio        Approx F       Num DF      Den DF      Pr > F
1          0.10182098       6.0742          32       226.5521     0.0001
2          0.34910951       3.7644          21       178.5806     0.0001
3          0.60694884       2.9776          12            126     0.0011
4          0.86345990       2.0241           5             64     0.0870

                    Total Canonical Structure
                    CAN1          CAN2           CAN3            CAN4
TDFRM           0.038469      0.384150       0.020255       -0.070203
TDCD            0.496960     -0.059493       0.548827       -0.234956
TDMLK          -0.552840      0.313902       0.460134        0.331502
TDSGR           0.587469      0.173795      -0.173077       -0.203818
TDCF            0.572577     -0.012674       0.183974        0.061963
TDTB            0.272146     -0.431450      -0.457377        0.384797
TDBN           -0.129356      0.179771      -0.296964        0.221236
TDFAM           0.129188      0.360798       0.350263       -0.374388
```

problem allows these questions to be answered. But because two pronounced dimensions are present in these two data sets, regression must be undertaken with the general linear model (GLM). Next, to assess the soundness of the regression results, diagnostics need to be applied to the multivariate set of residuals.

EXAMPLE PR-9.12. Calculate GLM results based upon regressing the prominent CVs of the USDA matrix on the matrix of CPR values.

Using the SAS procedure

```
PROC GLM DATA=CANVAR;
   MODEL V1-V2 = CPR1-CPR5;
   OUTPUT OUT=RESDMLK1 R=USDARCV1-USDARCV2;
```

RUN;

produced SAS output that includes the following two segments:

R-Square	C.V.	Root MSE	V1 Mean
0.994814	9999.99	0.0747	0.0000

Parameter	Estimate	T for H0: Parameter=0	Pr > |T|	Std Error of Estimate
INTERCEPT	-2.986423852	-100.02	0.0001	0.02985771
CPR1	0.238897336	6.95	0.0001	0.03437119
CPR2	0.151391335	2.90	0.0051	0.05226331
CPR3	0.241079724	4.31	0.0001	0.05588444
CPR4	0.280814287	5.33	0.0001	0.05267283
CPR5	0.117393989	3.60	0.0006	0.03262483

R-Square	C.V.	Root MSE	V2 Mean
0.940024	-9999.99	0.2539	-0.0000

Parameter	Estimate	T for H0: Parameter=0	Pr > |T|	Std Error of Estimate
INTERCEPT	0.852697683	8.40	0.0001	0.10153724
CPR1	-1.375247156	-11.77	0.0001	0.11688622
CPR2	-0.531356022	-2.99	0.0039	0.17773206
CPR3	0.233940814	1.23	0.2226	0.19004645
CPR4	1.061406295	5.93	0.0001	0.17912469
CPR5	0.260954133	2.35	0.0216	0.11094740

These results are consistent with those reported in Example PR-9.6. Primary differences between these two CCA components are that (1) CPR3 is significantly related to the V1 but not V2; and, (2) the contribution of all CPR variables to the statistical explanation of V1 is nearly equal whereas considerably differential contributions are made by these variables to the statistical explanation of V2.

Individual regressions yield the following variance homogeneity test results, using the SPEC option in the REG procedure of SAS:

```
Test of First and Second Moment Specification
DF:    20    Chisq Value: 14.723612416    Prob>Chisq:    0.7920

Test of First and Second Moment Specification
DF:    20    Chisq Value: 22.465026237    Prob>Chisq:    0.3158
```

Clearly there is no evidence to indicate that the regression residuals come from populations having heterogeneous variance. For the geographic distribution of the bivariate statistical distributions of these residuals, Bartlett's test statistic using the agricultural administrative regions classification yields

```
Test Chi-Square Value =      22.414496
with      12 DF         Prob > Chi-Sq = 0.0331
```

using the coastal lowlands/interior highlands classification yields

```
Test Chi-Square Value =      2.853087
with        3 DF        Prob > Chi-Sq = 0.4148
```

and, using the rural-urban dichotomy classification yields

```
Test Chi-Square Value =      0.935067
with        3 DF        Prob > Chi-Sq = 0.8170
```

Again, there is no evidence to indicate that the regression residuals come from bivariate populations having geographically heterogeneous variance. Furthermore, these results should be quite robust since the Q-Q plot correlation of 0.99197 is nearly unity.

Finally, the cross-MCs for the regression residuals from this CCA are

```
V1     -0.04618
V2     -0.01972      -0.06537
           V1            V2  .
```

The Geary Ratio (GR) for V1 is 0.98476, while its expected MC value is -0.042915; the GR for V2 is 1.08156, while its expected MC value is -0.042915. Based on the asymptotic standard error for MC of 0.076249, neither of these MC values is significant, implying the absence of residual spatial autocorrelation. Furthermore, because these canonical variates are orthogonal, the expected value of the cross-MC is zero. Consequently, there is no evidence to indicate that the regression residuals for these canonical variates contain spatial autocorrelation in the population. This is not a surprising result, because milk production is being predicted from milk production.

Overall, then, the CCA results appear to be on a sound statistical foundation. Part of this evaluation, particularly that of spatial autocorrelation, is not possible without rewriting CCA in its regression version.

9.5. Problems

P-9.1. Given the *Old Maps* (Table 10.1 in Andrews and Herzberg, 1985) data set, identify the canonical dimensions that span the actual latitude-longitude measures (this is **Y**), and the 11 map latitude-longitude measures (this is **X**). How do these results compare with those for P-4.1 and P-6.1?

P-9.2. Given the *Identifying Groundwater Populations* (Table 17.1 in Andrews and Herzberg, 1985) data set, identify the canonical dimensions that span the geo-referenced coordinates and the set of 12 interval/ratio soil characteristics.

9.6. References

Andrews, D. and H. Herzberg. 1985. *Data: A Collection of Problems from Many Fields for the Student and Research Worker.* Berlin: Springer Verlag.

Bailey, T. and A. Gatrall. 1995. *Interactive Spatial Data Analysis.* Harlow, U.K.: Longman Scientific Publishing.

de Jong, P., C. Sprenger and F. van Veen. 1984. "On Extreme Values of Moran's I and Geary's c." *Geographical Analysis* 16: 17–24.

Griffith, D. 1982. "Geometry and Spatial Interaction." *Annals of the Association of American Geographers* 72: 332–46.

—. 1988. *Advanced Spatial Statistics*, Dordrecht: Kluwer.

—. 1992. "Estimating Missing Values in Spatial Urban Census Data." *The Operational Geographer* 10 (2): 23–26 [also translated into French].

—. 1995. "The General Linear Model and Spatial Autoregressive Models." In *New Directions in Spatial Econometrics*, edited by L. Anselin and R. Florax, 273– 95. Berlin: Springer-Verlag.

—. 1996. "Spatial autocorrelation and eigenfunctions of the geographic weights matrix accompanying geo-referenced data." *The Canadian Geographer* 40: 351-67.

Griffith, D. and C. Amrhein. 1991. *Statistical Analysis for Geographers.* Englewood Cliffs, N.J. Prentice Hall.

Hair Jr., J., R. Anderson, R. Tatham, and W. Black. 1995. *Multivariate Data Analysis.* 4th ed. Englewood Cliffs, N.J.: Prentice Hall.

Johnson, R. and D. Wichern. 1992. *Applied Multivariate Statistical Analysis.* Englewood Cliffs, N.J.: Prentice Hall.

Tiefelsdorf, M. and B. Boots. 1995. "The Exact Distribution of Moran's I." *Environment and Planning A* 27: 985–99.

PART III

NONLINEAR AND CATEGORICAL DATA MODELING

CHAPTER 10

NONLINEAR REGRESSION ANALYSIS

Chapter Preview. In Chapter 1 and again in Chapter 4, we emphasized the linear nature of the relationships that are assumed to link the dependent and independent variables. However, as discussed in Chapter 1, there are many instances in which the nature of relationships is nonlinear. While nonlinear regression is certainly not for the faint-of-heart (note the label of advanced material attached to §10.2.1 and §10.3), current software and computing power no longer place it in the outer limits of statistical analysis.

In this chapter, we discuss how to manage nonlinearities in error terms as well as in relationships between the independent and dependent variables. No longer able to use ordinary least squares to solve this class of problems, this chapter introduces alternative estimation routines such as nonlinear least squares (NLS).

NLS often can be expressed as a sequence of OLS results, where ranges of nonlinear parameters are set, and then conventional regression is used to estimate the linear parameters (in other words, the nonlinear relationship is estimated as a series of piecewise linear segments). This conceptual viewpoint reinforces the workhorse label attached to linear regression in Chapter 4 and aids in acquiring a better understanding of spatial statistics in the next chapter.

10.0. Introduction

At times research problems involving the analysis of empirical data violate the assumption of linearity. Classical statistical models work with linear rather than nonlinear relationships (Chapter 1). But how should the analyst proceed when the assumption of linearity appears to be violated? How might models be respecified in light of apparently nonlinear relationships? Some of the answers to these and related questions refer to how the error term is written in a model, whereas others refer to variable transformations that improve the linear trend exhibited by pairwise plots of variables. All of the answers often require moving beyond the popular version of the general linear model (GLM), which is the hallmark of traditional statistical analysis, often into the realm of the generalized linear interactive model (GLIM) framework. The GLIM framework expresses the statistical model simply as

$$\mathbf{Y} = f(\mathbf{X}; \boldsymbol{\beta}) + \epsilon \, , \tag{10.1}$$

regardless of the functional form represented by f(☼).

In the presence of nonlinearities, problem statements and research questions previously posited for individual multivariate techniques remain unaltered. What does change is the functional form employed in the individual analysis, each of which needs to be rewritten again as a regression problem.

10.1. Error Structure Options

The GLM represents data by the multiple linear regression equation, which is stated in equation (4.2) as

$$\mathbf{Y} = \mathbf{X}\boldsymbol{\beta} + \epsilon \ .$$

This specification, which has been the focus of discussion so far in this text, implies that the error in an analysis is additive. But there are many cases in which error is either known or assumed to be nonadditive. For example, consider population undercounts in the U.S. census. In part, because of the tendency for immigrants to move to larger urban areas, statisticians recognize that the undercount error increases as the population of an urban area increases. In this case, the error structure is one of proportionality. For a bivariate regression in which X_1 is the population variable, the appropriate model is

$$\begin{aligned}\mathbf{Y} &= \mathbf{1}\beta_0 + \mathbf{X}_1(\mathbf{I}+\xi)\beta_1 + \epsilon \\ &= \mathbf{X}\boldsymbol{\beta} + (\mathbf{X}_1\xi\,\beta_1 + \epsilon) \ ,\end{aligned}$$

where ξ is the undercount measurement error. Alternatively, consider the relationship between population density and distance from a monocentric city center. The observed negative exponential decline of population with distance requires the regression equation to be rewritten as

$$LN(\mathbf{Y}) = \mathbf{X}\boldsymbol{\beta} + \epsilon \ , \qquad (10.2)$$

which, after manipulation (a back-transformation), becomes

$$\begin{aligned}\mathbf{Y} &= \beta_0 \prod_{j=1}^{p} \mathbf{X}_j^{\beta_j} \times EXP\ (\epsilon) \\ &= \hat{\mathbf{Y}} \times EXP\ (\epsilon) \ .\end{aligned} \qquad (10.3)$$

This pair of equations suffers from a number of problems. While equation (10.2) constitutes an attractive linearization of a nonlinear trend,[1] neither the quantity ($\mathbf{Y}$ - $\hat{\mathbf{Y}}$) nor the quantity $EXP(\epsilon)$ in equation (10.3) has a mean of 0, as required. In fact, if the parameter vector, $\boldsymbol{\beta}$, of equation (10.2) is estimated using OLS, then the only guarantee is that $EXP(\bar{\epsilon}) = 1$. Further, if ϵ is normally distributed, then $EXP(\epsilon)$ is not. And, if ϵ has constant variance, then $EXP(\epsilon)$ does not. Deviation

[1]Relationships where this type of linearization is possible sometimes are referred to as intrinsically or transformably linear.

of the sum $\sum_{i=1}^{n} (y_i - \hat{y}_i)$ from 0 sometimes is referred to as bias due to nonlinearity. One advantage of pursuing an analysis in this fashion arises from all of the traditional statistical theory that can be related to equation (10.2). The spirit of transforming to equation (10.2) is equivalent to that for undertaking a Box-Cox transformation (Chapter 1).

But equation (10.3) especially may produce unacceptable predictions, regardless of how well the error term in equation (10.2) behaves. In cases where the back-transformed results are unacceptable or unstable, an analyst may wish to respecify the equation as having an additive error structure. Accordingly, equation (10.3) could be rewritten as

$$\mathbf{Y} = \mathbf{1}\beta_{p+1} + \beta_0 \prod_{j=1}^{p} X_j^{\beta_j} + \epsilon \ . \qquad (10.4)$$

The term $\mathbf{1}\beta_{p+1}$ guarantees that the error term in equation (10.4) does have a mean of 0. Parameters are still estimated by minimizing the squared errors. This revised process is known as nonlinear least squares (NLS). Now the power of linear statistical theory no longer fully addresses the complexities encountered in a data analysis. Frequently, since an underlying normal distribution cannot be assumed, asymptotic statistical results are used to obtain approximate inferences.

Thus, transforming data involves transforming the error term. Transforming the error term in turn affects the assumptions that underlie it. As a result,

> *RULE OF THUMB 10.1:* The decision about whether to transform variable Y depends upon how the error behaves and how the error structure should be conceptualized.

Fortunately, the increasingly widespread availability of efficient nonlinear estimation algorithms, such as SAS's PROC NLIN, has greatly reduced the need for the linear transformation of Y simply for the sake of tractability.

In the special case of bivariate regression (when p = 1), the estimation of equation (10.4) is equivalent to undertaking a Box-Tidwell transformation (Chapter 1). Although the relationship between variables Y and X is nonlinear, the usual assumptions of normally and independently distributed errors with constant variance hold, at least approximately. The goal, then, is to select an appropriate transformation of variable X so that the relationship between Y and f(X) is as simple as possible. One distinguishing difference between this Box-Tidwell approach and practiced nonlinear regression analysis is that the former is sequential (like PCA, for instance), whereas NLS simultaneously estimates β_2, β_0, and β_1.

10.2. Principles of NLS

Just as is done in OLS, non-linear least squares employs first and second derivative tools from calculus but does so in a far less transparent fashion. The practical importance of these calculus results arises from a need to obtain parameter estimates and to establish confidence intervals for the unknown parameters **ß** as well as for the purpose of hypothesis testing.

Considering the general regression equation given by equation (10.1), let

$$\mathbf{F} = \frac{\partial f(\boldsymbol{\beta})}{\partial \boldsymbol{\beta}^T} , \qquad \text{and} \qquad \mathbf{e} = (\mathbf{Y} - \hat{\mathbf{Y}}) = [\mathbf{Y} - f(\boldsymbol{\beta})] .$$

The second expression, the residual vector, is familiar from the earlier discussion of OLS theory. Because both matrix **F** and vector **e** are functions of the parameter vector **ß**, a closed form solution to the estimation problem does not exist, as given by $(\mathbf{X}^T\mathbf{X})^{-1}\mathbf{X}^T\mathbf{Y}$ in OLS theory. Hence, estimation must proceed using an iterative procedure, which begins by taking an initial, usually ad hoc, guess of **b** provided by the analyst and then continuously improving upon this guess until the SSE reaches a minimum, that is, does not decrease markedly from one iteration to the next (e.g., by more than 10^{-8}). Moreover, a matrix of change, Δ, is sought at iterative step τ such that

$$\text{SSE}(\mathbf{b}_\tau + k\Delta) < \text{SSE}(\mathbf{b}_{\tau-1}) , \ \tau = 1, 2, ...,$$

where k is the size of the step, equaling 1 unless

$$\text{SSE}(\mathbf{b}_\tau + \Delta) \geq \text{SSE}(\mathbf{b}_{\tau-1}) ,$$

which triggers k to be reduced (sequentially halved in SAS) in an attempt to reverse the inequality. Failure to find a smaller SSE results in nonconvergence. Further, as is well known from calculus, one pitfall here is that a local rather than global optimum may be converged upon. And as is well known from numerical analysis, a second danger is that the initial guess is so far away from the optimum that the solution cannot be reached in a reasonable number of iterations. There are a number of techniques for defining Δ, each of which has strengths and weaknesses; the selection of the best single or combination of techniques may require the assistance of a statistical consultant.

Because NLS is plagued by complexities and the potential of being sidetracked by local minima, sometimes troubleshooting strategies need to be invoked. If convergence is not accomplished in a specified number of steps, then the maximum number of possible steps may need to be increased (the MAXITER default for SAS is 50, whereas experience suggests that many geo-referenced data analyses require at least 100 and sometimes as many as 500 steps). If the matrix **F** built from first derivatives becomes singular, then just as in the case with the presence of severe multicollinearity in OLS, the model may be overparameterized, and it may need to be respecified. But now the source of this problem also might be a poor ad hoc choice of $\mathbf{b}_{\tau=0}$ values, which values could lead to lack of movement of the trajectory of parameter estimates in the direction of improved minimization, often resulting in assumed convergence. Worse yet is movement in the direction of parameter estimate divergence. This situation, which is not all that uncommon in NLS, arises when functions such as LOG and SQRT encounter illegal arguments (e.g., negative values) or when two parameters begin mutual compensation, with one increasing toward ∞ while the other decreases toward $-\infty$. Finally, for many geo-referenced data analyses, there is no guarantee that a global minimum has been reached. One means of increasing confidence in the final results is to select a wide range of different, reasonable starting values, $\mathbf{b}_{\tau=0}$ and then to see whether or not their trajectories converge upon the same parameter estimates (except, perhaps, for convergence rounding error).

10.2.1. Methods for Defining Δ (*advanced material*)

One iterative SSE minimization procedure is known as ***steepest descent***. The operational criterion is that the trajectory of $\mathbf{b}_\tau$ should move in the direction of greatest change in parameter estimates, defined by the gradient vector. Precisely, movement is in the direction of a given gradient until SSE($\mathbf{b}_\tau$) no longer decreases. Then the next step, $\tau + 1$, begins movement in a new direction, along a new gradient. This method exploits gradient information contained in the first derivative matrix and defines the matrix of change as

$$\Delta = \mathbf{F}^{\mathbf{T}}\mathbf{e} = \left(\frac{\partial f(\mathbf{ß})}{\partial \mathbf{ß}^{\mathbf{T}}}\right)^{\mathrm{T}} [\mathbf{Y} - f(\mathbf{ß})] \ .$$

Depicting the parameter space as a landscape of all possible SSE hills, planes, and valleys, with the initial guess being a position somewhere in this landscape, the gradient direction is orthogonal to contours of iso-SSEs. Hence, proceeding to the minimum of a given gradient means reaching a point of tangency with a lower iso-SSE contour line. This method tends to make rapid progress during beginning iterations, increasingly slowing down by taking smaller and smaller steps along a zigzagging path and generally converging very slowly. Nevertheless, steepest descent is valuable for securing initial guesses of $\mathbf{b}$, namely $\mathbf{b}_{\tau=0}$.

A second popular iterative procedure is known as ***Gauss-Newton***, which is a special case of the more general Newton-Raphson method. The operational criterion is that the trajectory of $\mathbf{b}_\tau$ should be guided by a local quadratic approximation to the SSE objective function. This approximation is based upon the first two terms of a Taylor series expansion of f(**ß**) and defines the matrix of change as

$$\Delta = (\mathbf{F}^{\mathrm{T}}\mathbf{F})^{-1}\mathbf{F}^{\mathrm{T}}\mathbf{e}$$

$$= \left[\left(\frac{\partial f(\mathbf{ß})}{\partial \mathbf{ß}^{\mathbf{T}}}\right)^{\mathrm{T}} \left(\frac{\partial f(\mathbf{ß})}{\partial \mathbf{ß}^{\mathbf{T}}}\right)\right]^{-1} \left(\frac{\partial f(\mathbf{ß})}{\partial \mathbf{ß}^{\mathbf{T}}}\right)^{\mathrm{T}} [\mathbf{Y} - f(\mathbf{ß})] \ .$$

This method exploits gradient information as well as the second derivative matrix. In contrast to steepest descent, its performance tends to be best close to the final solution. One associated drawback is that $(\mathbf{F}^{\mathrm{T}}\mathbf{F})^{-1}$ may be undefined because matrix $\mathbf{F}^{\mathrm{T}}\mathbf{F}$ is singular (i.e., it contains serious multicollinearity; see §4.2.3); in part, this problem can be circumvented by the use of a generalized inverse when it is encountered.

A third popular iterative procedure is known as ***Marquardt***, which essentially is a compromise between the steepest descent and Gauss-Newton methods. Here the matrix of change is defined as

$$\Delta = [\mathbf{F}^T\mathbf{F} + \lambda(\mathbf{F}^T\mathbf{F})_{diag}]^{-1}\mathbf{F}^T\mathbf{e} \ ,$$

where $(\mathbf{F}^T\mathbf{F})_{diag}$ denotes a diagonal matrix whose diagonal elements are those making up the diagonal of matrix $\mathbf{F}^T\mathbf{F}$, with steepest descent being equivalent to $\lambda \to \infty$ and Gauss-Newton being equivalent to $\lambda \to 0$. In the Marquardt technique, the direction of movement potentially is guided by inflation of the diagonal term and follows steps intermediate between those given by steepest descent and Gauss-Newton. The operational criterion establishing the trajectory of $\mathbf{b}_\tau$ is that if SSE improves at each iteration, $\lambda \to 0$, increasingly exploiting the advantages of the Gauss-Newton method. Otherwise, if SSE does not improve at each iteration, $\lambda \to \infty$, increasingly (and, it is hoped, temporarily) exploiting the advantages of the steepest descent method. PROC NLIN in SAS initially sets $\lambda = 10^{-3}$, moves it toward 0 by dividing by 10 for each step, and moves it toward infinity by multiplying by 10 for each step.

Implementation of the three preceding procedures requires the ability to write down analytical derivatives of the regression equation with respect to the parameters. Obtaining these derivatives involves some knowledge of calculus, which can be easily bolstered by resorting to the aid of software such as MathCad, Maple V, or Mathematica. The task of obtaining and coding these derivatives is both tedious and error-prone, particularly when they are messy. Two important benefits that analytical derivatives offer, though, are additional information that tends to enhance the accuracy of estimation and speedier convergence. A convenient alternative is to employ a method utilizing numerical derivatives based upon finite differences, such as DUD (doesn't use derivatives) in NLIN. But DUD does not simply use numerical approximations to derivatives. Rather than a tangent plane approximation, it is based upon a secant plane approximation to the expectation surface. This method is similar to the Gauss-Newton one, except that derivatives are approximated from the history of iterations as opposed to being analytically supplied. This history is inaugurated by taking the vector of initial guesses, $\mathbf{b}_\tau$=0, and automatically displacing each of the P parameter estimates by either 10% or 0.1 (in SAS these iterative steps are denoted by negative integers). These displacements then are used to calculate the necessary secant plane, which in turn is used to compute parameter estimates at the next iterative step. Once launched, each iteration is based upon the current history summarized only in the previous (P+1) steps. One cost the researcher should keep in mind when resorting to the convenience of using DUD is that the propagation of error introduced by using numerical derivatives through to the second derivatives, upon which the asymptotic variance estimates are based, can render considerably inaccurate, unreliable confidence intervals.

Therefore,

> *RULE OF THUMB 10.2:* Begin an analysis using DUD. Once intuitively reasonable parameter estimates have been obtained with it, convert to Marquardt to obtain final estimates.

Of course, when analytical derivatives are virtually impossible or even just far too complicated to write down, then the Marquardt phase cannot be implemented. In addition, expressly when derivatives are complex or messy, the final estimates obtained with the Marquardt method should be double-checked with DUD. If the two sets of results do not agree (within rounding error), then the analytical derivatives should be reexamined, because incorrect analytical derivatives will introduce specification error into the estimation process.

10.3. Some Statistical Theory for NLS Estimation (*advanced material*)

One direct extension of linear statistical analysis is to assume that the errors are normally distributed. This assumption casts the nonlinear estimation and inference problem within the context of conventional maximum likelihood (ML) estimation, where the product of the individual observation probabilities is optimized with respect to the vector of parameters β being estimated. Again, least squares estimates are identical to ML estimates. This finding means that the NLS error variance estimate is given by

$$s^2 = \frac{SSE(\mathbf{b})}{(n - p)},$$

which is a familiar expression from OLS analysis, that vector **b** and s^2 are independent (as with OLS), and that **b** is multivariate normally distributed as MVN[**ß**, $\sigma^2(\mathbf{F}^T\mathbf{F})^{-1}$], with matrix **F** being solved by substituting vector **b** for **ß**.

An alternative to this ML approach is to employ conventional asymptotic arguments. Asymptotic variances are obtained by utilizing the second derivatives of the error sum-of-squares, and are given by

$$2s^2\left\{\partial^2[\mathbf{Y} - f(\mathbf{ß})]^T \sum \frac{[\mathbf{Y} - f(\mathbf{ß})]}{(\partial\mathbf{ß}\ \partial\mathbf{ß}^T)}\right\}^{-1},$$

evaluating this matrix at **b**. Again, this result is familiar from OLS analysis. The argument here is that as n increases to infinity, vector **b** converges to **ß**, s^2 converges to σ^2, and $(\mathbf{b} - \mathbf{ß})\sqrt{n}$ becomes distributed as MVN[**0**, $n\sigma^2(\mathbf{F}^T\mathbf{F})^{-1}$].

10.4. A Comparison of Various Models for Estimating Empirical Probabilities

Suppose an analyst is interested in estimating the probability that an individual areal unit belongs to one of two regions constituting some regionalization scheme. A simple way of obtaining a solution is to treat this as a two-groups discriminant function problem. The observed probability of group membership, then, is given by the 0-1 binary variable, and the predicted probabilities from a regression analysis (§7.7) yield the desired set of estimated probabilities. The equation used is (4.2), which is linear in functional form and has additive error. One serious flaw of this conceptualization is that the estimated probabilities are not guaranteed to be contained in the interval [0, 1]. A second, slightly more complicated, way of obtaining a solution is to solve this directly as a discriminant function problem and to compute the posterior probabilities of regional membership (§7.5). Now the equation used is

$$\Pr(\text{region } k|\mathbf{X}) = \frac{EXP[-.5D_k^2(\mathbf{X})]}{\sum_{j=1}^{2} EXP[-.5D_k^2(\mathbf{X})]}$$

where $D_j^2 = (\mathbf{X} - \bar{\mathbf{X}}_j)^T S(\mathbf{X} - \bar{\mathbf{X}}_j)$. Because the error term is present both in the numerator and denominator of this second probability estimation expression, it now is nonadditive. One serious flaw of this conceptualization is that the estimated probabilities, which now are guaranteed to fall into the interval [0, 1], are a sequential consequence of the parameter estimation DFA procedure. A third, even more elaborate way of obtaining a solution is to solve the nonlinear regression problem

$$\mathbf{Y} = \frac{1}{1 + EXP(\mathbf{X}\beta)} + \epsilon ,$$

whose error is additive, whose probabilities are guaranteed to fall between 0 and 1, and whose parameters are simultaneously estimated with the probabilities.

EXAMPLE PR-10.1. Estimate the probability of areal unit membership in the coastal lowlands/interior highlands regionalization of Puerto Rico using multiple linear regression, based upon the agricultural density variables of farms (DFRM), sugar cane production (DSGR), and farm families (DFAM).

Using the SAS procedure

```
PROC REG;
   MODEL LND = DFRM DSGR DFAM;
   OUTPUT OUT=OLSOUT PRED=YHTOLS;
RUN;
```

produced SAS output that includes

Analysis of Variance

Source	DF	Sum of Squares	Mean Square	F Value	Prob>F
Model	3	8.77148	2.92383	22.257	0.0001
Error	69	9.06413	0.13136		
C Total	72	17.83562			

Parameter Estimates

Variable	DF	Parameter Estimate	Standard Error	T for H0: Parameter=0	Prob > |T|
INTERCEP	1	-0.047109	0.10873220	-0.433	0.6662
DFRM	1	0.045995	0.00971353	4.735	0.0001
DSGR	1	-0.016114	0.00630057	-2.558	0.0127
DFAM	1	0.072852	0.03404678	2.140	0.0359

A cursory inspection indicates that each of these three variables is significantly related to a differentiation of the coastal lowlands and interior highlands municipios. As was anticipated, though, six of the estimated probabilities are negative (realizing a minimum of -0.44006), whereas three of the estimated probabilities exceed unity (realizing a maximum of 1.29751). One easy and intuitively appealing corrective strategy for modifying these infeasible estimates is to truncate the negative values at 0 and the excessively large values at 1; one statistically more appealing strategy is to better specify the model.

EXAMPLE PR-10.1 (continued) Estimate the probability of areal unit membership in the coastal lowlands/interior highlands regionalization of Puerto Rico using DFA.

Using the SAS procedure

```
PROC DISCRIM POOL=TEST LIST;
   CLASS LND;
   VAR DFRM DSGR DFAM;
RUN;
```

produced this SAS output:

```
Test Chi-Square Value =       5.793800
with       6 DF       Prob > Chi-Sq = 0.4467
Since the chi-square value is not significant at the  0.1 level,
a pooled covariance matrix will be used in the discriminant
function.

       Number of Observations and Percent Classified into LND:
From LND                  0                  1                Total

       0                 36                  6                   42
                      85.71              14.29               100.00
       1                  7                 24                   31
                      22.58              77.42               100.00
   Total                 43                 30                   73
Percent               58.90              41.10               100.00
```

This output implies that there is no evidence to indicate that the covariance matrices for the two regions are not equal in the population and that thirteen of the municipios would be misclassified by the estimated discriminant function. The comparable SSE for the associated probability estimates is 8.71416, which contains a nonlinear bias of at least 0.00385.

EXAMPLE PR-10.1 (continued) Estimate the probability of areal unit membership in the coastal lowlands/interior highlands of Puerto Rico using nonlinear regression employing numerical derivatives.

Because of ignorance of reasonable parameter values for a nonlinear model, ad hoc guesses are based upon the notion of regional membership being equally likely (probability of regional membership equals 0.5). Using the SAS procedure

```
PROC NLIN MAXITER=100 METHOD=DUD;
   PARMS B0=1 B1=0 B2=0 B3=0;
   EXPONENT = EXP(B1*DFRM + B2*DSGR + B3*DFAM);
   MODEL LND = 1/(1 + B0*EXPONENT);
RUN;
```

produced the SAS output on the following page.

Of note is the set of negative iterative steps, which appear because the DUD procedure is used and which involve displacements of 0.1 to inaugurate development of a history. Since the SSE value of 8.27151 attained here is similar to that achieved with OLS, and in fact is even smaller, one can be more confident in believing that a global minimum has been reached. Next, the parameter estimates obtained here (i.e., B0 = 24.06593314, B1 = -0.28655414, B2 = 0.17611345, B3 = -0.58623919) can serve as the initial guesses for the Marquardt method.

Iter	B0	B1	B2	B3	Sum of Squares
-5	1.000000	0	0	0	18.250000
-4	1.100000	0	0	0	18.029478
-3	1.000000	0.100000	0	0	24.245979
-2	1.000000	0	0.100000	0	15.009213
-1	1.000000	0	0	0.100000	19.250862
0	1.000000	0	0.100000	0	15.009213
1	3.325970	-0.230025	0.100554	-0.281416	12.671452
2	4.511189	-0.229745	0.126542	-0.490398	11.825492
3	6.117237	-0.257737	0.150789	-0.616598	11.511434
4	10.682584	-0.286372	0.248336	-0.805973	9.377661
5	13.223569	-0.318353	0.291825	-0.656719	8.954186
6	13.011049	-0.320644	0.290120	-0.597204	8.928422
7	11.648185	-0.314683	0.275348	-0.538170	8.906628
8	9.992206	-0.295154	0.219461	-0.321174	8.644586
9	24.711292	-0.292179	0.234140	-0.834446	8.433308
10	25.077163	-0.274641	0.201687	-0.797175	8.352433
11	34.859396	-0.300165	0.172256	-0.706960	8.304396
12	31.226638	-0.287660	0.182654	-0.723462	8.295711
13	30.816481	-0.300153	0.185516	-0.643780	8.284343
14	25.376316	-0.294184	0.179779	-0.589742	8.272548
15	24.919927	-0.292829	0.179780	-0.592069	8.272503
16	24.491288	-0.291272	0.178862	-0.590780	8.272437
17	24.354592	-0.290411	0.179590	-0.590152	8.272186
18	23.641891	-0.287262	0.176811	-0.579289	8.271691
19	23.778863	-0.286645	0.176491	-0.583387	8.271576
20	23.951366	-0.286143	0.175907	-0.586050	8.271521
21	24.042010	-0.286494	0.175934	-0.586003	8.271512
22	24.048882	-0.286486	0.176077	-0.586132	8.271507
23	24.065933	-0.286554	0.176113	-0.586239	8.271507

NOTE: Convergence criterion met.

Source	DF	Sum of Squares	Mean Square
Regression	4	22.728492955	5.682123239
Residual	69	8.271507045	0.119876914
Uncorrected Total	73	31.000000000	
(Corrected Total)	72	17.835616438	

Parameter	Estimate	Asymptotic Std. Error	Asymptotic 95 % Confidence Interval Lower	Upper
B0	24.06593314	22.841417761	-21.501536974	69.633403256
B1	-0.28655414	0.092170260	-0.470429140	-0.102679149
B2	0.17611345	0.057686999	0.061030809	0.291196084
B3	-0.58623919	0.293879536	-1.172513965	0.000035577

Asymptotic Correlation Matrix

Corr	B0	B1	B2	B3
B0	1	-0.748809331	0.3555038604	-0.567961202
B1	-0.748809331	1	-0.562622392	0.0556012086
B2	0.3555038604	-0.562622392	1	-0.298314465

EXAMPLE PR-10.1 (continued) Estimate the probability of areal unit membership in the coastal lowlands/interior highlands of Puerto Rico using nonlinear regression employing analytical derivatives.

Using the SAS procedure

```
PROC NLIN MAXITER=100 METHOD=MARQUARDT;
   PARMS B0=24.07 B1=-0.28 B2=0.18 B3=-0.59;
   EXPONENT = EXP(B1*DFRM + B2*DSGR + B3*DFAM);
   MODEL LND = 1/(1 + B0*EXPONENT);
   OUTPUT OUT=NLSOUT PRED=YHATNLS;
DER.B0=-EXPONENT/(1 + B0*EXPONENT)**2;
DER.B1=-DFRM*B0*EXPONENT/(1 + B0*EXPONENT)**2;
DER.B2=-DSGR*B0*EXPONENT/(1 + B0*EXPONENT)**2;
DER.B3=-DFAM*B0*EXPONENT/(1 + B0*EXPONENT)**2;
RUN;
```

produced SAS output that includes

```
Iter       B0          B1          B2          B3       Sum of Squares
 0    24.070000   -0.280000   0.180000   -0.590000         8.281528
 1    23.296575   -0.285367   0.174744   -0.570233         8.271746
 2    23.961252   -0.286341   0.175763   -0.582154         8.271510
 3    23.980587   -0.286469   0.175927   -0.583398         8.271500
 4    24.006709   -0.286542   0.175982   -0.583908         8.271500
 5    24.009716   -0.286549   0.175992   -0.584023         8.271500
NOTE: Convergence criterion met.

Source                  DF  Sum of Squares     Mean Square

Regression               4    22.728500366     5.682125092
Residual                69     8.271499634     0.119876806
Uncorrected Total       73    31.000000000
(Corrected Total)       72    17.835616438

Parameter       Estimate      Asymptotic              Asymptotic 95 %
                              Std. Error            Confidence Interval
                                                    Lower          Upper
         B0   24.00971617   22.821538301  -21.518095435   69.537527767
         B1   -0.28654915    0.092174526   -0.470432656   -0.102665645
         B2    0.17599226    0.057614562    0.061054129    0.290930389
         B3   -0.58402284    0.293542361   -1.169624965    0.001579281

                       Asymptotic Correlation Matrix
Corr                     B0                B1                B2                B3
-----------------------------------------------------------------------------------
        B0                1      -0.747347281      0.3515500769      -0.565752616
        B1     -0.747347281                 1      -0.561940569      0.0529465411
        B2     0.3515500769      -0.561940569                 1      -0.295456169
        B3     -0.565752616      0.0529465411      -0.295456169                 1
```

The SAS statements DER.B0, DER.B1, . . . are the coding of the analytical derivatives supplied by the analyst or his/her consultant. Because the SSE value of 8.27150 attained here is equivalent to that achieved with DUD, and in fact is slightly smaller, one can continue to be confident in believing that a global minimum has been reached. The detectable nonlinear bias here is roughly -0.00392. Differences occurring between the DUD and Marquardt output are attributable to inaccuracies that emerge when using methods based upon numerical derivatives.

> *RULE OF THUMB 10.3:* Virtually always use DUD to double-check parameter estimates obtained with Marquardt, using Marquardt results as the initial estimates for the DUD solution.

Some conspicuous differences between these three specifications merit review. First, the equivalent OLS and DFA results suggest that the coefficients attached to each of the three predictor variables are significantly different from 0 in the population (the DFA results can be confirmed with PROC STEPDISC; see §7.1.3). In contrast, NLS renders output suggesting that the coefficient attached to DFAM is not significantly different from 0. Second, the probability estimates produced by each specification usually are moderately to markedly different. As mentioned, in nine cases the OLS results are inadmissible; corresponding DFA and NLS results for these nine case are, however, perfectly acceptable. The most similar estimates are obtained for San Lorenzo:

method	OLS	NLS	DFA
probability estimates	0.9786	0.9798	0.9829 .

One of the biggest discrepancies occurs for Aguada:

method	OLS	NLS	DFA
probability estimates	0.5021	0.2814	0.7502 .

And, one of the closest pairwise results is for Yabucoa:

method	NLS	DFA
probability estimates	0.8084	0.8082 .

Overall, this example clearly illustrates that the accommodation of nonlinearities and that the decision regarding the error structure specification can have profound impacts on statistical results.

10.5. Problems

P-10.1. Given the *Number of Species in the Galápogas Islands* (Table 49.1 in Andrews and Herzberg, 1985) data set, predict the density of observed number of species (this is **Y**) from the distance variables (this is **X**), specifying each of these distance variables as negative exponential functions. (Note: Upton and Fingleton, 1985, pp. 293–97 analyze these data.)

P-10.2. Given the *Old Maps* (Table 10.1 in Andrews and Herzberg, 1985) data set, repeat P-4.1 assuming a multiplicative rather than an additive error structure.

10.6. References

Andrews, D. and H. Herzberg. 1985. *Data: A Collection of Problems from Many Fields for the Student and Research Worker.* Berlin: Springer-Verlag.

Upton, G. and B. Fingleton. 1985. *Spatial Data Analysis by Example: Volume 1.* New York: Wiley.

CHAPTER 11

SPATIAL AUTOREGRESSIVE ANALYSIS

Chapter Preview. Not surprisingly, most students find the study of statistics hard, and the study of spatial statistics very, very hard! This reaction is not because simple descriptive spatial statistics are so challenging, or even that elementary spatial statistics, such as the Geary Ratio and the Moran Coefficient, generate fear. Rather, the formidable reputation of spatial statistics, we think, lies in the very demanding adjustments that are required to accommodate the reality of spatial relationships (as defined by nonzero spatial autocorrelation) in these aspatial, classical techniques. In this chapter, we take a first step in this direction with an introduction to spatial autoregressive analysis. By the end of this chapter, it should be clear that nonlinear regression is the workhorse of spatial statistics.

Much of the advancement, perhaps nearly all, in the development and implementation of spatial autoregressive models (and the related spatial econometric analysis; see Anselin, 1988; Anselin and Florax, 1995) has occurred in the last ten years. When we pursued this methodology in the early 1980s, the development of models examined in this chapter was in its early stages, and progress was severely hampered by available computing power limitations.

A number of still-unresolved issues hamper the application of spatial autoregressive models. For example, construction of the connectivity matrix (matrix **C** or **W**) that quantifies the nature of the connections in geographic space is subject to arbitrary decisions by analysts. In addition, accounting for spatial dependence through either the mean response or the variance component (also referred to as first-order and second-order effects) is subject to arbitrariness on the part of analysts.

In this chapter we are unable to definitively resolve these issues. We provide only an introduction to complex material and ideas that still are evolving.

11.0. Introduction

Statistical methodology for modeling geo-referenced data, in which complexities due to the presence of spatial autocorrelation lie dormant, has only recently begun to reach the level of sophistication achieved early in this century in traditional multivariate statistical analysis and just several decades ago in time series analysis. Much of this has been accomplished by rewriting conventional statistical techniques as regression models and then estimating parameters with generalized least squares procedures. Just as the regression technique is the workhorse of data analysis in general, nonlinear regression is the workhorse of geo-referenced data analysis.

> *SA Question #1*: What is the rule of correspondence between variable Y and a battery of X variables that enables Y to be predicted from these Xs when nonzero spatial autocorrelation is present?

The approach to solve this problem leads to a model equation that is a special case of equation (10.1) and may be written as

$$\frac{\mathbf{L}_1(\rho)\mathbf{Y}}{EXP\,[J^{\frac{1}{2}}(\rho)]} = \frac{\mathbf{L}_2(\rho)\mathbf{X}\beta}{EXP\,[J^{\frac{1}{2}}(\rho)]} + \frac{\mathbf{L}_3(\rho)\epsilon}{EXP\,[J^{\frac{1}{2}}(\rho)]} \tag{11.1}$$

(see Griffith, 1988) where ρ is the spatial autocorrelation parameter, the n-by-n matrix $\mathbf{L}_j(\rho)$ is a geographic weights matrix that represents the geographic configuration of areal units channeling spatial autocorrelation, and $EXP[J^{1/2}(\rho)]$ is a normalizing constant that ensures that the total of the probabilities associated with the probability density function for this equation equals 1. The parameters of equation (11.1) must be estimated using nonlinear procedures, because the spatial autocorrelation parameter ρ appears in both the numerator and the denominator of this equation, as well as a multiplier of vector $\mathbf{\beta}$.

Instead of attempting to specify the operator matrix $\mathbf{L}$ directly in equation (11.1), we specify it indirectly through spatial interaction schemes. This specification is posited by including in the nonlinear regression model those relationships between each areal unit's attributes and its geographically contiguous areal unit attributes, introducing here only the additional parameter ρ (for spatial autocorrelation) that needs to be estimated. Such specifications do not necessarily address issues of heteroscedasticity. (Some impacts of the arbitrariness of these specifications are outlined in Griffith, 1996).

11.1. Specifying the n-by-n Matrix $\mathbf{L}_j(\rho)$

Several popular specifications of the triplet of matrices $\mathbf{L}_j(\rho)$ appear in the literature. As an introduction to this topic, the two most commonly employed ones will be reviewed here. If the error structure serves as the vehicle by which spatial autocorrelation appears in geo-referenced data, then one specification becomes

$$\mathbf{L}_1(\rho) = \mathbf{I}\ ,\ \mathbf{L}_2(\rho) = \mathbf{I}\ ,\ \text{and}\ \mathbf{L}_3(\rho) = (\mathbf{I} - \rho\mathbf{W})^{-1}\ ,$$

rendering the SA model specification, which focuses on spatial autocorrelation in the error term,

$$\frac{\mathbf{Y}}{EXP\,[J^{\frac{1}{2}}(\rho)]} = \frac{[\rho\mathbf{W}\mathbf{Y} + (\mathbf{I} - \rho\mathbf{W})\mathbf{X}\mathbf{\beta}]}{EXP\,[J^{\frac{1}{2}}(\rho)]} + \frac{\epsilon}{EXP\,[J^{\frac{1}{2}}(\rho)]}\ , \tag{11.2}$$

where the n-by-n matrix $\mathbf{W}$ is the row-standardized version of the geographic contiguity matrix $\mathbf{C}$ used to calculate the Moran Coefficient (MC) (§2.3.1). This specification is known as the ***simultaneous autoregressive*** (SAR) model and is common in the spatial statistics literature. Phenomena that appear to conform to its specification include multiple linear regression models misspecified because of missing variables and situations where measurement error exhibiting a geographic pattern is present (e.g., remotely sensed data reflectance). The argument for its normalizing constant is given by

$$J^{\frac{1}{2}}(\rho) = \sum_{i=1}^{n} LN(1-\rho\lambda_i) \ , \qquad (11.3)$$

where λ_i is the i-th of the n eigenvalues of matrix **W**. These eigenvalues relate to orthogonalization of the geo-referenced data (recall that spatial autocorrelation indicates that information redundancies exist in the data) can be obtained like those for matrix **C** (see §6.5).

If the response variable itself serves as the vehicle by which spatial autocorrelation appears in geo-referenced data, then another specification becomes

$$\mathbf{L}_1(\rho) = \mathbf{I} - \rho\mathbf{W} \ , \ \mathbf{L}_2(\rho) = \mathbf{I} \ , \ \text{and} \ \mathbf{L}_3(\rho) = \mathbf{I} \ .$$

rendering the SAR model specification, which focuses on spatial autocorrelation latent in variable Y's values:

$$\frac{\mathbf{Y}}{EXP\ [J^{\frac{1}{2}}(\rho)]} = \frac{(\rho\mathbf{WY} + \mathbf{X}\beta)}{EXP\ [J^{\frac{1}{2}}(\rho)]} + \frac{\epsilon}{EXP\ [J^{\frac{1}{2}}(\rho)]} \ . \qquad (11.4)$$

This specification is known as the ***autoregressive response*** (AR) model and is common in the spatial econometrics literature. Phenomena that appear to conform to its specification include housing prices and population density. The argument for its normalizing constant is given by equation (11.3).

One should keep in mind that the SAR model, equation (11.2), is more frequently utilized by spatial statisticians, whereas the AR model, equation (11.4), is more frequently used by spatial econometricians. Rarely will statistical diagnostics suggest a preference for either of these specifications. Selection should be based upon substantive conceptual or theoretical considerations. Attractive features of the SAR model include its furnishing a direct analogy with AR model specifications found in the time series literature. It has an unbiased OLS (ordinary least squares) **b**, a relatively simpler asymptotic covariance matrix, and an autoregressive error distribution. In contrast, the AR model specification is simpler to estimate.

11.2. Computing the Normalizing Constant

The first step in undertaking a spatial autoregressive analysis is to deal with the normalizing constant. The importance of this constant is illustrated in Figure 11.1a–c which has been constructed with hypothetical, simulated geo-referenced data. In Figure 11.1c, where 0 spatial autocorrelation prevails, the normalizing constant has a smooth quadratic-type plot that achieves a

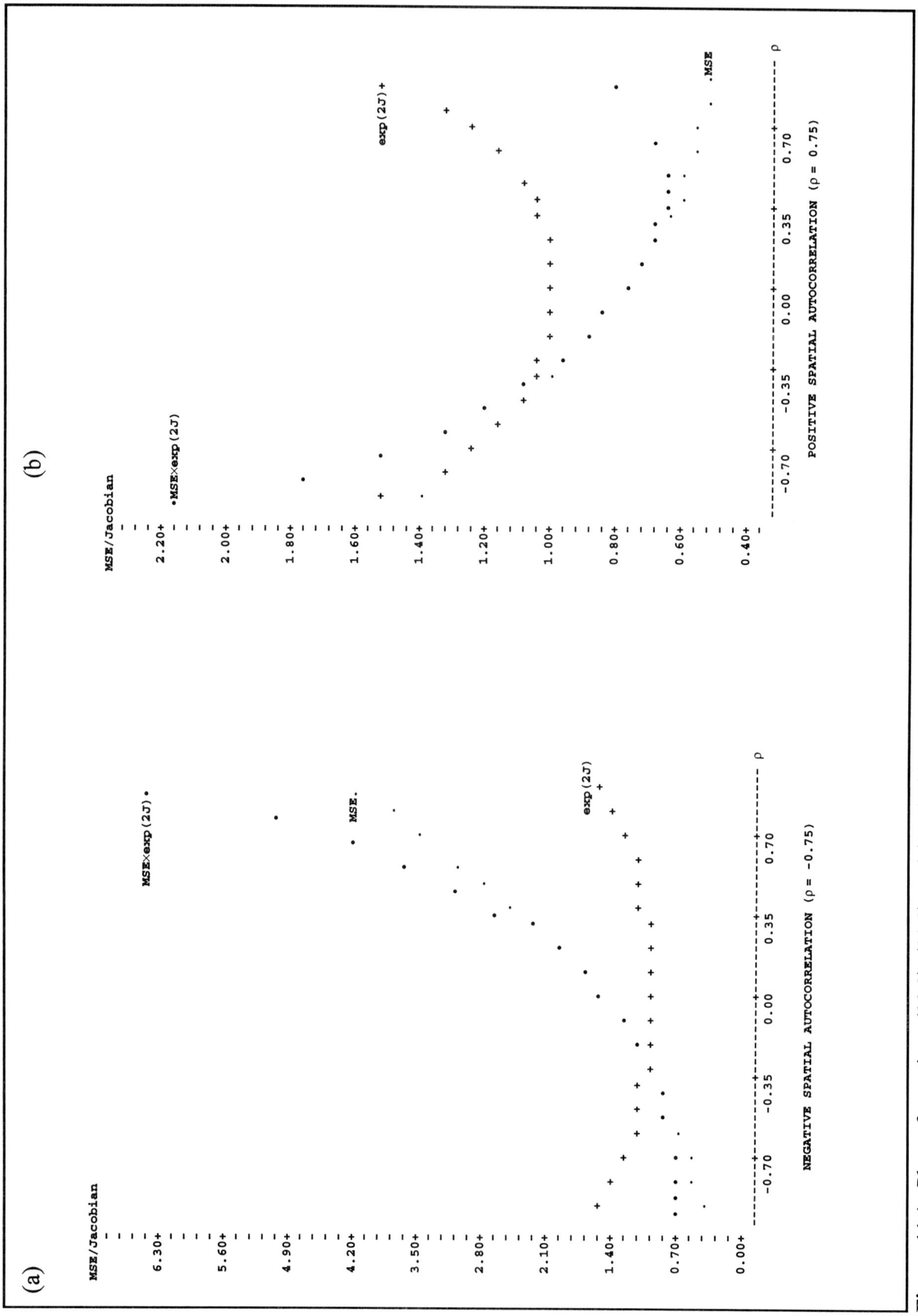

Figure 11.1. Plots of equations (11.2), (11.3), and (11.4): (a) positive spatial autocorrelation, (b) negative spatial autocorrelation.

minimum at $\rho = 0$. Minima for the MSE (mean squared error) from OLS and the $\frac{MSE}{EXP[J^{\frac{1}{2}}(\rho)]}$ coincide at roughly this ρ value, too. In Figure 11.1a, where moderately strong positive spatial autocorrelation prevails, the normalizing constant again has a smooth quadratic-type plot achieving a minimum at $\rho = 0$. But now the minimum for the MSE from OLS occurs beyond 1, while that for the $\frac{MSE}{EXP[J^{\frac{1}{2}}(\rho)]}$ occurs near 0.5. In Figure 11.1b, where moderately strong negative spatial autocorrelation prevails, the normalizing constant continues to have a smooth quadratic-type plot achieving a minimum at $\rho = 0$. But now the minimum for the MSE from OLS occurs well below -1, while that for the $\frac{MSE}{EXP[J^{\frac{1}{2}}(\rho)]}$ occurs near -0.75. Therefore, as mentioned, the normalizing constant acts as a constraint on $\hat{\rho}$, safeguarding against physically meaningless spatial autocorrelation coefficient values (recall that the value of a correlation coefficient cannot exceed 1). This normalizing constant, operating much like a weight in weighted least squares (WLS), also ensures that the correct MSE curve is being evaluated. The smooth, systematic nature of the normalizing constant between the limits of $\hat{\rho}$ is a property that can be exploited in spatial autoregressive analysis.

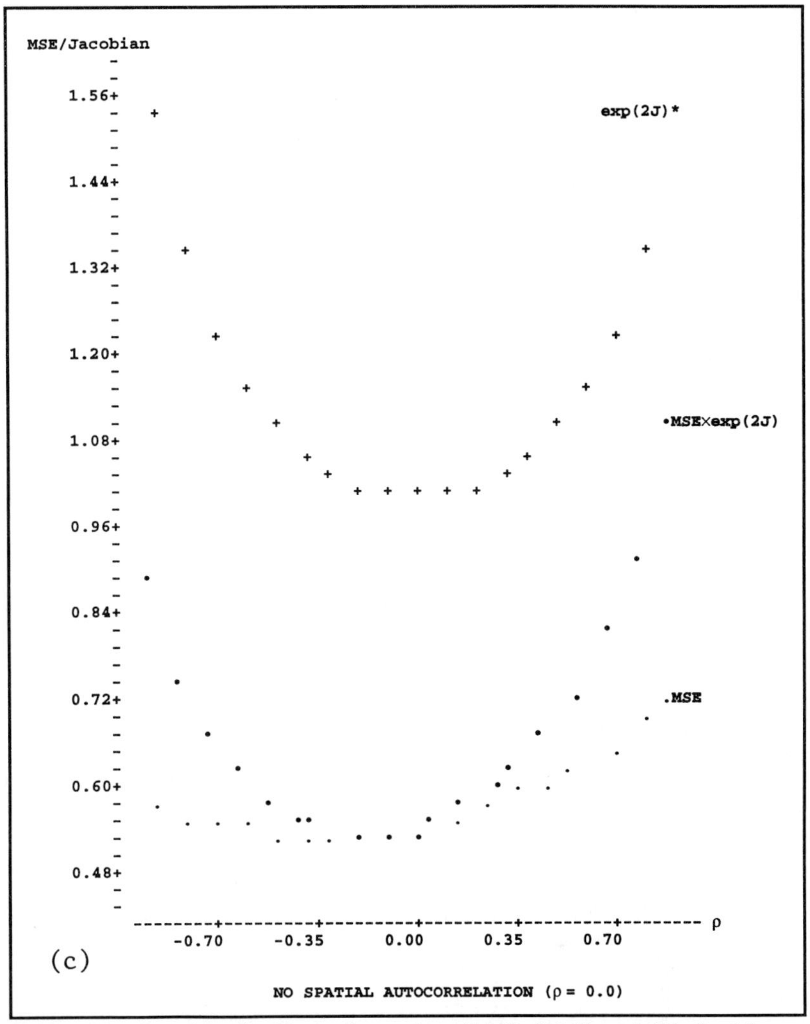

Figure 11.1 (continued). Plots of equations (11.2), (11.3), and (11.4): (c) zero spatial autocorrelation.

One instinctive and straightforward approach to the task of calculating equation (11.3) is to compute the n eigenvalues of matrix **W**.

EXAMPLE PR-11.1. (Example PR-6.5 continued). Compute the eigenvalues of the stochastic version of binary matrix **C** for the Puerto Rico municipio surface partitioning.

Using the SAS procedure

```
DATA STEP2;
   INFILE CONN;
   INPUT NUMCON C1-C73;
ARRAY CMTX{73} C1-C73;
RSUM=0;
DO I=1 TO 73;
   RSUM = RSUM + CMTX{I};
END;
DROP I;
RUN;
PROC TRANSPOSE DATA=STEP2 PREFIX=TRSUM OUT=TEMP1;
   VAR RSUM;
RUN;
DATA IW;
   SET STEP2;
   IF _N_=1 THEN SET TEMP1;
ARRAY CMTX{73} C1-C73;
ARRAY WMTX{73} W1-W73;
ARRAY RMTX{73} TRSUM1-TRSUM73;
DO I=1 TO 73;
   WMTX{I} = CMTX{I}/SQRT(RSUM*RMTX{I});
   IF NUMCON=I THEN WMTX{I} = 1;
END;
DROP I;
RUN;
DATA WCORREL(TYPE=CORR);
   SET IW;
DROP C1-C73 _NAME_ RSUM NUMCON TRSUM1-TRSUM73;
   _TYPE_='CORR';
RUN;
PROC FACTOR DATA=WCORREL METHOD=PRIN N=0
                    OUTSTAT=WEIGEN;
   VAR W1-W73;
RUN;
DATA WEIGEN (REPLACE=YES);
   SET WEIGEN (KEEP=W1-W73);
RUN;
PROC TRANSPOSE DATA=WEIGEN OUT=WEIGEN (REPLACE=YES);
RUN;
DATA WEIGEN(REPLACE=YES);
   SET WEIGEN (KEEP=COL1-COL78);
WEIGNVAL=COL78-1;
DROP COL1-COL78;
RUN;
```

produced the SAS output

1	1.00000	26	0.08577	51	-0.34363
2	0.98571	27	0.06371	52	-0.34705
3	0.94965	28	0.03800	53	-0.36092
4	0.88760	29	0.02731	54	-0.36490
5	0.87942	30	-0.02288	55	-0.37741
6	0.85317	31	-0.02708	56	-0.38172
7	0.79606	32	-0.02978	57	-0.38523
8	0.76476	33	-0.06821	58	-0.40251
9	0.72850	34	-0.09235	59	-0.41099
10	0.65969	35	-0.10851	60	-0.42238
11	0.61704	36	-0.12014	61	-0.43651
12	0.57737	37	-0.14230	62	-0.44630
13	0.55007	38	-0.15219	63	-0.45714
14	0.49944	39	-0.16779	64	-0.47330
15	0.47991	40	-0.19313	65	-0.48160
16	0.44719	41	-0.21830	66	-0.50119
17	0.41639	42	-0.23164	67	-0.51311
18	0.36135	43	-0.24193	68	-0.52543
19	0.31643	44	-0.24861	69	-0.53572
20	0.29932	45	-0.26679	70	-0.54708
21	0.27315	46	-0.27933	71	-0.56420
22	0.23419	47	-0.29488	72	-0.57407
23	0.19991	48	-0.30388	73	-0.60139
24	0.18656	49	-0.31441		
25	0.13862	50	-0.33840		

As in §6.5, this preceding computer code tricks the SAS software into thinking that it is dealing with a correlation matrix. Once again, the matrix that is being analyzed is (**I** + **W**), because each of the diagonal entries of a correlation matrix must equal 1; in other words, any variable must perfectly and positively correlate with itself. This modification simply adds 1 to each of the resulting eigenvalues, which is subtracted here with the SAS statement WEIGNVAL=COL78-1. The corresponding eigenvalue problem (see Chapter 3) has been rewritten as follows:

$$\begin{aligned}
DET\ [(\mathbf{I} + \mathbf{W} - \lambda\mathbf{I}] &= DET(I)DET[(\mathbf{I} + \mathbf{CD}) - \lambda\mathbf{I}] \\
&= DET(\mathbf{D}^{\frac{1}{2}} \times (\mathbf{D}^{-\frac{1}{2}} DET[(\mathbf{I} + \mathbf{CD}) - \lambda\mathbf{I}] \\
&= DET(\mathbf{D}^{\frac{1}{2}}) DET[(\mathbf{I} + \mathbf{CD} - \lambda\mathbf{I}] DET(\mathbf{D}^{-\frac{1}{2}}) \\
&= DET[(\mathbf{I} + \mathbf{D}^{\frac{1}{2}}\mathbf{CD}^{\frac{1}{2}}) - \lambda\mathbf{I}] \\
&= 0
\end{aligned}$$

This version yields the correct eigenvalues of matrix **W**, plus 1, but extracts them from an equivalent symmetric matrix. The resulting eigenvalues of matrix **W** are those used in calculating equation (11.3).

But the computation of eigenvalues for equation (11.3) is numerically intensive, and when n gets sufficiently large, this computational chore becomes infeasible even with a supercomputer. To circumvent this restriction, and in doing so exploit the smooth, systematic curvature displayed by the normalizing constant (see Figure 11.1) across the range of possible ρ values, the following approximation equation has been formulated as a substitute for equation (11.3):

$$J^{\frac{1}{2}}(\rho) \approx -\left\{\frac{\rho(\alpha_{2,n}\lambda_{min}^2 - \alpha_{1,n}\lambda_{max}^2)}{2} + \alpha_{2,n}|\lambda_{min}|LN\left(\frac{\gamma_{2,n}}{|\lambda_{min}|}\right) + \alpha_{1,n}\lambda_{max}LN\left(\frac{\gamma_{1,n}}{\lambda_{max}}\right) - \alpha_{2,n}|\lambda_{min}|LN\left(\frac{\gamma_{2,n}}{|\lambda_{min}|}+\rho\right) - \alpha_{1,n}\lambda_{max}LN\left(\frac{\gamma_{1,n}}{\lambda_{max}}-\rho\right)\right\} \times n^{-1}, \tag{11.5}$$

This equation can be calibrated directly with the eigenvalues of matrix **W**, to obtain estimates of its four unknown coefficients $\alpha_{1,n}$, $\alpha_{2,n}$, $\gamma_{1,n}$, and $\gamma_{2,n}$; λ_{max} and λ_{min} are the extreme eigenvalues obtained from the preceding eigenfunction problem. Principal benefits and costs of employing equation (11.5) are outlined in Griffith and Sone (1995).

EXAMPLE PR-11.2. Compute the normalizing constant approximation equation (11.5) for the Puerto Rico municipio surface partitioning, using the eigenvalues of matrix **W**.

Using the SAS procedure

```
DATA STEP2 (REPLACE=YES);
   SET WEIGEN;
N = 73;
LAMBDA=WEIGNVAL;
RUN;
PROC UNIVARIATE NOPRINT;
   VAR LAMBDA;
   OUTPUT OUT=EXTREMEL MAX=LMAX MIN=LMIN;
RUN; PROC PRINT DATA=EXTREMEL; VAR LMAX LMIN; RUN;
DATA STEP3;
   IF _N_=1 THEN SET EXTREMEL;
   SET STEP2;
FINISH = 0.999/LMAX;
START =  0.999/LMIN;
INC = (FINISH - START)/21;
ARRAY JACOB{22} JAC1-JAC22;
RHO = START;
DO I = 1 TO 22;
   JACOB{I} = -LOG(1 - RHO*LAMBDA);
   RHO = RHO + INC;
END;
RUN;
PROC MEANS NOPRINT;
   VAR JAC1-JAC22;
```

```
   OUTPUT OUT=JACOB1 MEAN=;
RUN;
PROC TRANSPOSE OUT=JACOB2;
   VAR JAC1-JAC22;
RUN;
DATA STEP3 (REPLACE=YES);
   SET STEP3;
   IF _N_=1 THEN SET EXTREMEL;
   DO RHO = START TO FINISH BY INC;
      OUTPUT;
   END;
DROP START FINISH INC;
RUN;
DATA STEP3 (REPLACE=YES);
   SET JACOB2;
   SET STEP3;
J = COL1;
DROP COL1;
RUN;
PROC PLOT;
   PLOT J*RHO;
RUN;
PROC NLIN MAXITER=500 METHOD=MARQUARDT;
   PARMS A1 =  9
      A2 = 30
      D1 = 1.05
      D2 = 1.11;
   BOUNDS 1.00001 < D1 < 2.0, 1.00001 < D2 < 2.0,
      0.001 < A1, 0.001 < A2;
   MODEL J = (RHO/(2*N))*(A2*(LMIN**2) - A1*(LMAX**2)) +
         (1/N)*(A2*ABS(LMIN)*LOG(D2/ABS(LMIN)) +
            A1*LMAX*LOG(D1/LMAX) -
            A2*ABS(LMIN)*LOG(D2/ABS(LMIN) + RHO) -
            A1*LMAX*LOG(D1/LMAX - RHO));
    DER.A1 = (RHO/(2*N))*(-(LMAX**2)) + (1/N)*(LMAX*LOG(D1/LMAX)
       - LMAX*LOG(D1/LMAX - RHO));
   DER.A2 = (RHO/(2*N))*((LMIN**2))+
       (1/N)*(ABS(LMIN)*LOG(D2/ABS(LMIN))-
       ABS(LMIN)*LOG(D2/ABS(LMIN) + RHO));
   DER.D1 = (1/N)*(A1*LMAX/D1 - A1/(D1/LMAX - RHO));
   DER.D2 = (1/N)*(A2*ABS(LMIN)/D2 - A2/(D2/ABS(LMIN) + RHO));
RUN;
```

produced SAS output that includes

Source	DF	Sum of Squares	Mean Square
Regression	4	0.53610118744	0.13402529686
Residual	18	0.00038759811	0.00002153323
Uncorrected Total	22	0.53648878555	
(Corrected Total)	21	0.30172941915	

Parameter	Estimate	Asymptotic Std. Error	Asymptotic 95 % Confidence Interval Lower	Upper
A1	9.28291114	0.34158569804	8.565270956	10.000551315
A2	30.27615771	0.87458174546	28.438741782	32.113573630
D1	1.04872104	0.00564606158	1.036859182	1.060582895
D2	1.10781400	0.00768457245	1.091669422	1.123958584

The relative SSE term of $\frac{0.00038759811}{0.30172941915} \approx 1.28 \times 10^{-3}$ is remarkably small and is typical of those achieved with equation (11.5). These four parameter estimates, combined with the extreme eigenvalues of 1 and -0.60139 reported in Example PR-11.1, can be substituted into equation (11.5) and then used to all but eliminate the computational burden in a spatial autoregressive analysis.

A plot of the normalizing constant for the Puerto Rico municipio surface partitioning appears in Figure 11.2. Once more the smooth, quadratic-type plot that achieves a minimum at $\rho = 0$ can be recognized. This plot fails to be symmetric because the surface partitioning is irregular (the one used for constructing Figure 11.1 was the regular square tessellation that underlies remotely sensed data).

Because matrix **W** is row-standardized, then theoretically λ_{max} equals 1 (property P12, Table 3.1). The minimum eigenvalue, λ_{min}, is contained in the open interval (-0.5, -1), with an average value apparently around -0.66542 (based upon a convenient, nonrandom sample of irregular surface partitionings), which is not too different from -0.60139. Preliminary findings suggest that the remaining coefficients may be estimated as follows:

$$\hat{\alpha}_{1,n} = 1.79861 + 0.09359n$$

$$\Rightarrow \frac{1.79861}{n} + 0.09359, \text{ incorporating the } \frac{1}{n} \text{ term appearing in equation (11.5),}$$

$$\hat{\alpha}_{2,n} = -1.22926 + 0.35397n$$

$$\Rightarrow \frac{-1.22926}{n} + 0.35397, \text{ incorporating the } \frac{1}{n} \text{ term appearing in equation (11.5),}$$

$$\hat{\gamma}_{1,n} = 1.10366 - 0.11719EXP(-0.01247n) \text{ , and}$$

$$\hat{\gamma}_{2,n} = 1.61079 - 0.59421EXP(-0.00165n) \ .$$

A tabulation of results for the Puerto Rico situation produces

$$\frac{\alpha_{1,n}}{n} = 0.12716 \qquad \frac{\alpha_{1,n}}{n} = 0.41474 \qquad \gamma_{1,n} = 1.04872 \ \ \gamma_{2,n} = 1.10781$$

$$\frac{\hat{\alpha}_{1,n}}{n} = 0.11823 \qquad \frac{\hat{\alpha}_{1,n}}{n} = 0.33713 \qquad \hat{\gamma}_{1,n} = 1.05650 \ \ \hat{\gamma}_{2,n} = 1.08401$$

These pairs are very similar, with the closer pairs being affiliated with the positive range for spatial autocorrelation. This latter feature is fortunate, because most empirical geographic distributions contain moderate, positive spatial autocorrelation.

> *RULE OF THUMB 11.1:* Begin a spatial autoregressive analysis utilizing the approximate normalizing constant.

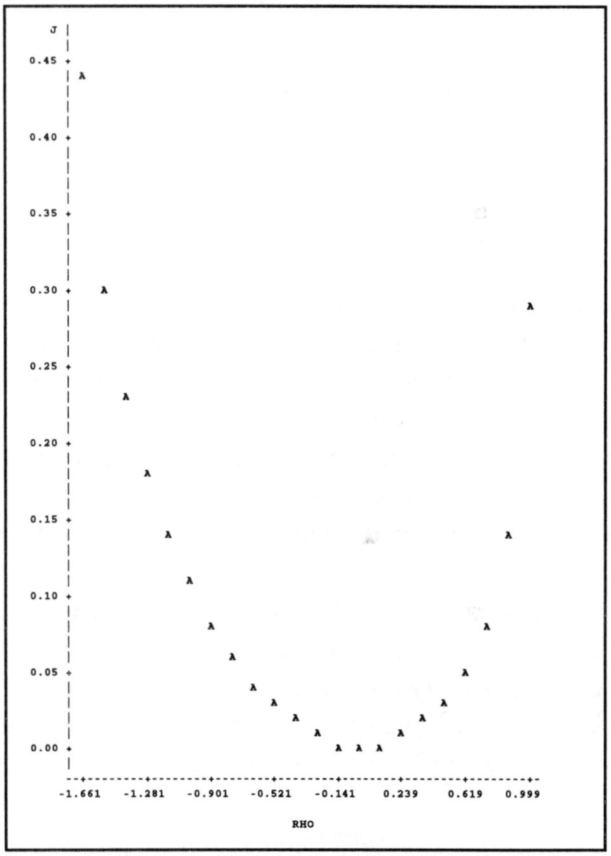

Figure 11.2. Plot of the normalizing constant across the feasible parameter space.

11.3. Estimating the Parameters of an SAR Model

Equation (11.2) can be specified for estimating a population mean, a customary multiple linear regression, a two-groups discriminant function model, bivariate correlation coefficients, and ANOVA models. (Extensive discussion of these reformulations of traditional statistical techniques appears in Griffith, 1993). Because the conventional OLS coefficients are unbiased, the estimates **b** obtained with an SAR model rarely differ markedly from their OLS counterparts. What do tend to differ substantially at times are the accompanying standard errors, because the presence of spatial autocorrelation compromises the statistical efficiency of OLS estimates (see Cordy and Griffith, 1993), sometimes causing a reversal of hypothesis testing decisions.

EXAMPLE PR-11.3. Estimate a constant mean SAR model specification for √(sugar cane production density) in Puerto Rico employing the exact normalizing constant.

Using the SAS procedure, based upon equation (11.3),

```
DATA STEP2;
   SET STEP1 (KEEP=Y);
   INFILE CONN;
   INPUT NUMCON C1-C73;
ARRAY CMTX{73} C1-C73;
ARRAY YMTX{73} CY1-CY73;
RSUM=0;
DO I=1 TO 73;
   RSUM = RSUM + CMTX{I};
   YMTX{I} = Y*CMTX{I};
END;
DROP I;
RUN;
⋮
PROC TRANSPOSE DATA=WEIGEN PREFIX=EIG OUT=EIGOUT;
RUN;
PROC MEANS DATA=STEP2 NOPRINT;
   VAR CY1-CY73;
   OUTPUT OUT=CYOUT1 SUM=CY1-CY73;
RUN;
PROC TRANSPOSE DATA=CYOUT1 PREFIX=WY OUT=CYOUT2;
   VAR CY1-CY73;
RUN;
DATA STEP3;
   IF _N_=1 THEN SET EIGOUT;
   SET STEP1;
   SET STEP2;
   SET CYOUT2;
YL = WY1/RSUM;
DROP C1-C73 CY1-CY73 _NAME_ WY1;
RUN;
PROC UNIVARIATE; VAR Y; RUN;
PROC NLIN DATA=STEP3 MAXITER=500 METHOD=MARQUARDT;
   PARMS RHO=0.500, B0= 7.76;
   BOUNDS -1.66667 < RHO < 1.0;
   ARRAY EIGEN{73} EIG1-EIG73;
   JACOB = 0;
   JDER = 0;
```

```
    DO I=1 TO 73;
       JACOB = JACOB + LOG(1 - RHO*EIGEN{I});
       JDER = JDER - EIGEN{I}/(1 - RHO*EIGEN{I});
    END;
    JACOB = EXP(-JACOB/73);
    JDER = -JDER/73;
    ZY = Y*JACOB;
    MODEL ZY = (RHO*YL + B0*(1 - RHO))*JACOB;
    DER.B0 = (1 - RHO)*JACOB;
    DER.RHO = ((RHO*YL + B0*(1 - RHO) - Y)*JDER
          + (YL - B0))*JACOB;
RUN;
```

produced SAS output that includes

Source	DF	Sum of Squares	Mean Square
Regression	2	1537.9930522	768.9965261
Residual	71	662.4808148	9.3307157
Uncorrected Total	73	2200.4738671	
(Corrected Total)	72	975.5044076	

Parameter	Estimate	Asymptotic Std. Error	95 % Confidence Interval Lower	Upper
RHO	0.623447814	0.16841654158	0.2876340029	0.9592616248
B0	3.985137932	0.90298388133	2.1846351101	5.7856407544

Of note is that the derivative $\frac{\partial ZY}{\partial RHO}$ is quite messy. This solution required PROC NLIN to execute 493 iterations in order to attain convergence, even though the initial estimates were quite close to this optimal solution. The OLS estimate b_0 is 3.8831, which is not markedly different from the value of 3.98514 computed here. The spatial autocorrelation estimate, $\hat{\rho} \approx 0.62345$, suggests the presence of moderate, positive spatial autocorrelation. Meanwhile, the OLS standard error estimate for b_0 is 0.40838, less than half of the value of 0.90298 yielded by the SAR model.

Example PR-11.3 involved the calculation of 73 eigenvalues and the recalculation of equation (11.3) for each of the 493 nonlinear regression iterations. This massive amount of computation can be reduced dramatically by using equation (11.5).

EXAMPLE PR-11.3. (continued). Estimate a constant mean SAR model specification for √DSGR in Puerto Rico, employing the approximate normalizing constant.

Replacing equation (11.3) with equation (11.5), and then using the revised SAS procedure

```
PROC NLIN DATA=STEP3 MAXITER=500 METHOD=MARQUARDT;
   PARMS RHO=0.100, B0=0.000;
   LMAX = 1.0; LMIN=-0.66542;
   A1 = 0.11823; A2 = 0.33713;
   D1 = 1.04872; D2 = 1.08401;
   BOUNDS -1.50281 < RHO < 1.0;
   JHAT = EXP((A2*(LMIN**2)-A1*(LMAX**2))*RHO/2
        + A1*LMAX*LOG(D1/LMAX)
        + A2*ABS(LMIN)*LOG(D2/ABS(LMIN))
        - (A1*LMAX*LOG(D1/LMAX-RHO)
        + A2*ABS(LMIN)*LOG(D2/ABS(LMIN)+RHO)));
   ZY = Y*JHAT;
   MODEL ZY = (RHO*YL + B0*(1 - RHO))*JHAT;
   DER.B0 = (1 - RHO)*JHAT;
   DER.RHO = ((RHO*YL + B0*(1 - RHO) - Y)*
         ((A2*(LMIN**2) - A1*(LMAX**2))/2 +
         (A1*LMAX/(D1/LMAX -RHO) -
         A2*ABS(LMIN)/(D2/ABS(LMIN)+RHO)))
         + (YL - B0))*JHAT;
   RUN;
```

produced SAS output that includes

```
Non-Linear Least Squares Summary Statistics     Dependent Variable ZY

  Source                 DF Sum of Squares      Mean Square
  Regression              2   1520.4006000      760.2003000
  Residual               71    649.5094839        9.1480209
  Uncorrected Total      73   2169.9100839

  (Corrected Total)      72    961.9550055

  Parameter      Estimate    Asymptotic              Asymptotic 95 %
                             Std. Error            Confidence Interval
                                                   Lower          Upper
       RHO    0.637414259 0.16744480628  0.3035380378  0.9712904806
        B0    3.991229495 0.93527077923  2.1263482806  5.8561107100
```

Of note is that the derivative $\frac{\partial \text{ZY}}{\partial \text{RHO}}$ is even messier than in the preceding case of the exact normalizing constant. This second solution required PROC NLIN to execute 11 iterations to attain convergence and did not require calculation of the 73 eigenvalues of matrix **W**. Moreover, the reduction in numerical intensity connected with implementing this SAR model is spectacular. The output, computed using the normalizing constant approximation, gives almost exactly the same estimates for β_0 and ρ and their estimators' respective standard errors, with the small differences being attributable to specification error.

11.4. Estimating the Parameters of an AR Model

Equation (11.4) also can be specified for estimating a population mean, a customary multiple linear regression, a two-groups discriminant function model, bivariate correlation coefficients, and ANOVA models. (Extensive discussion of these reformulations of traditional statistical techniques also appears in Griffith, 1993).

EXAMPLE PR-11.4. Estimate a constant mean AR model specification for √(sugar cane production density) in Puerto Rico.

Modifying the SAS procedure presented in Example PR-11.3, based upon equation (11.3), so that the model statement becomes

```
MODEL ZY = (RHO*YL + B0)*JHAT;
DER.B0 = JHAT;
DER.RHO = ((RHO*YL + B0 - Y)*
     ((A2*(LMIN**2) - A1*(LMAX**2))/2 +
     (A1*LMAX/(D1/LMAX -RHO) -
      A2*ABS(LMIN)/(D2/ABS(LMIN)+RHO))) + YL)*JHAT;
```

produced SAS output that includes

Source	DF	Sum of Squares	Mean Square
Regression	2	1520.3631117	760.1815559
Residual	71	649.5094845	9.1480209
Uncorrected Total	73	2169.8725962	
(Corrected Total)	72	961.9383866	

Parameter	Estimate	Asymptotic Std. Error	Asymptotic 95 % Confidence Interval Lower	Upper
RHO	0.637371911	0.16744769906	0.30348992150	0.9712539004
B0	1.447349438	0.72364968416	0.00442966306	2.8902692124

This solution required PROC NLIN to execute 11 iterations to attain convergence when the initial estimates were those obtained for the SAR model in Example PR-11.3. While $\hat{\rho}$ is essentially the same for both spatial statistical specifications, the estimate b_0 for the AR model virtually equals (1 - 0.63737) × b_0 obtained with the SAR model (i.e., 3.99123). In other words, in the constant mean case, these two models are equivalent. Recall that statistical criteria alone cannot indicate which of these two models best describes a given geo-referenced data set. Rather, selection of the specification must be based upon conceptual considerations.

RULE OF THUMB 11.2: A preliminary estimate of $\hat{\rho}$ can be secured with the AR model, which is easier to estimate. Nevertheless, final selection of either the SAR or AR specification must be based upon a sound geographic conceptual framework.

With *RULE OF THUMB 11.2* in mind, a researcher might argue that the AR specification best describes the geographic distribution of sugar cane production across the island of Puerto Rico, because nearby farms tend to produce the same crops. If one farmer chooses to grow a given crop, this judgment inspires neighboring farmers to make the same decision. Examples PR-4.3 through PR-4.8 supply a traditional regression analysis of this agricultural production, revealing that density of sugar cane production can be predicted from density of farms (DFRM), density of cultivated land (CDC), whether a municipio has any sugar cane production (I0), distance from San Juan (DISTSJ), and distance from Arecibo (DISTA). The estimated regression model displays homogeneity of variance, no aberrant municipios, normally distributed error, and significant spatial autocorrelation in the residuals.

EXAMPLE PR-11.5. Estimate an AR version of Example PR-4.3.

Using the SAS code

```
PROC NLIN DATA=STEP3 MAXITER=500 METHOD=MARQUARDT;
   PARMS RHO=0.5, B0=-0.8, B1=-0.1, B2=0.02, B3=1.5, B4=.2, B5=.3;
   LMAX = 1.0; LMIN=-0.66542;
   A1 = 0.11823; A2 = 0.33713;
   D1 = 1.04872; D2 = 1.08401;
   BOUNDS -1.50281 < RHO < 1.0;
   JHAT = EXP((A2*(LMIN**2)-A1*(LMAX**2))*RHO/2 +
     A1*LMAX*LOG(D1/LMAX) + A2*ABS(LMIN)*LOG(D2/ABS(LMIN))
     - (A1*LMAX*LOG(D1/LMAX-RHO) +
     A2*ABS(LMIN)*LOG(D2/ABS(LMIN)+RHO)));
   ZY = Y*JHAT;
   MODEL ZY = (RHO*YL + B0 + B1*DFRM + B2*DCD + B3*I0 +
           B4*DISTA + B5*DISTSJ)*JHAT;
   DER.B0=JHAT;
   DER.B1=DFRM*JHAT;
   DER.B2=DCD*JHAT;
   DER.B3=I0*JHAT;
   DER.B4=DISTA*JHAT;
   DER.B5=DISTSJ*JHAT;
   DER.RHO=((RHO*YL + B0 + B1*DFRM + B2*DCD + B3*I0 +
         B4*DISTA + B5*DISTSJ - Y)*
         ((A2*(LMIN**2) - A1*(LMAX**2))/2 +
         (A1*LMAX/(D1/LMAX -RHO) -
         A2*ABS(LMIN)/(D2/ABS(LMIN)+RHO))) + YL)*JHAT;
RUN;
```

produced SAS output that includes

```
Source                  DF  Sum of Squares     Mean Square

Regression               7    386.54085722      55.22012246
Residual                66     61.32749693       0.92920450
Uncorrected Total       73    447.86835415
(Corrected Total)       72    181.35671389

Parameter      Estimate     Asymptotic             Asymptotic 95 %
                            Std. Error           Confidence Interval
                                                 Lower          Upper
      RHO    0.313731534 0.23803461976 -0.1615205904  0.7889836581
       B0   -0.746401900 0.59132121872 -1.9270144742  0.4342106751
       B1   -0.082218674 0.02651070351 -0.1351490761 -0.0292882717
       B2    0.020876555 0.01130871753 -0.0017020604  0.0434551695
       B3    1.432360996 0.32277778853  0.7879134375  2.0768085554
       B4    0.129938460 0.10639156081 -0.0824794455  0.3423563647
       B5    0.185913679 0.13499750964 -0.0836179260  0.4554452849
```

This solution required PROC NLIN to execute 170 iterations to attain convergence. The value $\hat{\rho} \approx 0.31373$ indicates that weak to moderate, positive spatial autocorrelation is present in the residuals of Example PR-4.3, which is exactly what the MC test statistic reported in Example PR-4.8 suggests. The final estimates of the parameters do not differ greatly from their OLS counterparts. The most dramatic hypothesis testing decision switch here is for DISTSJ, whose standard error nearly doubled. Meanwhile, the coefficient for DISTA moved from just significant at the 5% level to not significant at this level. Unfortunately, according to these SAS results the estimate $\hat{\rho}$ is not significantly different from 0 which is inconsistent with the MC results reported in Example PR-4.8; however, the spatial statistical approximate standard error is $\sqrt{\frac{2}{344}}$, which indicates that $\hat{\rho}$ indeed is significant.

RULE OF THUMB 11.3: Because the OLS parameter estimates are unbiased, initialize nonlinear SAR or AR model estimation with them, setting $\hat{\rho}$ to 0.5 if the MC computed for the OLS residuals indicates the presence of positive spatial autocorrelation, and to -0.5 if the MC indicates the presence of negative spatial autocorrelation.

A researcher could consider these conflicting statistics to be indicative of model misspecification. In other words, the alternative argument that the SAR specification best describes the geographic distribution of sugar cane production across the island of Puerto Rico might be put forth, within the context of missing variables, because nearby farms tend to encounter similar soil types and climatic conditions, two variables not included in the original regression equation.

EXAMPLE PR-11.6. Estimate an SAR version of Example PR-4.3.

Using the SAS code

```
ARRAY X1MTX{73} CDFRM1-CDFRM73;
ARRAY X2MTX{73} CDCD1-CDCD73;
ARRAY X3MTX{73} CIO1-CIO73;
ARRAY X4MTX{73} CDSTA1-CDSTA73;
ARRAY X5MTX{73} CDSTSJ1-CDSTSJ73;
DO I=1 TO 73;
   X1MTX{I} = DFRM*CMTX{I};
   X2MTX{I} = DCD*CMTX{I};
   X3MTX{I} = I0*CMTX{I};
   X4MTX{I} = DISTA*CMTX{I};
   X5MTX{I} = DISTSJ*CMTX{I};
END;
⋮
PROC MEANS DATA=STEP2 NOPRINT;
   VAR CDFRM1-CDFRM73;
   OUTPUT OUT=CYOUT1(REPLACE=YES) SUM=CX1-CX73;
RUN;
PROC TRANSPOSE DATA=CYOUT1 PREFIX=CDFRM OUT=CXOUT1;
   VAR CX1-CX73;
RUN;
PROC MEANS DATA=STEP2 NOPRINT;
   VAR CDCD1-CDCD73;
   OUTPUT OUT=CYOUT1(REPLACE=YES) SUM=CX1-CX73;
RUN;
PROC TRANSPOSE DATA=CYOUT1 PREFIX=CDCD OUT=CXOUT2;
   VAR CX1-CX73;
RUN;
PROC MEANS DATA=STEP2 NOPRINT;
   VAR CIO1-CIO73;
   OUTPUT OUT=CYOUT1(REPLACE=YES) SUM=CX1-CX73;
RUN;
PROC TRANSPOSE DATA=CYOUT1 PREFIX=CIO OUT=CXOUT3;
   VAR CX1-CX73;
RUN;
PROC MEANS DATA=STEP2 NOPRINT;
   VAR CDSTA1-CDSTA73;
   OUTPUT OUT=CYOUT1(REPLACE=YES) SUM=CX1-CX73;
RUN;
PROC TRANSPOSE DATA=CYOUT1 PREFIX=CDSTA OUT=CXOUT4;
   VAR CX1-CX73;
RUN;
PROC MEANS DATA=STEP2 NOPRINT;
```

```
   VAR CDSTSJ1-CDSTSJ73;
   OUTPUT OUT=CYOUT1(REPLACE=YES) SUM=CX1-CX73;
RUN;
PROC TRANSPOSE DATA=CYOUT1 PREFIX=CDSTSJ OUT=CXOUT5;
   VAR CX1-CX73;
RUN;
DATA STEP3 (REPLACE=YES);
   SET STEP3;
   SET CXOUT1;
   SET CXOUT2;
   SET CXOUT3;
   SET CXOUT4;
   SET CXOUT5;
WDFRM=CDFRM1/RSUM;
WDCD=CDCD1/RSUM;
WI0=CIO1/RSUM;
WDSTA=CDSTA1/RSUM;
WDSTSJ=CDSTSJ1/RSUM;
RUN;
PROC NLIN DATA=STEP3 MAXITER=500 METHOD=MARQUARDT;
   PARMS RHO=0.31, B0=-1.1, B1=-0.1, B2=0.02, B3=1.4, B4=.1, B5=.2;
   LMAX = 1.0; LMIN=-0.66542;
   A1 = 0.11823; A2 = 0.33713;
   D1 = 1.04872; D2 = 1.08401;
   BOUNDS -1.50281 < RHO < 1.0;
   JHAT = EXP((A2*(LMIN**2)-A1*(LMAX**2))*RHO/2
      + A1*LMAX*LOG(D1/LMAX) +
        A2*ABS(LMIN)*LOG(D2/ABS(LMIN))
      - (A1*LMAX*LOG(D1/LMAX-RHO) +
      A2*ABS(LMIN)*LOG(D2/ABS(LMIN)+RHO)));
   ZY = Y*JHAT;
   MODEL ZY = (RHO*YL + B0*(1-RHO) + B1*(DFRM-RHO*WDFRM)
         + B2*(DCD-RHO*WDCD) + B3*(I0-RHO*WI0) +
         B4*(DISTA-RHO*WDSTA) +
         B5*(DISTSJ-RHO*WDSTSJ))*JHAT;
   DER.B0=(1-RHO)*JHAT;
   DER.B1=(DFRM-RHO*WDFRM)*JHAT;
   DER.B2=(DCD-RHO*WDCD)*JHAT;
   DER.B3=(I0-RHO*WI0)*JHAT;
   DER.B4=(DISTA-RHO*WDSTA)*JHAT;
   DER.B5=(DISTSJ-RHO*WDSTSJ)*JHAT;
   DER.RHO=((RHO*YL + B0*(1-RHO) + B1*(DFRM-RHO*WDFRM) +
       B2*(DCD-RHO*WDCD) + B3*(I0-RHO*WI0) +
       B4*(DISTA-RHO*WDSTA) + B5*(DISTSJ-RHO*WDSTSJ) - Y)*
```

```
        ((A2*(LMIN**2) - A1*(LMAX**2))/2 +
        (A1*LMAX/(D1/LMAX -RHO) -
         A2*ABS(LMIN)/(D2/ABS(LMIN)+RHO))) +
         YL - (B0 + B1*WDFRM + B2*WDCD + B3*WI0 +
         B4*WDSTA + B5*WDSTSJ))*JHAT;
RUN;
```

produced SAS output that includes

```
Source                   DF  Sum of Squares       Mean Square

Regression                7    389.94606095       55.70658014
Residual                 66     60.56332163        0.91762609
Uncorrected Total        73    450.50938258
(Corrected Total)        72    182.42615368

Parameter     Estimate      Asymptotic              Asymptotic 95 %
                            Std. Error           Confidence Interval
                                                 Lower           Upper
      RHO    0.357490147 0.22774111236 -0.0972102980  0.8121905910
       B0   -1.141608589 0.73453032553 -2.6081477755  0.3249305965
       B1   -0.082948650 0.02825544350 -0.1393625427 -0.0265347569
       B2    0.025857041 0.01253733257  0.0008254133  0.0508886684
       B3    1.556554960 0.33067688945  0.8963363145  2.2167736047
       B4    0.233729560 0.13238672788 -0.0305894442  0.4980485640
       B5    0.308284166 0.10210935702  0.1044159683  0.5121523629
```

This solution required PROC NLIN to execute 241 iterations to attain convergence. The value $\hat{\rho} \approx 0.35749$ also indicates that weak to moderate, positive spatial autocorrelation is present in the residuals of Example PR-4.3. But it still is not significant, perhaps implying that yet another spatial statistical model specification is needed. The final estimates of the parameters do not differ greatly from their AR counterparts (recall the aforementioned relationship between the intercept term, b_0, in these two models). Significance of coefficients has changed, primarily because the AR standard errors are far more complicated to calculate than are their SAR counterparts. Across all three of these models, coefficients for both DFRM and I0 remain significant. Hence, there is strong evidence to indicate that at least these two variables are related to the density of sugar cane production in the population.

11.5. A Repeated Regressions Interpretation of the Spatial Statistical Analysis

Because of the smooth curvature exhibited by the normalizing constant and its operating as a weighting in a regression equation, a spatial statistical analysis may be viewed as an execution of the following simple steps;

Step 1: select some feasible value for $\hat{\rho}_{\tau=0}$, say 0, and then estimate equation (11.1) using a regular regression technique (i.e., begin with an OLS solution);

Step 2: at iteration τ, increment $\hat{\rho}_\tau$ by some given amount δ, and then again estimate equation (11.1) using a regular regression technique, choosing the increment δ so that the SSE decreases;

Step 3: repeat Step 2 until the SSE reaches a minimum.

This procedure emphasizes that a spatial statistical analysis is equivalent to OLS for which a variable transformation is introduced. In this case, the variable transformation is one that filters out latent spatial autocorrelation, removing data information redundancies (analogous to removing multicollinearity effects with PCA). Finding the best transformation involves computing numerous regression estimates, a source of numerical intensity. One convenience of PROC NLIN is that it automatically searches for the optimal transformation, with each of its iterations actually being one of these regression analyses.

EXAMPLE PR-11.7. Sequentially estimate AR regression equations having the selected values of $\hat{\rho}$ equal to $0.1\times k$, k=0, 1, ..., 6, for Example PR-11.5.

Using the SAS code

```
REG1=LOG(1-.1*WEIGNVAL);
REG2=LOG(1-.2*WEIGNVAL);
REG3=LOG(1-.3*WEIGNVAL);
REG4=LOG(1-.4*WEIGNVAL);
REG5=LOG(1-.5*WEIGNVAL);
REG6=LOG(1-.6*WEIGNVAL);
⋮
PROC MEANS DATA=WEIGEN NOPRINT;
   VAR REG1-REG6;
   OUTPUT OUT=SAREGOUT SUM=SREG1-SREG6;
RUN;
⋮
Y1=(Y-.1*YL)/EXP(SREG1/73); DFRM1=DFRM/EXP(SREG1/73);
   DCD1=DCD/EXP(SREG1/73); I01=I0/EXP(SREG1/73);
   DISTA1=DISTA/EXP(SREG1/73); DISTSJ1=DISTSJ/EXP(SREG1/73);
Y2=(Y-.2*YL)/EXP(SREG2/73); DFRM2=DFRM/EXP(SREG2/73);
   DCD2=DCD/EXP(SREG2/73); I02=I0/EXP(SREG2/73);
   DISTA2=DISTA/EXP(SREG2/73); DISTSJ2=DISTSJ/EXP(SREG2/73);
Y3=(Y-.3*YL)/EXP(SREG3/73); DFRM3=DFRM/EXP(SREG3/73);
   DCD3=DCD/EXP(SREG3/73); I03=I0/EXP(SREG3/73);
   DISTA3=DISTA/EXP(SREG3/73); DISTSJ3=DISTSJ/EXP(SREG3/73);
Y4=(Y-.4*YL)/EXP(SREG4/74); DFRM4=DFRM/EXP(SREG4/74);
   DCD4=DCD/EXP(SREG4/74); I04=I0/EXP(SREG4/74);
   DISTA4=DISTA/EXP(SREG4/74); DISTSJ4=DISTSJ/EXP(SREG4/74);
Y5=(Y-.5*YL)/EXP(SREG5/75); DFRM5=DFRM/EXP(SREG5/75);
```

```
   DCD5=DCD/EXP(SREG5/75); I05=I0/EXP(SREG5/75);
   DISTA5=DISTA/EXP(SREG5/75); DISTSJ5=DISTSJ/EXP(SREG5/75);
Y6=(Y-.6*YL)/EXP(SREG6/76); DFRM6=DFRM/EXP(SREG6/76);
   DCD6=DCD/EXP(SREG6/76); I06=I0/EXP(SREG6/76);
   DISTA6=DISTA/EXP(SREG6/76); DISTSJ6=DISTSJ/EXP(SREG6/76);
⋮
PROC REG;
   MODEL Y = DFRM DCD I0 DISTA DISTSJ;
   MODEL Y1 = DFRM1 DCD1 I01 DISTA1 DISTSJ1;
   MODEL Y2 = DFRM2 DCD2 I02 DISTA2 DISTSJ2;
   MODEL Y3 = DFRM3 DCD3 I03 DISTA3 DISTSJ3;
   MODEL Y4 = DFRM4 DCD4 I04 DISTA4 DISTSJ4;
   MODEL Y5 = DFRM5 DCD5 I05 DISTA5 DISTSJ5;
   MODEL Y6 = DFRM6 DCD6 I06 DISTA6 DISTSJ6;
RUN;
```

produced SAS output that includes

Source	DF	Squares	Square	F Value	Prob>F
Model	5	114.35578	22.87116	23.816	0.0001
Error	67	64.34119	0.96032		
C Total	72	178.69697			
Model	5	101.56289	20.31258	21.643	0.0001
Error	67	62.88102	0.93852		
C Total	72	164.44392			
Model	5	90.30807	18.06161	19.509	0.0001
Error	67	62.02935	0.92581		
C Total	72	152.33742			
Model	5	80.49893	16.09979	17.451	0.0001
Error	67	61.81283	0.92258		
C Total	72	142.31176			
Model	5	72.03945	14.40789	15.506	0.0001
Error	67	62.25465	0.92917		
C Total	72	134.29410			
Model	5	64.91381	12.98276	13.710	0.0001
Error	67	63.44421	0.94693		
C Total	72	128.35802			
Model	5	59.14676	11.82935	12.095	0.0001
Error	67	65.52653	0.97801		
C Total	72	124.67328			

These sequential SSEs indicate that $\hat{\rho}$ is near 0.3, which the results reported in Examples PR-11.5 and PR-11.6 confirm. This output also helps underscore the danger of relying upon numerical rather than analytical derivatives. Replacing METHOD=MARQUARDT in Example PR-11.5 with METHOD=DUD and initiating PROC NLIN with estimates obtained using the analytical derivatives, produced this SAS output:

```
Source                    DF Sum of Squares      Mean Square

Regression                 5   403.80616558      80.76123312
Residual                  68    63.17174771       0.92899629
Uncorrected Total         73   466.97791329
(Corrected Total)         72   189.09480661

Parameter      Estimate     Asymptotic               Asymptotic 95 %
                            Std. Error            Confidence Interval
                                                  Lower           Upper
         RHO   0.530070575 0.00000000000 0.53007057511  0.5300705751
          B0  -0.692093694 0.00000000000 -.69209369445 -0.6920936944
          B1  -0.075776488 0.01771257679 -.11112144747 -0.0404315290
          B2   0.021811469 0.01057260082 0.00071413471  0.0429088023
          B3   1.363215214 0.30624980120 0.75210216948  1.9743282587
          B4   0.081534624 0.01618514001 0.04923762278  0.1138316251
          B5   0.076860849 0.11320381296 -.14903424196  0.3027559406
```

The most deviant estimate here is for $\hat{\rho}$, which has increased by nearly 100%. Given the preceding sequential regressions analysis, this value of 0.53007 clearly is incorrect. The principal culprit here stems from $\frac{\partial(Y^{*}JHAT)}{\partial RHO}$, which renders

$$\frac{(\partial Y)^{*}JHAT}{\partial RHO} + \frac{(\partial JHAT)^{*}Y}{\partial RHO} = [\partial(WY + X\beta)]^{*}\frac{JHAT}{\partial RHO} + (\partial JHAT)^{*}\frac{(WY + X\beta)}{\partial RHO} .$$

The estimation procedure requires the derivative to be the first term in the right-hand side of this equation, namely (∂Y)*JHAT/∂RHO, whereas the numerical derivatives are being calculated for the entire right-hand side of this equation. Subtracting (∂Y)*JHAT/∂RHO from both sides of this equation is why the term (- Y) appears in the analytical derivative.

Finally, Anselin and Griffith (1988) demonstrate and argue that spatial dependency effects can obscure the true meaning of standard regression diagnostics. In keeping with the approach employed in weighted least squares regression analysis and given the foregoing discussion,

> *RULE OF THUMB 11.4*: For more meaningful results, diagnostics should be performed on residuals generated by regressing the filtered response variable, whose values have been corrected for the presence of spatial autocorrelation, on the set of predictor variables.

The spatial analyst should be mindful of the approximate nature of at least some of these diagnostic results. For example, Cliff and Ord (1981, p. 240) note that "... standard tests for autocorrelation cease to be valid when the model contains an autoregressive component," with "the greater complexity of the estimation procedures for spatial models [being] such that no satisfactory technique has been developed to handle the problem to date."

EXAMPLE PR-11.8. Calculate regression diagnostics for the AR model estimated in Example PR-11.5.

Using the SAS statement

```
TY = Y-0.31373*WY1/RSUM;
```

and then substituting TY for SQDSGR in a replication of Examples PR-4.7 and PR-4.8, produced SAS output shown on the following page.

Since the matrix **X** has not changed, the multicollinearity diagnostics remain the same. Improvement has been achieved with both the homogeneity of variance and the Shapiro-Wilk test statistics. Little has changed in terms of the homogeneity of variance plot, or outlier and influential points diagnostics, either. Of course, those based solely on the **X** matrix are exactly the same. Others, such as Cook's D, have shrunk (e.g., Aguada's value decreased from 0.264 to 0.254). Finally, the MC value for these residuals has been reduced by nearly 50%, decreasing to $0.35949/4.739726 \approx 0.07585$. In other words, once spatial autocorrelation has been accounted for, behavior of regression residuals is even closer to that required by the model assumptions. Differences exposed here are attributable to the possible presence of weak to moderate spatial autocorrelation.

11.6. Problems

P-11.1. Given the *Soil Data from the Province of Murcia, Spain* (Table 18.1 in Andrews and Herzberg, 1985) data set, estimate the nature and degree of spatial autocorrelation present in the silt content measures (this is **Y**) using a constant mean model specification (a vector of ones is **X**). Repeat this exercise for the clay content measures.

P-11.2. Given the *Insurance Availability in Chicago* (Table 67.1 in Andrews and Herzberg, 1985) data set, predict the voluntary market activity rate from the socioeconomic variables of racial composition and income, supplemented with the fire and theft rates and the housing age variable, using an autoregressive response model specification.

```
              Test of First and Second Moment Specification
     DF:    19    Chisq Value:       20.0307    Prob>Chisq:     0.3927
```

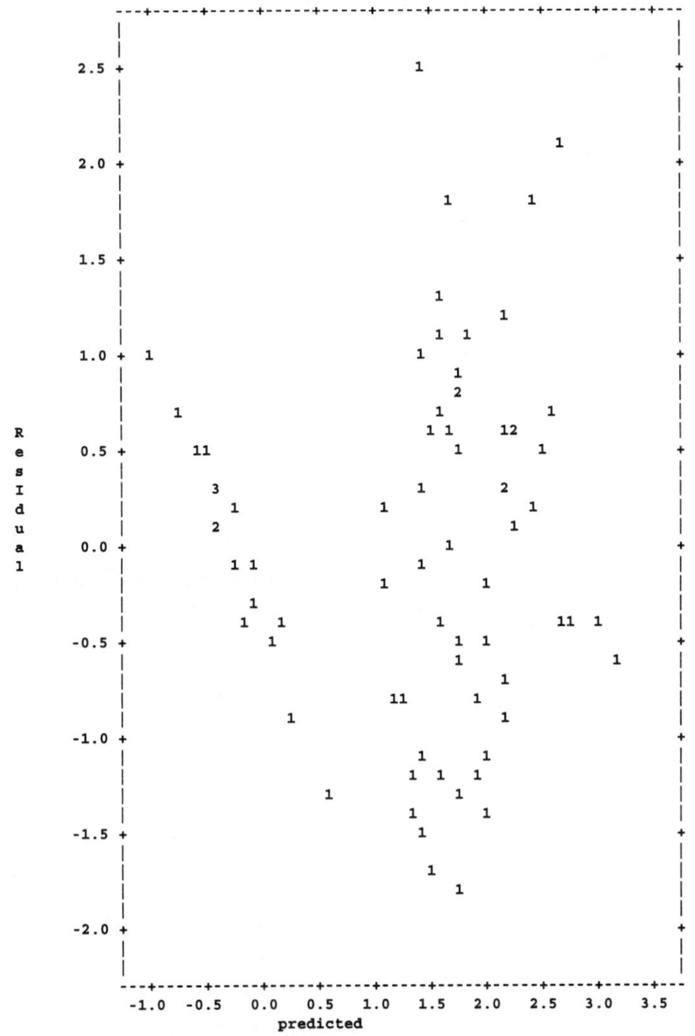


```
              W:Normal    0.980195   Pr<W            0.6283
```

11.7. References

Andrews, D. and H. Herzberg. 1985. *Data: A Collection of Problems from Many Fields for the Student and Research Worker.* Berlin: Springer-Verlag.

Anselin, L. and R. Florax. eds. 1995. *New Directions in Spatial Econometrics.* Berlin: Springer-Verlag.

Anselin, L. 1988. *Spatial Econometrics.* Dordrecht: Kluwer.

Anselin, L. and D. Griffith. 1988. "Do Spatial Effects Really Matter in Regression Analysis?" *Papers of the Regional Science Association* 65: 11–34.

Cliff, A. and J. Ord. 1981. *Spatial Processes.* London: Pion.

Cordy, C. and D. Griffith. 1993. "Efficiency of Least Squares Estimators in the Presence of Spatial Autocorrelation." *Communications in Statistics* 22 B: 1161–79.

Griffith, D. 1988. "Estimating Spatial Autoregressive Model Parameters with Commercial Statistical Packages." *Geographical Analysis* 20: 176–86.

—. 1993. *Spatial Regression Analysis on the PC: Spatial Statistics Using SAS.* Washington, D.C.: Association of American Geographers.

—. 1996. "Some Guidelines for Specifying the Geographic Weights Matrix Contained in Spatial Statistical Models." In *Practical Handbook of Spatial Statistics*, edited by S. Arlinghaus and D. Griffith, 65–82. Boca Raton, Fla.: CRC Press.

Griffith, D. and A. Sone. 1995. "Trade-offs Associated with Normalizing Constant Computational Simplifications for Estimating Spatial Statistical Models." *Journal of Statistical Computation and Simulation* 51: 165–83.

CHAPTER 12

SPECIAL NONLINEAR REGRESSION APPLICATIONS IN SPATIAL ANALYSIS

Chapter Preview. In a sense, here we combine the extensions of Chapter 10 (nonlinear regression) with Chapter 11 (spatially adjusted regression models) to form a chapter in which we introduce (very briefly) a set of special, nonlinear regression models used in explicitly spatial settings.

The first special model, logistic regression, has experienced explosive increase in use over the past decade. Entire sessions at professional meetings are devoted to its applications in transportation, migration, housing studies, and many other topics. We first introduced the term logistic regression in Chapter 7, in the context of discriminant function analysis (DFA). Here, we develop similarities between these two techniques in some detail. Initially dealing with the case in which the dependent variable is binary in nature (often used to analyze the history of some event taking place, such as a decision to move or stay, be employed or unemployed, and so on), we then extent our treatment to the polychotomous case in which the dependent variable can be assigned to a number of categories (style of house, choice of destinations, and so on).

We conclude with a brief introduction to Poisson regression. This technique is still not widely employed, although some researchers have argued that it is the most appropriate model in many more applications than analysts think (Flowerdew and Aitkin, 1982). Poisson regression is well suited to analyses of the distribution of rare discrete events in space. For example, in a matrix of origins and destinations (such as a county-level migration matrix for the United States), most of the cells for any year will contain 0, while proportionately few will have entries with large frequencies (for example, the row-column intersections that pair northeast counties with southeast counties). Such situations are described well by the Poisson distribution.

12.0. Introduction

Throughout this text, discussion has focused almost exclusively on either the univariate or multivariate normal distribution. Elementary statistics courses often stress that other distributions can prove very useful as well (Griffith and Amrhein, 1991, Chapter 7). For example, in general the uniform distribution supports many sampling notions. As has been emphasized in preceding chapters, the normal distribution generally supplies the basis for drawing inferences. The binomial distribution and its polychotomous generalization to the multinomial distribution support classification analysis (Chapter 10). And the Poisson distribution supports analysis of events that are both discrete and rare, and often counts (CNT), such as points on a surface in point pattern analysis (Griffith and Amrhein, 1991, Chapter 5).

> *CNT Question #1*: What is the rule of correspondence between variable Y and a battery of X variables that enables Y to be predicted from these Xs when Y is measured as counts or represents classification schemes?[1]

Answers to this question are rendered by generalized linear models.

12.1. The Binomial Distribution and Logistic Regression

Binary response variables are treated first by relating them to the DFA problem (which is shown to have model coefficients proportional to those appearing in its regression equation counterpart; see §7.4) and second by estimating an exponential function using nonlinear least squares (NLS) (Chapter 10). This exponential function approach supplies a link between the NLS results, which assumed a normally distributed error term, and logistic regression, which assumes a binomially distributed error term. The problem statement associated with logistic regression may be stated as:

> What is the relationship between a response probability and a set of predictor variables?

Expressing this problem in terms of the multiple linear regression equation (4.2),

$$< LN\left[\frac{(1 - p)}{p}\right] > \; = \mathbf{X}\beta + \epsilon , \tag{12.1}$$

where the vector $< LN\left[\frac{(1 - p)}{p}\right] >$ is composed of the logarithm of odds ratios and p is the binomial probability of the event under study occurring.

Taking the antilogarithm of both sides of equation (12.1) yields individual terms of the form

$$\frac{(1 - p_i)}{p_i} = EXP\,(\mathbf{X}_i^T\beta + \epsilon_i) .$$

Multiplying both sides of this equation by $(1-p_i)$ and then solving for p_i yields

[1]Multiple classification schemes are described either as polytomous or polychotomous. Although technically polytomous is the correct Greek word, we prefer to use polychotomous to maintain a parallel descriptor with dichotomous.

$$p_i = \frac{1}{[1 + EXP(\mathbf{X}_i^T\beta + \epsilon_i)]} , \qquad (12.2)$$

which is the logistic regression equation specification. The probability of the event under study not occurring is given by $(1\text{-}p_i)$, which equals

$$\frac{EXP(\mathbf{X}_i^T\beta + \epsilon_i)}{[1 + EXP(\mathbf{X}_i^T\beta + \epsilon_i)]} . \qquad (12.3)$$

Equation (12.2) is the specification employed in Example PR-10.1 to illustrate NLS estimation.

Several properties of equation (12.2) are worth mentioning at this point. First, as is obvious from its algebraic derivation, the error structure is not additive, as was the case for NLS estimation in Example PR-10.1. Second, the linearized version of equation (12.1) can be estimated using weighted least squares (WLS), where the unequal variance is given by

$$VAR(p_i) = \frac{1}{[n_i p_i (1 - p_i)]} .$$

This result is familiar, with the denominator being the variance of a binomial random variable (Griffith and Amrhein, 1991, p. 173), which is equivalent to n_i independent realizations of a Bernoulli random variable (Burt and Barber, 1996, p. 180). One caveat associated with using WLS is that none of the p_i values can be either 0 or 1.[2] Otherwise, maximum likelihood (ML) estimation should be used. Third, as a result, analyses such as Example PR-10.1, where individual observations are being classified and hence $n_i = 1$, require ML estimation.

EXAMPLE PR-12.1 (Example PR-10.1 continued) Estimate the probability of areal unit membership in the coastal lowlands/interior highlands regionalization of Puerto Rico using logistic regression, based upon the variables DFRM, DSGR, and DFAM.

Using the SAS procedure

```
PROC LOGISTIC;
   MODEL LND = DFRM DSGR DFAM;
   OUTPUT OUT=LOGSTOUT P=PROBHAT;
RUN;
```

---

[2]When this occurs, standard practice is to add a constant (usually 0.5). This modification should be used with caution, as it can overwhelm the subsequent statistical analysis.

produced this SAS output:

```
                    Criteria for Assessing Model Fit
                                 Intercept
                  Intercept         and
  Criterion          Only        Covariates    Chi-Square for Covariates

  -2 LOG L          99.536         53.272         46.264 with 3 DF (p=0.0001)

                     Analysis of Maximum Likelihood Estimates
                  Parameter  Standard      Wald          Pr >       Standardized      Odds
Variable    DF    Estimate     Error    Chi-Square   Chi-Square       Estimate        Ratio

INTERCPT    1       3.6085    0.9734      13.7436       0.0002             .          36.912
DFRM        1      -0.3186    0.0901      12.4969       0.0004       -0.898588         0.727
DSGR        1       0.1427    0.0564       6.3967       0.0114        0.552312         1.153
DFAM        1      -0.5464    0.3003       3.3108       0.0688       -0.441493         0.579

          Association of Predicted Probabilities and Observed Res
                     Concordant = 91.2%          Somers' D = 0.824
                     Discordant =  8.8%          Gamma     = 0.825
                     Tied       =  0.1%          Tau-a     = 0.408
                     (1302 pairs)                c         = 0.912
```

The overall model test statistic, related to the F-ratio in linear regression analysis, is the chi square statistic, which equals $\{-2LN(\text{Likelihood}_{\text{intercept only}}) - [-2LN(\text{Likelihood}_{\text{full model}})]\}$, and accordingly has degrees of freedom (df) equal to the number of variable coefficients only. Because the assumption of normality in linear regression causes the residual sum-of-squares for a fitted regression model to be proportional to the log-likelihood for that model, the computational results reported here can be used to calculate the following pseudo-R^2 value (this quantity typically underestimates the percent of variance accounted for, in part reflecting measurement scale and nonlinear constraints that are present):

$$\text{pseudo-}R^2 = \frac{[-2LN(\text{Likelihood}_{\text{intercept only}})] - [-2LN(\text{Likelihood}_{\text{full model}})]}{[-2LN(\text{Likelihood}_{\text{intercept only}})]},$$

which equals $\frac{46.264}{99.536} \approx 0.4648$ for this analysis. The point biserial correlation between Y and $\hat{p}$ equals 0.36130, whose squared counterpart equals 0.1305. This last number, 0.1305, is more directly comparable to the proportion of variance accounted for value of 0.4918 for OLS and 1 minus the relative error sum-of-squares value 0.5362 for NLS in Example 10.1 . The squared correlation value yielded by logistic regression often is lower than its OLS or NLS counterparts mainly because it is so highly constrained (i.e., all values must lie between 0 and 1). As was found in Example PR-10.1, the coefficient for DFAM is not significantly different from 0 at the 5% level. The variable coefficient estimates obtained with the logistic regression procedure are very similar to those obtained with NLIN, except for the intercept term [which is *EXP*(3.6085) = 36.9106].

These differences can be attributed to differences in the treatment of error, underlying assumptions, and non-linear estimation techniques.

Predictive effectiveness of the model is assessed through measures of ordinal association between the binary response variable and the estimated conditional probabilities of whether or not an areal unit belongs to the interior highlands region. All possible pairs of observations having different responses are considered (i.e., $n_1 \times n_2 = 42 \times 31 = 1302$ responses, where there are n_1 areal units in the coastal lowlands and n_2 areal units in the interior highlands). Those pairs with different values for Y are called concordant if the observation for which Y = 1 has a larger estimated conditional probability than does the observation for which Y = 0. If the observation for which Y = 1 has a smaller estimated conditional probability than does the observation for which Y = 0, then the pair is marked discordant. Otherwise the pair is labeled a tie. Three of the four measures of association are based upon differences between the number of concordant and discordant pairs. The (Goodman-Kruskal) gamma statistic suggests that roughly 82.5% fewer prediction errors are made by using the model, rather than chance alone, to predict regional membership. One drawback of this statistic is that it tends to exaggerate the strength of this relationship. The Somers' D statistic gives a measure of how well the estimated conditional probabilities can be predicted from the binary response variable. The (Kendall's) tau-a statistic is the ratio of the difference between concordant and discordant pairs in the numerator, and the maximum number of pairs in the denominator [i.e., $\frac{n(n-1)}{2}$; Griffith and Amrhein, 1991, p. 153]. The distinction between this statistic and the gamma statistic is that gamma has the total number of pairs with different responses in its denominator whereas the tau-a has the total number of possible pairs in its denominator [i.e., $\frac{n(n-1)}{2}$]. Hence, here tau-a is roughly $\frac{(1302-1)}{2628} \times \gamma \approx 0.4951 \times$ gamma. Finally, the statistic denoted with "c" is the adjusted percentage of subset pairs analyzed for con-/discordance that are labeled concordant, with an adjustment being made for the presence of ties. Consequently, given the reported statistics, a respectable model fit is attained here.

Output here that is analogous to the selected estimated conditional probabilities reported for Example PR-10.1 includes

San Lorenzo: $1 - 0.02738 \approx 0.9726$
Aguada: $1 - 0.46716 \approx 0.5328$
Yabucoa: $1 - 0.22307 \approx 0.7769$

This first probability estimate is consistent with those obtained using OLS, NLS, and DFA. This second probability estimate is most similar to that computed using OLS (recall the relationship between the multiple linear and logistic regression models). This third probability estimate is similar to that obtained using NLS and DFA and is a substantial improvement over the negative OLS estimate.

Exactly the same output[3] can be obtained by using the more general procedure

```
PROC CATMOD;
   DIRECT DFRM DSGR DFAM;
   MODEL LND = DFRM DSGR DFAM;
RUN;
```

The DIRECT statement instructs SAS that the variables are measured at the interval/ratio rather than the nominal scale. The advantage of CATMOD is that it enables numerous log-linear models to be implemented, including the polychotomous extension of logistic regression.

12.2. The Binomial Distribution and Measurement Error

Often a binary response variable is the outcome of a change of measurement scale, moving from a more detailed, finer ratio to a coarser nominal measurement. For example, the rural-urban classification variable (R-U) in the Puerto Rico data set is defined as an indicator variable whose value equals 1 if the 1980 ratio of urban to total municipio population is greater than 50% and 0 if this ratio lies in the interval [0, 0.5]. Although collapsing a ratio-level measurement may prove convenient in some analyses, it involves a loss of attribute information (Griffith and Amrhein, 1991, pp. 36–42). However, it is representative of many dichotomous variables employed in research that do nothing more than serve as proxies or surrogates for underlying constructs for which existing inadequate instruments prevent precise measurement on the interval/ratio scale. The notion of an urban municipio is a case in point here. When the more detailed measurement is available, it constitutes an empirical probability and becomes the response variable Y of the logistic regression equation. Comparison of results for these two different levels of measurement (nominal and ratio) is enlightening.

First consider the dichotomous classification of each of the 73 municipios as being either urban or rural (i.e., nonurban). The point biserial correlation coefficient for these two variables equals 0.868. The binomial probability, p, refers to the probability of a municipio being classified as urban.

EXAMPLE PR-12.2. Estimate the probability of areal unit membership in the urban grouping of Puerto Rico, using the binary classification variable RU, based upon the agricultural density variables of cultivated land (DCD), sugar cane production (DSGR), and milk (DMLK).

Using the SAS procedure

---

[3]When the response variable Y is binary, the parameter estimates from CATMOD and LOGISTIC are the same, except it is possible the signs may be reversed. This reversal arises because CATMOD models the probability of the first sorted value of the dependent variable, whereas LOGISTIC models the probability of the maximum value.

```
PROC LOGISTIC DATA=STEP1;
   MODEL RU = DCD DMLK DSGR;
   OUTPUT OUT=OUT1 P=PROBHAT1;
RUN;
```

produced SAS output that includes

```
                    Criteria for Assessing Model Fit
                                 Intercept
                 Intercept          and
  Criterion        Only         Covariates    Chi-Square for Covariates

  -2 LOG L        98.095          59.894       38.201 with 3 DF (p=0.0001)

                  Analysis of Maximum Likelihood Estimates
                Parameter  Standard       Wald          Pr >      Standardized   Odds
Variable   DF    Estimate     Error   Chi-Square   Chi-Square      Estimate     Ratio

INTERCPT   1     -3.5125     1.2639       7.7235       0.0055           .        0.030
DCD        1      0.1576     0.0404      15.1959       0.0001       1.038789    1.171
DMLK       1     -0.0335     0.0122       7.5612       0.0060      -0.921342    0.967
DSGR       1     -0.0905     0.0533       2.8866       0.0893      -0.350435    0.913

          Association of Predicted Probabilities and Observed Responses
                  Concordant = 89.0%             Somers' D = 0.781
                  Discordant = 10.9%             Gamma     = 0.782
                  Tied       =  0.1%             Tau-a     = 0.379
                  (1276 pairs)                   c         = 0.891
```

The pseudo-R^2 here is 0.3894. The point biserial correlation between Y and $\hat{p}$ equals 0.66920, whose squared counterpart equals 0.4478, a value more directly comparable with the R^2 value rendered by OLS. According to the gamma statistic, this logistic equation considerably enhances one's ability to predict membership in the urban and rural groups. Of particular note is the suggestion (using chi square probabilities reported by SAS) based on these results that DSGR may or may not help predict the conditional probability of membership in the urban and rural groups of municipios on the island.

Next, consider the actual percentage of urban population residing in each of the 73 municipios (see Appendix E). The dependent variable appearing in the SAS MODEL statement becomes, for each municipio, (1980 urban population)/(1980 total population).

EXAMPLE PR-12.3 (Example PR-12.2 continued) Estimate the probability of areal unit membership in the urban grouping of Puerto Rico, using the percentage of urban population, based upon DCD, DMLK, and DSGR.

Using the SAS procedure

```
PROC LOGISTIC DATA=STEP1;
   MODEL URB80/POP80 = DCD DMLK DSGR;
```

```
    OUTPUT OUT=OUT2 P=PROBHAT2;
RUN;
```

produced SAS output that includes

```
                    Criteria for Assessing Model Fit
                                  Intercept
                    Intercept        and
   Criterion          Only       Covariates    Chi-Square for Covariates

   -2 LOG L        4044877.2      3262201.2    782675.97 with 3 DF (p=0.0001)

                     Analysis of Maximum Likelihood Estimates
                    Parameter   Standard      Wald          Pr >      Standardized   Odds
  Variable   DF      Estimate      Error    Chi-Square   Chi-Square     Estimate     Ratio

INTERCPT   1        3.2400   0.00414   612047.862       0.0001          .          25.535
DCD        1       -0.0900   0.000126  511029.671       0.0001     -0.754289     0.914
DMLK       1       0.00588   0.00003   38878.2437       0.0001      0.141617     1.006
DSGR       1        0.0236   0.000237    9971.6968       0.0001      0.077403     1.024

       Association of Predicted Probabilities and Observed Responses
                     Concordant = 78.0%          Somers' D = 0.571
                     Discordant = 20.9%          Gamma     = 0.577
                     Tied       =  1.1%          Tau-a     = 0.253
                     (2.247973E12 pairs)         c         = 0.786
```

The pseudo-R^2 changes to 0.1935. Now the product moment correlation between Y and $\hat{p}$ has decreased to 0.60154, whose squared counterpart equals 0.3619. According to the Gamma statistic, the ability of this logistic equation to predict membership in the urban and rural groups has diminished by about a third. On the other hand, apparently DSGR really does help predict the conditional probability of membership in the urban and rural groups of municipios on the island (note the chi square probabilities reported by SAS).

> *RULE OF THUMB 12.1*: As the denominator of the Y variable increases, significance levels will increase, regardless of information uncovered by the model, and hence significance levels should not be relied upon as heavily in cases of large denominators when interpreting results.

The estimated conditional probabilities also disagree considerably. The correlation between the two sets is 0.567. Hatillo displays the closest agreement, while Rincon displays the greatest discrepancy. As might be anticipated, then, that dichotomizing the urban propensity variable results in a loss of attribute information. It also results in a concomitant loss of statistical power, especially in terms of DSGR. Comparing the estimated coefficients reveals that they are roughly of the same order of magnitude, but in each case the sign is reversed. This reversal of signs is due to PROC LOGISTIC estimating (1-p) when the binary dependent variable Y takes on the values of 0 and 1, where the probability p refers to a 1 occurring, while the corresponding rates version of Y has p, itself, estimated. Moreover, equation (12.3) is altered as follows:

$$\left\{\frac{EXP[-(\mathbf{X}_i^T\beta + \epsilon_i)]}{EXP[-(\mathbf{X}_i^T\beta + \epsilon_i)]}\right\}\left\{\frac{EXP[(\mathbf{X}_i^T\beta + \epsilon_i)]}{EXP[1 + (\mathbf{X}_i^T\beta + \epsilon_i)]}\right\} = \frac{1}{1 + EXP[-(\mathbf{X}_i^T\beta + \epsilon_i)]},$$

which is of the same form as equation (12.2). Cross-classification tabulations for these results may be summarized as:

		binary response prediction		ratio response prediction	
		0	1	0	1
empirical	0	38	6	26	18
classification	1	6	23	3	26

The phi correlation coefficient for this left-hand tabulation equals 0.657, whereas that for this right-hand tabulation equals 0.487.

12.3. The Multinomial Distribution and Polychotomous Logistic Regression

When the response variable Y is categorical, representing three or more mutually exclusive and collectively exhaustive, unordered categories, a set of binary logistic regressions can be formed by arbitrarily selecting any one of the categories to construct each denominator of the odds ratio. Then the parameters of each equation are estimated using ML. An additional, critical assumption must be invoked here, namely independence from irrelevant alternatives. This contention maintains that the ratio of alternative probabilities of any two alternatives for a specific areal unit is not systematically influenced by any other alternatives.

The procedure begins by generating the set of binary indicator variables $\{Y_k; k=1, 2, ..., K\}$ from the K-regions polychotomous response variable and then arbitrarily choosing one to omit, say Y_K. The set of odds ratios to be modeled may be summarized as $\{LN[\Pr(Y_k=1|\mathbf{X})/\Pr(Y_K=1|\mathbf{X})]; k=1, 2, ..., K-1\}$. Following the same algebraic logic to derive the logistic regression model, the K-1 conditional probabilities are given by

$$\Pr(Y_{k,i} = 1 | \mathbf{X}_i) = \frac{EXP\,[g_k(\mathbf{X}_i^T\beta + \epsilon_{k,i})]}{\{1 + \sum_{k=1}^{K-1} EXP\,[g_k(\mathbf{X}_i^T\beta + \epsilon_{k,i})]\}}, \tag{12.4}$$

with the conditional probability of the arbitrarily selected K-th region being given by

$$\Pr(Y_{k,i} = 1 | \mathbf{X}_i) = \frac{1}{\{1 + \sum_{k=1}^{K-1} EXP\,[g_k(\mathbf{X}_i^T\beta + \epsilon_{k,i})]\}}.$$

Thus, the conditional log-likelihood function for a sample of n independent areal units may be stated as follows:

$$LN\,[L(\beta)] = \sum_{i=1}^{n}\left(\sum_{k=1}^{K-1} Y_{k,i}\,g_k(X_i^t\beta) - LN\left\{1 + \sum_{k=1}^{K-1} EXP\,[g_k(X_i^T\beta)]\right\}\right) .$$

As before, the ML estimators are obtained by setting the partial derivatives of this equation equal to 0 and solving for the parameters. The results obtained here frequently are more interpretable than their discriminant function counterparts.

EXAMPLE PR-12.4. Estimate the probability of areal unit membership in each of the five agricultural administrative regions of Puerto Rico, based upon the agricultural density variables of cultivated land (DCD), sugar cane production (DSGR), and milk (DMLK).

Using the SAS procedure

```
PROC CATMOD;
   DIRECT DCD DSGR DMLK;
   MODEL ADM = DCD DSGR DMLK/PRED;
RUN;
```

produced the SAS output on the next page.

Model assessment is more difficult here than in the binary response case, because four, and by default five (one for each agricultural administrative region), equations are fitted. Certainly at the 10% level of significance, the three density variables of cultivated land, sugar cane production, and milk production are significant predictors of regional membership. DCD is a significant predictor for membership in the San Juan region (parameter 5 in the SAS output) when compared with membership in the Caguas region. DSGR is a significant predictor for membership in the Mayaguez region (parameter 11), and DMLK is a significant predictor for membership in the Mayaguez and Ponce regions (parameters 15 and 16).

These logistic regression functions produce conditional probability estimates that sum to 1 and that often deviate from those produced by DFA. The closest pairwise correspondence, overall, for the two sets of five assignment probabilities is for Bayamon/Catano. The largest overall difference in these pairwise comparisons is for San Sebastian. Quite dramatic discrepancies between these

MAXIMUM-LIKELIHOOD ANALYSIS-OF-VARIANCE TABLE

Source	DF	Chi-Square	Prob
INTERCEPT	4	12.51	0.0139
DCD	4	11.28	0.0236
DSGR	4	11.40	0.0224
DMLK	4	9.47	0.0505
LIKELIHOOD RATIO	276	160.64	1.0000

ANALYSIS OF MAXIMUM-LIKELIHOOD ESTIMATES

Effect	Parameter	Estimate	Standard Error	Chi-Square	Prob
INTERCEPT	1	4.1878	1.6998	6.07	0.0138
	2	-2.3412	1.9649	1.42	0.2334
	3	0.3838	1.9108	0.04	0.8408
	4	-1.3378	2.1228	0.40	0.5286
DCD	5	-0.0977	0.0456	4.58	0.0323
	6	0.0476	0.0479	0.98	0.3211
	7	-0.00428	0.0468	0.01	0.9270
	8	0.0603	0.0497	1.47	0.2255
DSGR	9	-0.1419	0.1122	1.60	0.2061
	10	0.0383	0.0985	0.15	0.6976
	11	0.2128	0.0940	5.13	0.0236
	12	0.1319	0.0928	2.02	0.1556
DMLK	13	-0.00842	0.0119	0.50	0.4786
	14	0.00473	0.00758	0.39	0.5328
	15	-0.0937	0.0397	5.56	0.0184
	16	-0.1506	0.0660	5.21	0.022

pairwise comparisons occur within each region, too. An overall summary of differences is provided by the following cross-classification tabulations:

		from discriminant analysis					from logistic regression				
		SJ	A	M	P	C	SJ	A	M	P	C
observed	SJ	10	1	0	1	6	14	0	4	0	1
regional	A	1	2	0	0	1	5	0	1	0	5
member-	M	0	0	7	0	2	2	2	6	3	3
ship	P	4	3	8	13	1	0	1	5	5	3
	C	3	5	1	0	3	0	2	4	7	0

The DFA equations correctly classify 35 of the areal units and misclassify 23 of the areal units into adjacent regions (suggesting the presence of spatial autocorrelation). In contrast, the logistic regression equations correctly classify 25 of the areal units and misclassify 28 of the areal units into adjacent regions (again suggesting the presence of spatial autocorrelation).

> *RULE OF THUMB 12.2*: DFA is useful when the emphasis is on regional means and variances for collections of areal units, whereas logistic regression is useful when the emphasis is on collections of items within individual areal units.

12.4. A Case Study of Children's Elevated Blood Lead Levels in Syracuse, New York

Because the goal of nonlinear statistical analysis is the same as conventional linear statistical analysis—that is, to find the best fitting and most parsimonious, yet conceptually reasonable equation to describe some phenomenon—models such as those used in logistic regression still require diagnostic treatment. These diagnostics follow the same general principles used in linear regression, except that in this nonlinear case they tend to be based upon the chi square rather than a t or normal distribution. Furthermore, visual assessment of these statistics must be relied upon in logistic regression analysis, because general statistical distribution theory of these diagnostics has yet to be developed. And, of course, the presence of spatial autocorrelation is still a concern when dealing with geo-referenced data.

> *RULE OF THUMB 12.3:* An individual areal unit value of a diagnostic statistic should be considered extreme if it appears as an outlier in the graphical display of these statistics, appearing to lie away from the balance of the other diagnostic statistics.

The preceding sections focus almost exclusively on issues of parameter and conditional probability estimation in various contexts. Attention now will be turned to application of logistic regression analysis to the important practical problem of estimating the conditional probability of elevated pediatric blood lead levels.

Johnson, McDade and Griffith (1996) studied the association between elevated blood lead levels ($\geq 10\mu g\ dL^{-1}$) and census tract characteristics for Syracuse, New York, using 1990 U.S. Bureau of the Census data aggregated into census tracts, coupled with April 1992 to March 1993 Onondaga County Health Department venous blood lead tested data for the 0–4 age cohort (Appendix F). Census tracts were included that comprise the city of Syracuse and its immediate suburban ring, with those having less than 10 elevated cases being eliminated (Map 12.1). These researchers found that the relative incidence of elevated blood lead is significantly associated with percentage of housing built before 1960 and percentage of population of Hispanic origin. In this present investigation, these two predictor variables are supplemented with population density and census tract centroid Cartesian coordinates (u_i, v_i).

The first step in this analysis involved estimation of equation (12.1) using weighted least squares (WLS). Using the SAS procedure

```
PROC REG;
   MODEL LOGY = PCTHIS DENSITY PCTH60 U V;
   WEIGHT WGT;
RUN;
```

produced SAS this output.

Analysis of Variance

Source	DF	Sum of Squares	Mean Square	F Value	Prob>F
Model	5	168.10162	33.62032	10.368	0.0001
Error	48	155.65072	3.24272		
C Total	53	323.75234			

Root MSE	1.80076	R-square	0.5192
Dep Mean	-0.06870	Adj R-sq	0.4691
C.V.	-2621.14505		

Parameter Estimates

Variable	DF	Parameter Estimate	Standard Error	T for H0: Parameter=0	Prob > |T|
INTERCEP	1	111.946416	34.58948098	3.236	0.0022
PCTHIS	1	4.549918	2.33017400	1.953	0.0567
DENSITY	1	0.000129	0.00006683	1.923	0.0604
PCTH60	1	1.969180	0.59832088	3.291	0.0019
U	1	-0.688938	0.52771886	-1.306	0.1979
V	1	-1.126725	0.39097836	-2.882	0.0059

The weighting factor, WGT, is the inverse of the in dividual rate binomial variance for each census tract. Computationally, this weight acts the same as dividing each areal unit odds ratio by its standard deviation, making all variances equal. WLS is possible to employ without adjustment here because none of the rates exactly equals 0 or 1. This output corroborates McDade's finding regarding percentage of housing stock built prior to 1960. And V being significant and U not suggests that these data display a linear (north-south) geographic trend.

The second step in this analysis involved estimation of equation (12.2) using the LOGISTIC procedure, which produced this output:

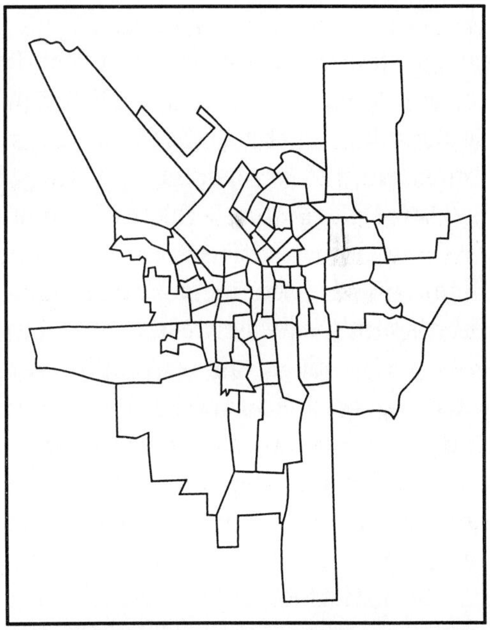

Map 12.1. 1990 census tract boundary map for Syracruse, New York.

Criterion	Intercept Only	Intercept and Covariates	Chi-Square for Covariates
-2 LOG L	3522.615	3302.735	219.881 with 5 DF (p=0.0001)

Analysis of Maximum Likelihood Estimates

Variable	DF	Parameter Estimate	Standard Error	Wald Chi-Square	Pr > Chi-Square	Standardized Estimate	Odds Ratio
INTERCPT	1	113.9	17.6788	41.4802	0.0001	.	999.000
PCTHIS	1	5.0664	1.2568	16.2500	0.0001	0.111125	158.595
DENSITY	1	0.000136	0.000036	14.2789	0.0002	0.104037	1.000
PCTH60	1	2.1132	0.3291	41.2292	0.0001	0.186493	8.275
U	1	-0.6357	0.2612	5.9229	0.0149	-0.064550	0.530
V	1	-1.1825	0.2024	34.1385	0.0001	-0.136363	0.307

Association of Predicted Probabilities and Observed Responses

Concordant = 64.9%	Somers' D = 0.334
Discordant = 31.6%	Gamma = 0.346
Tied = 3.5%	Tau-a = 0.166
(1617704 pairs)	c = 0.667

This output corroborates both of the Johnson's et al. findings and further suggests that these data relate to population density. The product moment correlation between Y and $\hat{p}$ is 0.60838, whose squared value equals 0.3701, which is in line with the gamma statistic. This result illustrates the unreliability of R^2 rendered by WLS, which is exaggerated here. Meanwhile, the pseudo-R^2 value is 0.0624, suggesting a relatively poor model fit. The estimated coefficients for these two models are very similar, differing primarily because of error structure and nonlinear estimation procedure distinctions. Also, the significance of both U and V indicates a more complex, two dimensional (north-south and east-west) spatial trend than was observed before.

One diagnostic utilized in linear regression concerns leverage, or the potential influence, areal units exert on model fit by virtue of their positions in the predictor variable mathematical space. This notion can be extended to logistic regression through the WLS version of the model. But logistic regression deviates substantially from linear regression because of the upper and lower bounds on probability p. These bounds result in leverage first increasing as the estimated probability gets further from 0.5, and then decreasing, rapidly approaching 0, as the estimated conditional probability converges on 0 or 1. Therefore, properly interpreting individual leverage values in the context of logistic regression requires knowledge about whether the estimated conditional probability is small (< 0.1) or large (> 0.9). As long as a given estimated probability falls within the interval (0.1, 0.9), its leverage value can be interpreted as before. With regard to the Syracuse blood lead level data, all estimated probabilities lie in the interval (0.1, 0.9). According to the leverage statistics, tracts #38 and #39 exhibit abundant influence potential, while tracts #53 and #61 exhibit modest influential potential. Tracts #38 and #39 are juxtaposed and are located in an area that historically has been subjected to massive industrial pollution. Tracts #43, #53, and #61 are close to the interstate highway that bisects the city.

Another conceptually appealing diagnostic statistic is the Pearson residual, which needs to be divided by √(1 - leverage) to parallel the studentized residual of linear regression. This statistic helps identify areal unit rates that are not well explained by the fitted logistic regression model. Two groups of such rates exist for the Syracuse blood lead data: (1) tract #52, which is the most aberrant and (2) tracts #28, #45, #54, #58, and #162, which are moderately anomalous. Tracts #52, #54, #58, and #162 embrace a major north-south transportation artery, historically a principal source of lead pollution.

Two useful diagnostics for detecting ill-fitted empirical rates, flagging areal units that contribute heavily to the disagreement between the data and the estimated conditional probabilities, are the deviance difference and the chi square difference. This pair of statistics is in keeping with the spirit of Cook's D statistic for linear regression (Chapters 4, 11, 12) and examines the effect that deleting all areal units with a particular covariate pattern has on the value of the parameter estimates. The deviance difference statistic reveals peculiar behavior for tracts #28, #35, #45, #52, #54, #58, and #162, with tract #52 again being the extreme. Meanwhile, the chi square difference statistic reveals idiosyncratic behavior for tracts #35, #52, #54, #58, and #162, once more with tract #52 being the extreme. These statistics tag tract #35, which encompasses several major east-west transportation arteries, historically sources of automobile emission contamination.

Evaluating the residuals obtained as the difference ($\hat{p}_i$ - p_i) discloses a mean of -0.04708, the nonlinear bias. The MC for these residuals is 0.15248, down from 0.35603 for the original empirical rates, Y, while the GR is 0.73894, which is closer to 1 than the value of 0.62689 for the empirical rates. Thus, discernible positive spatial autocorrelation appears to be present. This result is consistent with the geographic cluster of tracts #38, #39, and #52 being highlighted by the preceding diagnostic statistics. Unfortunately, as the recent work by le Cessie and van Houwelingen (1994) symbolizes, formulating a proper spatial statistical companion for the logistic regression model is virtually impossible.

12.5. Poisson Regression

One last special nonlinear regression topic to be presented in this chapter entails relatively rare events. Besides arising in geo-referenced data analysis that involves point pattern analysis, flows between areal units (see, for example Flowerdew and Aitkin, 1982) and disease incidences within areal units (see, for instance, Cressie, 1991, §7.6) should be viewed as rare events. In each case, the Poisson probability model is far more appropriate than the normal probability model. The model equation here is

$$\Pr(Y_i = n_i) = EXP \frac{(-\mu_i)\mu_i^{n_i}}{(n_i)!}$$

(Griffith and Amrhein, 1991, §7.3), where $\mu_i = \sigma_i^2$. The mean for areal unit i, as $LN(\mu_i)$, becomes the response variable, Y, of the regression equation, which is of the form

$$\langle LN(\mu_i)\rangle = \langle LN(\mathrm{N})\rangle + \mathbf{X}\beta + \epsilon .$$

Within this context the concept of rare signifies that the maximum probability lies in the interval (0.05, 0.01).

For such events with small probabilities of occurrence, logistic regression theoretically yields the same parameter estimates and standard errors as does Poisson regression.[4] In practice, using appropriately dedicated procedures to perform either Poisson or logistic regression is preferable, though, when possible. Resorting to logistic regression to implement Poisson regression tends to affect the intercept estimate only. The remaining parameter estimates and all standard errors are practically the same as those obtained with a dedicated Poisson routine.

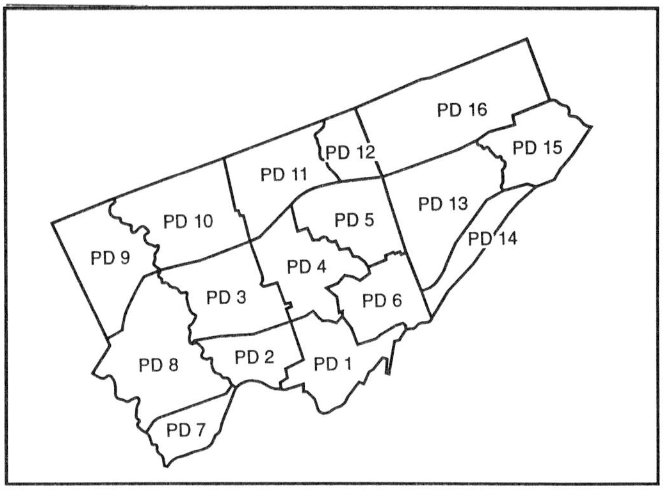

Map 12.2. 1986 planning district boundary map for Metropolitan Toronto, Ontario.

Implementation of Poisson regression will be demonstrated with fall 1986 planning district origin-destination travel data for metropolitan Toronto. The city has been divided into sixteen districts (Map 12.2). A 5% random sample of households in the study area was selected from Bell Canada's residential billing files. A total of 61,700 out of 1,470,000 households were successfully interviewed (a 4.2% sample). A trip was defined as a one-way movement between two districts for a single purpose during the 6 - 9 AM period. Trip information was collected for all persons at least six years of age, over a 24-hour period. This example uses only the 6 - 9 AM flows matrix to avoid the dependency between the AM and PM flows imbedded in the 24 hour flows matrix. Only trips made Monday through Friday were recorded. Finally, a set of expansion factors was used to inflate the information contained in the completed surveys to represent the total population residing in the Greater Toronto Area. The estimated total number of trips for all purposes made within metropolitan Toronto during the 6 - 9 AM period is 1,972,800, which becomes the denominator in the empirical probability calculations. These data appear in Appendix G. Using PROC UNIVARIATE on the empirical probabilities of flows generated the following summary information:

```
                     Quantiles(Def=5)
  100% Max   0.017944            99%   0.016271
   75% Q3    0.001825            95%   0.009327
   50% Med    0.00071            90%   0.003954
   25% Q1    0.000253            10%   0.000101
    0% Min          0             5%   0.000051
                                  1%          0
```

[4]For example, the observed number of deaths in any area, although strictly binomial, is only a very small proportion of the total population and hence can be considered to follow the Poisson approximation to a binomial distribution.

Clearly these probabilities are in keeping with the notion of a rare event.

EXAMPLE 12.5. Estimate a Poisson regression gravity-type inter-areal units flows model, based upon the variables of *LN*(origin size), *LN*(destination size), and interdistrict centroid distance.

Using the SAS procedure

```
PROC LOGISTIC;
   MODEL FLOWS/TOTAL=LOGOI LOGDJ DIST;
RUN;
```

where the origin and destination totals, OI and DJ, have been transformed as follows:

```
LOGOI=LOG(OI);
LOGDJ=LOG(DJ);
```

produced the SAS output

```
                                Intercept
                 Intercept         and
 Criterion          Only        Covariates    Chi-Square for Covariates
 -2 LOG L         13666710        12183445      1483264.3 with 3 DF (p=0.0001)

                  Analysis of Maximum Likelihood Estimates
                 Parameter  Standard       Wald          Pr >      Standardized      Odds
Variable   DF    Estimate     Error     Chi-Square   Chi-Square      Estimate        Ratio

INTERCPT   1     -21.7999    0.0358 370375.742        0.0001            .            0.000
LOGOI      1       0.5846   0.00299 38302.4557        0.0001          0.143325       1.794
LOGDJ      1       0.9423   0.00135 487942.017        0.0001          0.374690       2.566
DISTIJ     1      -0.0333 0.000041 673627.694         0.0001         -0.662489       0.967

     Association of Predicted Probabilities and Observed Responses
                Concordant = 66.5%          Somers' D = 0.557
                Discordant = 10.8%          Gamma     = 0.721
                Tied       = 22.7%          Tau-a     = 0.002
                (4.7264341E14 pairs)        c         = 0.779
```

The pseudo-R^2 here equals 0.1085, which is unimpressive. The squared correlation between the expected and observed flows is 0.8127. In contrast, the gamma statistic suggests that improved predictability is substantial when this model is used. And all parameter estimates are significant, which is not very surprising given the spatial interaction literature. The intercept term must be increased by LN(1,972,800), the logarithm of the total number of O-D flows, changing it to -7.3049.[5]

Comparing this solution with the standard log-normal estimation results can be informative. Using the SAS procedure

```
PROC REG;
   MODEL LOGTIJ = LOGOI LOGDJ DIST;
RUN;
```

where LOGTIJ = LOG(FLOWS) generated the following summary information:

Analysis of Variance

Source	DF	Sum of Squares	Mean Square	F Value	Prob>F
Model	3	644.20144	214.73381	350.414	0.0001
Error	252	154.42565	0.61280		
C Total	255	798.62709			

Root MSE	0.78282	R-square	0.8066
Dep Mean	7.10104	Adj R-sq	0.8043
C.V.	11.02395		

Parameter Estimates

Variable	DF	Parameter Estimate	Standard Error	T for H0: Parameter=0	Prob > |T|
INTERCEP	1	-11.547984	1.48794575	-7.761	0.0001
LOGOI	1	0.689462	0.11198959	6.156	0.0001
LOGDJ	1	1.196480	0.06877535	17.397	0.0001
DISTIJ	1	-0.030899	0.00140010	-22.069	0.0001

That the distance decay coefficient appearing in the term $EXP(-\beta_3 \text{DIST})$ is quite similar to that obtained with the Poisson regression. On the other hand, the origin and destination size exponent OLS estimates are distinctly larger than their Poisson regression counterparts. All parameter estimates unquestionably remain significant.[6]

---

[5]Since the SAS package provides an approximation to Poisson regression using logistic regression, these results were verified using the Poisson regression procedures in the GLIM and STATA packages. The parameter estimates were accurate to at least two, and often three, decimal places, whereas the standard errors essentially were identical. One caveat to keep in mind here is that PROC LOGISTIC and PROC CATMOD were designed to perform nonparametric analyses on categorical data (ignoring information captured by available parameters), whereas Poisson regression is a parametric procedure (exploiting information captured by available parameters). Accordingly, when probabilities are suitably small, SAS merely can be tricked into performing Poisson regression with nearly negligible specification error. This specification error grows as probabilities increase.

[6]*LN*(0) for zero flows was set to 0, with an antilogarithm equaling 1. This fudging value of 1 still is markedly smaller than the minimum value of 100 found in the O-D table.

Part of the observed discrepancies between the OLS and Poisson regression estimates is attributable to the posited error structures. This idea can be illustrated by comparing these preceding two outputs with that for the gravity model parameter estimates obtained with PROC NLIN. Using the SAS procedure,

```
PROC NLIN MAXITER=200 METHOD=MARQUARDT;
   PARMS A=0.00001 B=0.69, C=1.20, D=0.03;
   MODEL FLOWS = A*(OI**B)*(DJ**C)*EXP(-D*DIST);
DER.A=(OI**B)*(DJ**C)*EXP(-D*DIST);
DER.B=A*LOG(OI)*(OI**B)*(DJ**C)*EXP(-D*DIST);
DER.C=A*LOG(DJ)*(OI**B)*(DJ**C)*EXP(-D*DIST);
DER.D=-A*DIST*(OI**B)*(DJ**C)*EXP(-D*DIST);
RUN;
```

generated the following summary information:

Source	DF	Sum of Squares	Mean Square
Regression	4	12459084814	3114771204
Residual	252	1463115186	5806013
Uncorrected Total	256	13922200000	
(Corrected Total)	255	10488240000	

Parameter	Estimate	Asymptotic Std. Error	Asymptotic 95 % Confidence Interval Lower	Upper
A	0.000029484	0.00003734632	-.00004406764	0.0001030356
B	1.058078107	0.10321163627	0.85480818194	1.2613480325
C	0.750390373	0.02890165043	0.69347008278	0.8073106639
D	0.032103804	0.00113961366	0.02985939396	0.0343482131

Asymptotic Correlation Matrix

Corr	A	B	C	D
A	1	-0.964338061	-0.426036103	-0.174238858
B	-0.964338061	1	0.1723808487	0.1666245172
C	-0.426036103	0.1723808487	1	0.1307649535
D	-0.174238858	0.1666245172	0.1307649535	1

Now the intercept has decreased to -10.431666 (the logarithm of 0.000029484), and the origin and destination size exponent estimates are roughly the same as their logistic regression counterparts; but each attached to the opposite areal unit factor, and the distance decay exponent remains very similar.

> *RULE OF THUMB 12.3:* Because measurement scale transformations of data impact upon results, when possible, data should be analyzed at the scale upon which they originally were collected and error structure assumptions should be carefully selected.

This is reminiscent of *Rule of Thumb 8.2*, which pertains to the same notion with respect to geographic scale.

The OLS results include an R^2 value of 0.8066, which suggests that the gravity-type flows model describes these Toronto data reasonably well. The failure to expose this same degree of description for the logistic regression model offers one argument for also computing the product moment correlation between Y and $\hat{p}$ as standard practice. This contention is further corroborated by inspection of 1 minus the relative error sum-of-squares from NLIN, which equals a quite respectable 0.8605.

It is important to recognize that this data set does not fit the assumptions of Poisson regression for a number of reasons. The set of origin and destination attributes used in Example 12.1 is incomplete because only origin and destination total flows are used, ignoring information on other attribute information, such as parking space, office space, and employment totals. In addition, there is spatial dependence in the origin flows (positive spatial autocorrelation) arising from multiple related trip-makers residing in the same location (families of various sorts). This condition is related to a common problem encountered in migration research, in which family members are constrained to move as a group, rather than independently of one another (Flowerdew and Aitkin, 1982).

12.6. Problems

P-12.1. Given the *Particle Size Distribution of Soil Profiles* (Table 16.1 in Andrews and Herzberg, 1985) data set, predict the percentage of sand (this is **Y**) from depth (this is **X**) across the 20 plots using a logit model. What cautions about the statistical analysis results should be emphasized due to the use of 12 repeated measures on each of the 20 plots? Repeat this analysis, predicting the set of sand, silt, and clay percentages, using a multinomial logit model.

P-12.1. Given *The Spatio-temporal Spread of Fox Rabies* (Table 50.1 in Andrews and Herzberg, 1985) data set, construct a trend surface description of the number of case occurrences using a Poisson regression model specification.

12.7. References

Andrews, D. and H. Herzberg. 1985. *Data: A Collection of Problems from Many Fields for the Student and Research Worker*. Berlin: Springer-Verlag.

Burt, J. and G. Barber. 1996. *Elementary Statistics for Geographers*. 2nd ed. New York: Guilford Press.

Cressie, N. (1991). *Statistics for Spatial Data*. New York: Wiley.

Flowerdew, R., and M. Aitkin. (1982). "A Method of Fitting the Gravity Model Based on the Poisson Distribution." *Journal of Regional Science* 22: 191–202.

Griffith, D. and C. Amrhein. 1991. *Statistical Analysis for Geographers*. Englewood Cliffs, N.J.: Prentice Hall.

le Cessie, S., and J. van Houwelingen. 1994. "Logistic Regression for Correlated Binary Data." *Applied Statistics* 43: 95–108.

Johnson, D., K. McDade, and D. Griffith. 1996. "Seasonal variation in paediatric blood lead levels in Syracuse, NY, USA." *Environmental Geochemistry and Health*. 18: 81-88.

EPILOGUE

Chapters 4 through 12 contain what we believe to be one of the few thorough treatments of spatial analysis in a multivariate context—a treatment in which specifically spatial forms of multivariate analysis are presented, geo-referenced data are the focus of attention, and spatial autocorrelation is used as a key concept to understand nuances and idiosyncrasies peculiar to geo-referenced data (rather than distractions that need to be expunged from such data). We distinguish our approach from a spatial econometrics approach (see Anselin, 1988, for an excellent and comprehensive discussion of this topic), in part because we are interested in statistical description and hypothesis testing rather than econometric model specification issues.

As is evident from the technical complexities encountered throughout this book, especially those outlined in the last three chapters, explicitly modifying statistical techniques to accommodate the presence of latent spatial relations is not an easy task. But the rewards are well worth the effort, especially in terms of gains in statistical efficiency and improvements in statistical explanation of geo-referenced phenomena. We believe that we understand what we have done well enough to explain it largely in ordinary English, with a minimum of technical language. Of course, some technical jargon, such as the notion of eigenfunctions, cannot be avoided in most discussions of multivariate analysis.

We see two main pursuits as possible next steps in the acquisition of spatial statistical skills. The first is to continue an in-depth study of the theory of spatial statistics, as well as of linkages between this theory and the temporal realm (note that there is no mention in this book of time series analysis adjusted for spatial dependencies). Simultaneous space-time statistics have been formulated, but few serious applications have been undertaken (Christensen, 1991). Second, there is a pressing need to become sufficiently familiar with spatial statistical techniques, without requiring more than a modest technical ability, while still having a thorough understanding of the data analysis issues involved (after all, most of us want to concentrate on the science/substantive problem, not the methodology!). This familiarity is necessary so that standard commercial software packages (including GISs) can be used for applied spatial statistical analysis (Griffith, 1993). In other words, until firms such as MINITAB, SAS, and SPSS include spatial analysis procedures in their new releases, like S+ has done with its recent S+SpatialStats module, spatial analysts need to possess the knowledge and skills necessary to trick these packages into computing correct spatial statistical results. Alternatively, the discipline will have to undertake the development of specialized packages in a manner similar to that of SpaceStat (Anselin, 1992) in the spatial econometrics discipline.

Aware that no single text will be error free or answer all questions, we conclude by listing some additional volumes to which the interested reader might turn for supplementary assistance. Our list includes a few references to geostatistical techniques (e.g., Carr, 1995; Isaaks and Srivastava, 1989). There is substantial overlap between these books and ours, especially with topics such as trend surface modeling and cluster analysis, although we feel that our treatment of these subjects places a much stronger emphasis on the spatial dimension of geo-referenced data. The reader should be aware that there are a number of important spatial techniques used in geostatistics that we have not mentioned in our book. Some of these, such as kriging, are special applications of the general linear model and can be viewed as another topic for a chapter subsequent to our Chapter 12. Other topics, such as semi-variograms and their modelings, essentially are extensions of variance-covariance analysis. While there are initial efforts underway to formally link these approaches with those of spatial autoregression (Griffith and Csillag, 1993), there is much additional conceptual and theoretical work that needs to be completed yet. Finally, there are important special considerations that still require modifications to classical statistics just as we have undertaken for our purposes.

Additional References

Anselin, L. 1988. *Spatial Econometrics: Methods and Models.* Dordrecht: Kluwer Academic Publishers.

——. 1992. *SPACESTAT Tutorial: A Workbook for Using SpaceStat in the Analysis of Spatial Data.* Technical Software Series S-92-1. Santa Barbara, Calif.: National Center for Geographic Information and Analysis.

Arbia, G. 1989. *Spatial Data Configuration in Statistical Analysis of Regional Economic and Related Problems.* Dordrecht: Kluwer Academic Publishers.

Bailey, T. and A. Gatrell. 1995. *Interactive Spatial Data Analysis.* Harlow: Longman Scientific Publishers.

Carr, J. 1995. *Numerical Analysis for the Geological Sciences.* Englewood Cliffs, N.J.: Prentice Hall.

Christensen, R. 1991. *Linear Models for Multivariate, Time Series, and Spatial Data.* Berlin: Springer-Verlag.

Cliff, A. and J. Ord. 1981. *Spatial Processes: Models and Applications.* London: Pion.

Cressie, N. 1991. *Statistics for Spatial Data.* New York: Wiley.

Getis, A. 1995. *The Tyranny of Data.* San Diego, Calif.: San Diego State University Press.

Ghosh, A. and G. Rushton. 1987. *Spatial Analysis and Location-Allocation Models.* New York: Van Nostrand Reinhold.

Griffith, D. 1987. *Spatial Autocorrelation: A Primer.* Washington D.C.: Association of American Geographers.

——. 1988. *Advanced Spatial Statistics*. Dordrecht: Kluwer Academic Publishers.

——. 1993. *Spatial Regression Analysis on the PC: Spatial Statistics Using SAS*. Washington D.C.: Association of American Geographers.

Griffith, D. and F. Csillag. 1993. "Exploring Relationships between Semi-variogram and Spatial Autoregressive Models." *Papers in Regional Science* 72: 283–95.

Haining, R. 1990. *Spatial Data Analysis in the Social and Environmental Sciences*. Cambridge: Cambridge University Press.

Isaaks, E. and R. Srivastava. 1989. *An Introduction to Applied Geostatistics*. Oxford: Oxford University Press.

Okabe, A., B. Boots, and K. Sugihara. 1992. *Spatial Tessellations: Concepts and Applications of Voronoi Diagrams*. New York: Wiley.

Ripley, B. 1981. *Spatial Statistics*. New York: Wiley.

Upton, G. and B. Fingleton. 1985. *Spatial Data Analysis by Example: Volume 1—Point Pattern and Quantitative Data*. New York: Wiley.

——. 1989. *Spatial Data Analysis by Example: Volume 2—Categorical and Directional Data*. New York: Wiley.

Wrigley, N. 1985. *Categorical Data Analysis for Geographers and Environmental Scientists*. Harlow: Longman Scientific Publishers.

APPENDICES

APPENDIX A

Puerto Rico (PR) Agricultural Data Set

No.	Name	ADM	LND	R-U	DFRM	DCD	DMLK	DSGR	DCF	DTB	DBN	DFAM	u-co-ordinate	v-co-ordinate	area
	SAN JUAN REGION (1)														
10	Barranquitas	1	1	0	16.1	38.0	2.0	0.0	3279.7	32.2	5.4	4.3	7.2654	4.65032	28.760
11	Bayamón/Cataño	1	0	1	5.2	9.4	2.2	0.0	758.2	0.5	0.9	0.9	8.0910	5.68378	42.033
15	Carolina	1	0	1	5.5	22.7	43.0	4.0	11.0	0.0	0.2	1.1	9.3291	5.83102	46.644
17	Ceiba	1	0	1	3.0	13.2	20.8	2.1	23.7	0.0	0.2	0.7	11.0857	5.12181	25.021
21	Comerío	1	1	0	14.9	52.6	1.7	0.0	708.0	153.7	3.3	2.4	7.7923	4.81340	23.947
22	Corozal	1	1	0	13.7	40.7	28.6	0.0	2816.2	0.3	5.2	2.4	7.1367	5.30614	35.778
23	Dorado	1	0	1	2.7	26.2	66.6	7.6	8.5	0.5	0.2	2.5	7.4061	6.15557	20.604
24	Fajardo	1	0	1	3.3	24.5	12.0	10.0	15.6	0.0	0.3	0.5	11.0796	5.55172	27.574
28	Guaynabo	1	0	1	7.5	15.0	9.6	0.0	193.7	0.0	1.2	1.0	8.4028	5.61831	22.953
41	Loíza/Canóvanas	1	0	1	3.8	23.7	19.2	4.2	720.3	0.0	0.8	2.5	9.7248	5.82213	48.459
42	Luquillo	1	0	0	4.2	23.3	45.1	1.8	22.7	0.0	0.4	1.1	10.7439	5.69578	21.230
50	Naranjito	1	1	0	18.6	34.2	3.1	0.0	1591.7	7.5	6.2	1.7	7.5852	5.20299	24.393
51	Orocovis	1	1	0	17.6	53.6	10.8	2.1	13265.8	19.7	19.3	4.4	6.5317	4.71552	53.390
57	Río Grande	1	0	1	2.8	14.3	6.9	0.8	14.0	0.0	0.4	0.1	10.2143	5.69530	53.090
61	San Juan/Río Piedras	1	0	1	2.7	2.9	0.0	0.0	31.1	0.0	0.2	0.1	8.7020	5.90144	44.283
65	Toa Alta	1	1	1	7.4	27.7	94.0	0.0	16.0	0.0	0.8	1.5	7.6053	5.69592	23.095
66	Toa Baja	1	0	1	0.7	28.2	91.1	8.3	0.0	0.0	0.0	1.4	7.8017	6.15770	21.244
67	Trujillo Alto	1	1	1	9.2	20.4	27.4	0.0	7.2	0.0	0.3	0.7	9.1106	5.58670	19.610
69	Vega Alta	1	0	1	6.6	31.7	49.1	2.7	59.0	0.0	0.3	1.0	7.0459	5.96160	23.994
	ARECIBO REGION (2)														
7	Arecibo	2	0	1	4.5	25.6	46.7	6.1	678.7	1.5	0.3	1.9	5.0114	5.87870	108.924
9	Barceloneta/Florida	2	0	0	5.6	32.5	32.3	10.9	268.6	0.3	0.6	1.5	5.7017	5.93904	29.078
14	Camuy	2	0	0	12.1	39.9	68.1	3.6	413.1	1.7	0.4	1.1	3.8933	5.92480	38.420
18	Ciales	2	1	0	8.5	35.0	4.9	0.1	12070.0	10.0	9.4	2.7	6.0015	5.17091	57.390
30	Hatillo	2	0	1	8.9	53.4	368.6	1.0	45.1	0.0	0.9	4.9	4.2885	5.86894	35.575
38	Lares	2	1	0	17.9	47.5	0.0	0.2	48874.9	0.0	25.6	7.2	3.9092	4.94337	54.120

APPENDIX A (continued)

							ARECIBO REGION (2)								
43	Manatí	2	0	1	3.0	34.8	89.2	3.1	80.5	0.0	0.1	1.5	6.1300	6.01849	39.191
48	Morovis	2	1	0	12.9	50.9	50.4	0.0	3158.9	23.7	3.4	4.5	6.5648	5.38675	33.600
55	Quebradillas	2	0	0	10.7	36.9	40.9	9.8	610.4	11.0	0.4	0.5	3.4896	6.04380	18.372
68	Utuado	2	1	0	18.3	46.8	2.7	0.0	26677.2	10.0	9.0	2.8	4.8815	5.00825	96.313
70	Vega Baja	2	0	1	4.8	32.7	39.1	9.1	227.9	0.3	0.3	1.1	6.6805	6.07988	40.399
							MAYAGUEZ REGION (3)								
2	Aguada	3	0	0	21.2	41.7	0.1	26.1	10286.9	0.0	2.2	1.3	2.0297	5.45670	26.953
3	Aguadilla	3	0	1	5.3	28.1	5.7	14.3	199.6	0.8	0.2	2.3	2.3154	6.11329	30.995
6	Anasco	3	0	0	9.3	37.0	0.0	14.7	11811.4	0.0	2.1	2.1	2.3814	5.02043	33.357
12	Cabo Rojo	3	0	0	5.5	48.3	22.5	14.8	7.0	0.0	0.2	3.6	2.2667	3.44951	60.642
25	Guánica	3	0	1	2.4	39.0	4.4	9.9	147.9	0.0	0.0	1.0	3.6730	3.09100	35.086
31	Hormigueros	3	1	1	9.8	44.5	0.1	27.7	4130.0	0.0	1.3	2.0	2.4778	4.05086	8.125
33	Isabela	3	0	0	11.0	32.3	39.6	5.8	683.6	8.3	0.5	2.0	3.0253	6.07891	41.831
37	Lajas	3	0	0	5.3	62.5	6.1	13.1	31.5	0.1	0.2	2.0	2.9428	3.28687	49.608
39	Las Marías	3	1	0	18.2	40.3	0.0	0.4	34270.9	3.2	9.3	2.5	3.2054	4.72949	37.547
44	Maricao	3	1	0	14.0	42.0	0.0	0.0	48140.1	0.0	14.3	4.8	3.4891	4.31602	31.802
46	Mayaguez	3	0	1	9.2	26.4	3.8	7.6	9763.0	0.3	2.9	1.6	2.4779	4.51769	46.166
47	Moca	3	1	0	14.3	37.9	2.1	15.3	6482.3	0.1	1.2	2.6	2.5783	5.60533	42.395
56	Rincón	3	0	0	8.4	21.6	0.0	7.5	2391.8	0.0	0.4	0.9	1.7038	5.28606	12.971
58	Sabana Grande	3	1	0	4.9	24.3	3.5	4.9	4898.7	0.0	0.1	1.7	3.4965	3.76124	29.204
60	San Germán	3	1	0	11.9	49.4	0.5	14.1	10993.9	5.8	1.7	4.3	2.9189	3.89939	45.117
63	San Sebastián	3	1	0	14.7	45.9	6.9	17.4	11689.8	1.8	1.8	2.7	3.2641	5.31295	60.169
							PONCE REGION (4)								
1	Adjuntas	4	1	0	15.6	41.1	2.1	0.0	35710.4	0.0	10.8	3.2	4.6135	4.40692	55.074
8	Arroyo	4	0	0	9.3	41.4	0.0	13.4	2510.8	0.0	1.3	0.8	8.8282	3.41202	13.900
20	Coamo	4	1	0	5.3	43.6	10.9	0.0	2253.0	1.0	2.7	1.4	6.9811	3.96919	62.282
26	Guayama	4	0	1	3.7	36.8	5.3	9.7	1228.8	0.6	1.2	0.8	8.3673	3.44502	54.640
27	Guayanilla	4	0	0	6.5	62.9	0.0	7.6	9031.0	0.4	2.9	1.9	4.4257	3.51843	33.962
34	Jayuya	4	1	0	14.1	57.1	5.0	0.1	44081.6	0.1	32.4	5.2	5.5786	4.64993	38.224
35	Juana Díaz	4	0	0	2.0	49.5	8.6	12.9	3289.3	0.0	1.9	1.1	6.1901	3.66652	52.186
52	Patillas	4	0	0	8.3	26.7	0.0	1.5	485.9	1.5	1.2	2.0	9.0951	3.62582	41.547
53	Peñuelas	4	0	0	5.5	44.6	5.3	8.8	5978.4	0.0	3.2	4.0	4.8407	3.66012	39.171
54	Ponce	4	0	1	4.0	37.5	4.6	8.2	7788.3	0.0	2.8	1.3	5.4917	3.68603	98.289
59	Salinas	4	0	0	1.3	51.7	7.8	15.2	117.5	0.4	0.0	0.6	7.6215	3.44041	57.965
64	Santa Isabel	4	0	0	0.7	48.6	1.7	30.0	0.0	0.0	0.1	0.8	6.8331	3.31941	29.791
71	Villalba	4	1	0	10.9	39.4	0.0	0.8	8650.7	0.0	2.6	2.8	6.3026	4.15897	30.396
73	Yauco	4	0	0	9.9	39.1	2.3	2.6	15325.7	0.3	7.4	2.9	4.0069	3.74738	57.442

APPENDIX A (continued)

CAGUAS REGION (5)

4	Aguas Buenas	5	1	0	15.4	30.4	10.7	0.0	4505.6	19.4	4.5	1.8	8.3684	5.01915	27.638
5	Aibonito	5	1	0	8.3	37.3	16.6	0.0	2437.4	49.0	1.1	1.3	7.5508	4.20180	26.464
13	Caguas	5	1	1	10.0	36.5	82.8	0.0	304.7	9.6	1.7	4.7	8.8136	4.77268	51.928
16	Cayey	5	1	1	11.1	43.1	18.0	0.0	3660.0	6.7	10.0	1.4	8.2407	4.05145	43.829
19	Cidra	5	1	0	11.5	36.2	23.7	0.2	493.9	63.0	1.6	2.5	8.1605	4.49655	29.966
29	Gurabo	5	1	1	6.9	39.5	75.2	6.1	42.0	3.6	0.4	2.3	9.2308	5.12426	22.523
32	Humacao	5	0	1	5.0	18.5	15.0	10.3	13.0	0.2	0.3	3.6	10.2724	4.41366	36.778
36	Juncos	5	1	1	7.8	39.8	134.4	0.2	87.4	0.8	0.1	5.7	9.6773	4.87350	24.556
40	Las Piedras	5	1	0	15.2	36.8	60.0	0.9	494.4	3.0	1.8	3.0	9.8992	4.65550	31.494
45	Maunabo	5	0	0	5.5	37.5	0.0	16.5	14.5	0.2	2.1	1.5	9.6340	3.57284	19.959
49	Naguabo	5	0	0	4.3	27.2	68.2	4.0	346.9	0.0	0.4	0.9	10.5900	4.97730	44.397
62	San Lorenzo	5	1	1	14.8	35.4	32.5	1.5	225.9	8.2	1.3	4.9	9.2703	4.36967	44.966
72	Yabucoa	5	0	0	15.3	43.0	15.2	9.7	402.3	1.5	2.0	2.5	9.7715	3.90243	47.658

NOTE: "No." refers to the alphabetical order of the municipio names.
Sources: (1974) *U.S. Census of Agriculture* and Commonwealth of Puerto Rico (1974–75), *Facts and Figures on Puerto Rico's Agriculture.*

APPENDIX B

Puerto Rico Connectivity Matrix, **C**, Data Set

1	0	0	0	0	0	0	0	0	0	0	0	0	0	0	0	0	0	0	0	0	0	0	0	0	0	0	1	0	0	0	0	0	0	0	0	0	0	1	0	0	0	0	0	0	0	0	0	0	0	0	0	0	1	1	0	0	0	0	0	0	0	0	0	0	0	0	0	1	0	0	0	0	1
2	0	0	1	0	0	1	0	0	0	0	0	0	0	0	0	0	0	0	0	0	0	0	0	0	0	0	0	0	0	0	0	0	0	0	0	0	0	0	0	0	0	0	0	0	0	0	1	0	0	0	0	0	0	0	0	1	0	0	0	0	0	0	0	0	0	0	0	0	0	0	0	0	0
3	0	1	0	0	0	0	0	0	0	0	0	0	0	0	0	0	0	0	0	0	0	0	0	0	0	0	0	0	0	0	0	0	1	0	0	0	0	0	0	0	0	0	0	0	0	0	1	0	0	0	0	0	0	0	0	0	0	0	0	0	0	0	0	0	0	0	0	0	0	0	0	0	0
4	0	0	0	0	0	0	0	0	0	0	1	0	1	0	0	0	0	0	1	0	1	0	0	0	0	0	0	1	0	0	0	0	0	0	0	0	0	0	0	0	0	0	0	0	0	0	0	0	0	0	0	0	0	0	0	0	0	0	0	0	1	0	0	0	0	0	0	0	0	0	0	0	0
5	0	0	0	0	0	0	0	0	0	1	0	0	0	0	0	1	0	0	1	1	0	0	0	0	0	0	0	0	0	0	0	0	0	0	0	0	0	0	0	0	0	0	0	0	0	0	0	0	0	0	0	0	0	0	0	0	0	0	1	0	0	0	0	0	0	0	0	0	0	0	0	0	0
6	0	1	0	0	0	0	0	0	0	0	0	0	0	0	0	0	0	0	0	0	0	0	0	0	0	0	0	0	0	0	0	0	0	0	0	0	0	0	1	0	0	0	0	0	0	1	1	0	0	0	0	0	0	0	0	1	0	0	0	0	0	0	1	0	0	0	0	0	0	0	0	0	0
7	0	0	0	0	0	0	0	0	1	0	0	0	0	0	0	0	0	1	0	0	0	0	0	0	0	0	0	0	0	1	0	0	0	0	0	0	0	0	0	0	0	0	0	0	0	0	0	0	0	0	0	0	0	0	0	0	0	0	0	0	0	0	0	0	0	0	0	1	0	0	0	0	0
8	0	0	0	0	0	0	0	0	0	0	0	0	0	0	0	0	0	0	0	0	0	0	0	0	0	1	0	0	0	0	0	0	0	0	0	0	0	0	0	0	0	0	0	0	0	0	0	0	0	0	0	1	0	0	0	0	0	0	0	0	0	0	0	0	0	0	0	0	0	0	0	0	0
9	0	0	0	0	0	0	1	0	0	0	0	0	0	0	0	0	0	1	0	0	0	0	0	0	0	0	0	0	0	0	0	0	0	0	0	0	0	0	0	0	0	0	1	0	0	0	0	0	0	0	0	0	0	0	0	0	0	0	0	0	0	0	0	0	0	0	0	0	0	0	0	0	0
10	0	0	0	0	1	0	0	0	0	0	0	0	0	0	0	0	0	0	1	1	1	1	0	0	0	0	0	0	0	0	0	0	0	0	0	0	0	0	0	0	0	0	0	0	0	0	0	0	0	1	1	0	0	0	0	0	0	0	0	0	0	0	0	0	0	0	0	0	0	0	0	0	0
11	0	0	0	1	0	0	0	0	0	0	0	0	0	0	0	0	0	0	0	0	1	0	0	0	0	0	0	1	0	0	0	0	0	0	0	0	0	0	0	0	0	0	0	0	0	0	0	0	0	1	0	0	0	0	0	0	0	0	0	0	0	0	0	0	1	1	0	0	0	0	0	0	0
12	0	0	0	0	0	0	0	0	0	0	0	0	0	0	0	0	0	0	0	0	0	0	0	0	0	0	0	0	0	0	1	0	0	0	0	0	1	0	0	0	0	0	0	0	0	1	0	0	0	0	0	0	0	0	0	0	0	0	0	1	0	0	0	0	0	0	0	0	0	0	0	0	0
13	0	0	0	1	0	0	0	0	0	0	0	0	0	0	0	1	0	0	1	0	0	0	0	0	0	0	0	0	1	0	0	0	0	0	0	0	0	0	0	0	0	0	0	0	0	0	0	0	0	0	0	0	0	0	0	0	0	0	0	0	1	1	0	0	0	0	1	0	0	0	0	0	0
14	0	0	0	0	0	0	0	0	0	0	0	0	0	0	0	0	0	0	0	0	0	0	0	0	0	0	0	0	0	1	0	0	0	0	0	0	0	1	0	0	0	0	0	0	0	0	0	0	0	0	0	0	0	0	1	0	0	0	0	0	0	0	1	0	0	0	0	0	0	0	0	0	0
15	0	0	0	0	0	0	0	0	0	0	0	0	0	0	0	0	0	0	0	0	0	0	0	0	0	0	0	0	1	0	0	0	0	0	0	0	0	0	0	0	1	0	0	0	0	0	0	0	0	0	0	0	0	0	0	0	0	0	0	0	1	0	0	0	0	0	1	0	0	0	0	0	0
16	0	0	0	0	1	0	0	0	0	0	0	0	1	0	0	0	0	0	1	0	0	0	0	0	0	1	0	0	0	0	0	0	0	0	0	0	0	0	0	0	0	0	0	0	0	0	0	0	0	0	0	1	0	0	0	0	0	0	1	0	0	1	0	0	0	0	0	0	0	0	0	0	0
17	0	0	0	0	0	0	0	0	0	0	0	0	0	0	0	0	0	0	0	0	0	0	0	1	0	0	0	0	0	0	0	0	0	0	0	0	0	0	0	0	0	0	0	0	0	0	0	0	1	0	0	0	0	0	0	0	1	0	0	0	0	0	0	0	0	0	0	0	0	0	0	0	0
18	0	0	0	0	0	0	1	0	1	0	0	0	0	0	0	0	0	0	0	0	0	0	0	0	0	0	0	0	0	0	0	0	0	1	0	0	0	0	0	0	0	0	1	0	0	0	0	1	0	0	1	0	0	0	0	0	0	0	0	0	0	0	0	0	0	0	0	1	0	0	0	0	0
19	0	0	0	1	1	0	0	0	0	1	0	0	1	0	0	1	0	0	0	0	1	0	0	0	0	0	0	0	0	0	0	0	0	0	0	0	0	0	0	0	0	0	0	0	0	0	0	0	0	0	0	0	0	0	0	0	0	0	0	0	0	0	0	0	0	0	0	0	0	0	0	0	0
20	0	0	0	0	1	0	0	0	0	1	0	0	0	0	0	0	0	0	0	0	0	0	0	0	0	0	0	0	0	0	0	0	0	0	1	0	0	0	0	0	0	0	0	0	0	0	0	0	0	0	1	0	0	0	0	0	0	0	1	0	0	0	0	1	0	0	0	0	0	0	1	0	0
21	0	0	0	1	0	0	0	0	0	1	1	0	0	0	0	0	0	0	1	0	0	0	0	0	0	0	0	0	0	0	0	0	0	0	0	0	0	0	0	0	0	0	0	0	0	0	0	0	0	1	0	0	0	0	0	0	0	0	0	0	0	0	0	0	0	0	0	0	0	0	0	0	0
22	0	0	0	0	0	0	0	0	0	1	0	0	0	0	0	0	0	0	0	0	0	0	0	0	0	0	0	0	0	0	0	0	0	0	0	0	0	0	0	0	0	0	0	0	0	0	0	1	0	1	1	0	0	0	0	0	0	0	0	0	0	0	0	0	1	0	0	0	1	0	0	0	0
23	0	0	0	0	0	0	0	0	0	0	0	0	0	0	0	0	0	0	0	0	0	0	0	0	0	0	0	0	0	0	0	0	0	0	0	0	0	0	0	0	0	0	0	0	0	0	0	0	0	0	0	0	0	0	0	0	0	0	0	0	0	0	0	0	1	1	0	0	1	0	0	0	0
24	0	0	0	0	0	0	0	0	0	0	0	0	0	0	0	0	1	0	0	0	0	0	0	0	0	0	0	0	0	0	0	0	0	0	0	0	0	0	0	0	0	1	0	0	0	0	0	0	0	0	0	0	0	0	0	0	0	0	0	0	0	0	0	0	0	0	0	0	0	0	0	0	0
25	0	0	0	0	0	0	0	0	0	0	0	0	0	0	0	0	0	0	0	0	0	0	0	0	0	0	0	0	0	0	0	0	0	0	0	0	1	0	0	0	0	0	0	0	0	0	0	0	0	0	0	0	0	0	0	0	0	1	0	0	0	0	0	0	0	0	0	0	0	0	0	0	1
26	0	0	0	0	0	0	0	1	0	0	0	0	0	0	0	1	0	0	0	0	0	0	0	0	0	0	0	0	0	0	0	0	0	0	0	0	0	0	0	0	0	0	0	0	0	0	0	0	0	0	0	1	0	0	0	0	0	0	1	0	0	0	0	0	0	0	0	0	0	0	0	0	0
27	1	0	0	0	0	0	0	0	0	0	0	0	0	0	0	0	0	0	0	0	0	0	0	0	0	0	0	0	0	0	0	0	0	0	0	0	0	0	0	0	0	0	0	0	0	0	0	0	0	0	0	0	1	0	0	0	0	0	0	0	0	0	0	0	0	0	0	0	0	0	0	0	1
28	0	0	0	1	0	0	0	0	0	0	1	0	0	0	0	0	0	0	0	0	0	0	0	0	0	0	0	0	0	0	0	0	0	0	0	0	0	0	0	0	0	0	0	0	0	0	0	0	0	0	0	0	0	0	0	0	0	0	0	0	1	0	0	0	0	0	0	0	0	0	0	0	0
29	0	0	0	0	0	0	0	0	0	0	0	0	1	0	1	0	0	0	0	0	0	0	0	0	0	0	0	0	0	0	0	0	0	0	0	1	0	0	0	0	0	0	0	0	0	0	0	0	0	0	0	0	0	0	0	0	0	0	0	0	0	1	0	0	0	0	1	0	0	0	0	0	0
30	0	0	0	0	0	0	1	0	0	0	0	0	0	1	0	0	0	0	0	0	0	0	0	0	0	0	0	0	0	0	0	0	0	0	0	0	0	1	0	0	0	0	0	0	0	0	0	0	0	0	0	0	0	0	0	0	0	0	0	0	0	0	0	0	0	0	0	1	0	0	0	0	0
31	0	0	0	0	0	0	0	0	0	0	0	1	0	0	0	0	0	0	0	0	0	0	0	0	0	0	0	0	0	0	0	0	0	0	0	0	0	0	0	0	0	0	0	0	0	1	0	0	0	0	0	0	0	0	0	0	0	0	0	1	0	0	0	0	0	0	0	0	0	0	0	0	0
32	0	0	0	0	0	0	0	0	0	0	0	0	0	0	0	0	0	0	0	0	0	0	0	0	0	0	0	0	0	0	0	0	0	0	0	0	0	0	0	1	0	0	0	0	0	0	0	0	1	0	0	0	0	0	0	0	0	0	0	0	0	0	0	0	0	0	0	0	0	0	0	1	0
33	0	0	1	0	0	0	0	0	0	0	0	0	0	0	0	0	0	0	0	0	0	0	0	0	0	0	0	0	0	0	0	0	0	0	0	0	0	0	0	0	0	0	0	0	0	0	1	0	0	0	0	0	0	0	1	0	0	0	0	0	0	0	1	0	0	0	0	0	0	0	0	0	0
34	0	0	0	0	0	0	0	0	0	0	0	0	0	0	0	0	0	1	0	0	0	0	0	0	0	0	0	0	0	0	0	0	0	0	1	0	0	0	0	0	0	0	0	0	0	0	0	0	0	0	1	0	0	1	0	0	0	0	0	0	0	0	0	0	0	0	0	1	0	0	0	0	0
35	0	0	0	0	0	0	0	0	0	0	0	0	0	0	0	0	0	0	0	1	0	0	0	0	0	0	0	0	0	0	0	0	0	1	0	0	0	0	0	0	0	0	0	0	0	0	0	0	0	0	1	0	0	1	0	0	0	0	0	0	0	0	0	1	0	0	0	0	0	0	1	0	0
36	0	0	0	0	0	0	0	0	0	0	0	0	0	0	0	0	0	0	0	0	0	0	0	0	0	0	0	0	1	0	0	0	0	0	0	0	0	0	0	1	1	0	0	0	0	0	0	0	0	0	0	0	0	0	0	0	0	0	0	0	0	1	0	0	0	0	0	0	0	0	0	0	0
37	0	0	0	0	0	0	0	0	0	0	0	1	0	0	0	0	0	0	0	0	0	0	0	0	1	0	0	0	0	0	0	0	0	0	0	0	0	0	0	0	0	0	0	0	0	0	0	0	0	0	0	0	0	0	0	0	0	1	0	1	0	0	0	0	0	0	0	0	0	0	0	0	0
38	1	0	0	0	0	0	0	0	0	0	0	0	0	1	0	0	0	0	0	0	0	0	0	0	0	0	0	0	0	1	0	0	0	0	0	0	0	0	1	0	0	0	0	1	0	0	0	0	0	0	0	0	0	0	0	0	0	0	0	0	0	0	1	0	0	0	0	1	0	0	0	0	1
39	0	0	0	0	0	1	0	0	0	0	0	0	0	0	0	0	0	0	0	0	0	0	0	0	0	0	0	0	0	0	0	0	0	0	0	0	0	1	0	0	0	0	0	1	0	1	0	0	0	0	0	0	0	0	0	0	0	0	0	0	0	0	1	0	0	0	0	0	0	0	0	0	0
40	0	0	0	0	0	0	0	0	0	0	0	0	0	0	0	0	0	0	0	0	0	0	0	0	0	0	0	0	0	0	0	1	0	0	0	1	0	0	0	0	1	0	0	0	0	0	0	0	1	0	0	0	0	0	0	0	1	0	0	0	0	1	0	0	0	0	0	0	0	0	0	1	0
41	0	0	0	0	0	0	0	0	0	0	0	0	0	0	1	0	0	0	0	0	0	0	0	0	0	0	0	0	0	0	0	0	0	0	0	1	0	0	0	1	0	0	0	0	0	0	0	0	0	0	0	0	0	0	0	0	1	0	0	0	0	0	0	0	0	0	0	0	0	0	0	0	0
42	0	0	0	0	0	0	0	0	0	0	0	0	0	0	0	0	0	0	0	0	0	0	0	1	0	0	0	0	0	0	0	0	0	0	0	0	0	0	0	0	0	0	0	0	0	0	0	0	0	0	0	0	0	0	0	0	1	0	0	0	0	0	0	0	0	0	0	0	0	0	0	0	0
43	0	0	0	0	0	0	0	0	1	0	0	0	0	0	0	0	0	1	0	0	0	0	0	0	0	0	0	0	0	0	0	0	0	0	0	0	0	0	0	0	0	0	0	0	0	0	0	1	0	0	0	0	0	0	0	0	0	0	0	0	0	0	0	0	0	0	0	0	0	1	0	0	0
44	0	0	0	0	0	0	0	0	0	0	0	0	0	0	0	0	0	0	0	0	0	0	0	0	0	0	0	0	0	0	0	0	0	0	0	0	0	1	1	0	0	0	0	0	0	1	0	0	0	0	0	0	0	0	0	0	0	1	0	1	0	0	0	0	0	0	0	0	0	0	0	0	1
45	0	0	0	0	0	0	0	0	0	0	0	0	0	0	0	0	0	0	0	0	0	0	0	0	0	0	0	0	0	0	0	0	0	0	0	0	0	0	0	0	0	0	0	0	0	0	0	0	0	0	0	1	0	0	0	0	0	0	0	0	0	0	0	0	0	0	0	0	0	0	0	1	0
46	0	0	0	0	0	1	0	0	0	0	0	1	0	0	0	0	0	0	0	0	0	0	0	0	0	0	0	0	0	0	1	0	0	0	0	0	0	0	1	0	0	0	0	1	0	0	0	0	0	0	0	0	0	0	0	0	0	0	0	1	0	0	0	0	0	0	0	0	0	0	0	0	0
47	0	1	1	0	0	1	0	0	0	0	0	0	0	0	0	0	0	0	0	0	0	0	0	0	0	0	0	0	0	0	0	0	1	0	0	0	0	0	0	0	0	0	0	0	0	0	0	0	0	0	0	0	0	0	0	0	0	0	0	0	0	0	1	0	0	0	0	0	0	0	0	0	0
48	0	0	0	0	0	0	0	0	0	0	0	0	0	0	0	0	0	1	0	0	0	1	0	0	0	0	0	0	0	0	0	0	0	0	0	0	0	0	0	0	0	0	1	0	0	0	0	0	0	0	1	0	0	0	0	0	0	0	0	0	0	0	0	0	0	0	0	0	1	1	0	0	0
49	0	0	0	0	0	0	0	0	0	0	0	0	0	0	0	0	1	0	0	0	0	0	0	0	0	0	0	0	0	0	0	1	0	0	0	0	0	0	0	1	0	0	0	0	0	0	0	0	0	0	0	0	0	0	0	0	1	0	0	0	0	0	0	0	0	0	0	0	0	0	0	0	0
50	0	0	0	0	0	0	0	0	0	1	1	0	0	0	0	0	0	0	0	0	1	1	0	0	0	0	0	0	0	0	0	0	0	0	0	0	0	0	0	0	0	0	0	0	0	0	0	0	0	0	0	0	0	0	0	0	0	0	0	0	0	0	0	0	1	0	0	0	0	0	0	0	0
51	0	0	0	0	0	0	0	0	0	1	0	0	0	0	0	0	0	1	0	1	0	1	0	0	0	0	0	0	0	0	0	0	0	1	1	0	0	0	0	0	0	0	0	0	0	0	0	1	0	0	0	0	0	0	0	0	0	0	0	0	0	0	0	0	0	0	0	0	0	0	1	0	0
52	0	0	0	0	0	0	0	1	0	0	0	0	0	0	0	1	0	0	0	0	0	0	0	0	0	1	0	0	0	0	0	0	0	0	0	0	0	0	0	0	0	0	0	0	1	0	0	0	0	0	0	0	0	0	0	0	0	0	0	0	0	1	0	0	0	0	0	0	0	0	0	1	0
53	1	0	0	0	0	0	0	0	0	0	0	0	0	0	0	0	0	0	0	0	0	0	0	0	0	0	1	0	0	0	0	0	0	0	0	0	0	0	0	0	0	0	0	0	0	0	0	0	0	0	0	0	0	1	0	0	0	0	0	0	0	0	0	0	0	0	0	0	0	0	0	0	0
54	1	0	0	0	0	0	0	0	0	0	0	0	0	0	0	0	0	0	0	0	0	0	0	0	0	0	0	0	0	0	0	0	0	1	1	0	0	0	0	0	0	0	0	0	0	0	0	0	0	0	0	0	1	0	0	0	0	0	0	0	0	0	0	0	0	0	0	1	0	0	0	0	0
55	0	0	0	0	0	0	0	0	0	0	0	0	0	1	0	0	0	0	0	0	0	0	0	0	0	0	0	0	0	0	0	0	1	0	0	0	0	0	0	0	0	0	0	0	0	0	0	0	0	0	0	0	0	0	0	0	0	0	0	0	0	0	1	0	0	0	0	0	0	0	0	0	0
56	0	1	0	0	0	1	0	0	0	0	0	0	0	0	0	0	0	0	0	0	0	0	0	0	0	0	0	0	0	0	0	0	0	0	0	0	0	0	0	0	0	0	0	0	0	0	0	0	0	0	0	0	0	0	0	0	0	0	0	0	0	0	0	0	0	0	0	0	0	0	0	0	0
57	0	0	0	0	0	0	0	0	0	0	0	0	0	0	0	0	1	0	0	0	0	0	0	0	0	0	0	0	0	0	0	0	0	0	0	0	0	0	0	1	1	1	0	0	0	0	0	0	1	0	0	0	0	0	0	0	0	0	0	0	0	0	0	0	0	0	0	0	0	0	0	0	0
58	0	0	0	0	0	0	0	0	0	0	0	0	0	0	0	0	0	0	0	0	0	0	0	0	1	0	0	0	0	0	0	0	0	0	0	0	1	0	0	0	0	0	0	1	0	0	0	0	0	0	0	0	0	0	0	0	0	0	0	1	0	0	0	0	0	0	0	0	0	0	0	0	1
59	0	0	0	0	1	0	0	0	0	0	0	0	0	0	0	1	0	0	0	1	0	0	0	0	0	1	0	0	0	0	0	0	0	0	0	0	0	0	0	0	0	0	0	0	0	0	0	0	0	0	0	0	0	0	0	0	0	0	0	0	0	0	0	1	0	0	0	0	0	0	0	0	0
60	0	0	0	0	0	0	0	0	0	0	0	1	0	0	0	0	0	0	0	0	0	0	0	0	0	0	0	0	0	0	1	0	0	0	0	0	1	0	0	0	0	0	0	1	0	1	0	0	0	0	0	0	0	0	0	0	0	1	0	0	0	0	0	0	0	0	0	0	0	0	0	0	0

APPENDIX B (continued)

61	0	0	0	1	0	0	0	0	0	0	0	0	1	0	1	0	0	0	0	0	0	0	0	0	0	0	0	1	0	0	0	0	0	0	0	0	0	0	0	0	0	0	0	0	0	0	0	0	0	0	0	0	0	0	0	0	0	0	0	0	0	0	0	0	0	0	1	0	0	0	0	0	0
62	0	0	0	0	0	0	0	0	0	0	0	0	1	0	0	1	0	0	0	0	0	0	0	0	0	0	0	0	1	0	0	0	0	0	0	1	0	0	0	1	0	0	0	0	0	0	0	0	0	0	0	1	0	0	0	0	0	0	0	0	0	0	0	0	0	0	0	0	0	0	0	1	0
63	0	0	0	0	0	1	0	0	0	0	0	0	0	1	0	0	0	0	0	0	0	0	0	0	0	0	0	0	0	0	0	0	1	0	0	0	0	1	1	0	0	0	0	0	0	0	1	0	0	0	0	0	0	0	1	0	0	0	0	0	0	0	0	0	0	0	0	0	0	0	0	0	0
64	0	0	0	0	0	0	0	0	0	0	0	0	0	0	0	0	0	0	0	1	0	0	0	0	0	0	0	0	0	0	0	0	0	0	1	0	0	0	0	0	0	0	0	0	0	0	0	0	0	0	0	0	0	0	0	0	0	0	1	0	0	0	0	0	0	0	0	0	0	0	0	0	0
65	0	0	0	0	0	0	0	0	0	0	1	0	0	0	0	0	0	0	0	0	0	1	1	0	0	0	0	0	0	0	0	0	0	0	0	0	0	0	0	0	0	0	0	0	0	0	0	0	0	1	0	0	0	0	0	0	0	0	0	0	0	0	0	0	0	1	0	0	1	0	0	0	0
66	0	0	0	0	0	0	0	0	0	0	1	0	0	0	0	0	0	0	0	0	0	0	1	0	0	0	0	0	0	0	0	0	0	0	0	0	0	0	0	0	0	0	0	0	0	0	0	0	0	0	0	0	0	0	0	0	0	0	0	0	0	0	0	0	1	0	0	0	0	0	0	0	0
67	0	0	0	0	0	0	0	0	0	0	0	0	1	0	1	0	0	0	0	0	0	0	0	0	0	0	0	0	1	0	0	0	0	0	0	0	0	0	0	0	0	0	0	0	0	0	0	0	0	0	0	0	0	0	0	0	0	0	0	0	1	0	0	0	0	0	0	0	0	0	0	0	0
68	1	0	0	0	0	0	1	0	0	0	0	0	0	0	0	0	0	1	0	0	0	0	0	0	0	0	0	0	0	1	0	0	0	1	0	0	0	1	0	0	0	0	0	0	0	0	0	0	0	0	0	0	0	1	0	0	0	0	0	0	0	0	0	0	0	0	0	0	0	0	0	0	0
69	0	0	0	0	0	0	0	0	0	0	0	0	0	0	0	0	0	0	0	0	0	1	1	0	0	0	0	0	0	0	0	0	0	0	0	0	0	0	0	0	0	0	0	0	0	0	0	1	0	0	0	0	0	0	0	0	0	0	0	0	0	0	0	0	1	0	0	0	0	1	0	0	0
70	0	0	0	0	0	0	0	0	0	0	0	0	0	0	0	0	0	0	0	0	0	0	0	0	0	0	0	0	0	0	0	0	0	0	0	0	0	0	0	0	0	0	1	0	0	0	0	1	0	0	0	0	0	0	0	0	0	0	0	0	0	0	0	0	0	0	0	0	1	0	0	0	0
71	0	0	0	0	0	0	0	0	0	0	0	0	0	0	0	0	0	0	0	1	0	0	0	0	0	0	0	0	0	0	0	0	0	0	1	0	0	0	0	0	0	0	0	0	0	0	0	0	0	0	1	0	0	0	0	0	0	0	0	0	0	0	0	0	0	0	0	0	0	0	0	0	0
72	0	0	0	0	0	0	0	0	0	0	0	0	0	0	0	0	0	0	0	0	0	0	0	0	0	0	0	0	0	0	0	1	0	0	0	0	0	0	0	1	0	0	0	0	1	0	0	0	0	0	0	1	0	0	0	0	0	0	0	0	0	1	0	0	0	0	0	0	0	0	0	0	0
73	1	0	0	0	0	0	0	0	0	0	0	0	0	0	0	0	0	0	0	0	0	0	0	0	1	0	1	0	0	0	0	0	0	0	0	0	0	1	0	0	0	0	0	1	0	0	0	0	0	0	0	0	0	0	0	0	0	1	0	0	0	0	0	0	0	0	0	0	0	0	0	0	0

APPENDIX C

Samples Drawn for Use in Examples PR-4.10 and PR-4.11

municipio (areal unit) number	name	region	LND	R-U	2-way sample	3-way sample
1	ADJUNTAS	4	1	0	0	1
2	AGUADA	3	0	0	0	0
3	AGUADILLA	3	0	1	1	0
4	AGUAS-BUENAS	5	1	0	0	1
5	AIBONITO	5	1	0	0	0
6	AÑASCO	3	0	0	0	0
7	ARECIBO	2	0	1	1	1
8	ARROYO	4	0	0	0	0
9	BARCELONETA/FLORIDA	2	0	0	0	2
10	BARRANQUITAS	1	1	0	0	0
11	BAYAMÓN/CATAÑO	1	0	1	0	0
12	CABO-ROJO	3	0	0	0	1
13	CAGUAS	5	1	1	0	1
14	CAMUY	2	0	0	1	0
15	CAROLINA	1	0	1	1	0
16	CAYEY	5	1	1	0	0
17	CEIBA	1	0	1	0	0
18	CIALES	2	1	0	1	0
19	CIDRA	5	1	0	1	0
20	COAMO	4	1	0	1	0
21	COMERÍO	1	1	0	0	0
22	COROZAL	1	1	0	0	2
23	DORADO	1	0	1	0	0
24	FAJARDO	1	0	1	0	0
25	GUÁNICA	3	0	1	0	1
26	GUAYAMA	4	0	1	0	0
27	GUAYANILLA	4	0	0	0	0
28	GUAYNABO	1	0	1	1	0
29	GURABO	5	1	1	0	0
30	HATILLO	2	0	1	0	0
31	HORMIGUEROS	3	1	1	0	2
32	HUMACAO	5	0	1	0	1
33	ISABELA	3	0	0	1	0
34	JAYUYA	4	1	0	0	0
35	JUANA-DÍAZ	4	0	0	1	0
36	JUNCOS	5	1	1	0	0
37	LAJAS	3	0	0	0	1
38	LARES	2	1	0	0	0
39	LAS-MARÍAS	3	1	0	1	0
40	LAS-PIEDRAS	5	1	0	1	0
41	LOÍZA/CANOVANAS	1	0	1	0	0
42	LUQUILLO	1	0	0	0	0
43	MANATÍ	2	0	1	0	0
44	MARICAO	3	1	0	0	1
45	MAUNABO	5	0	0	0	0
46	MAYAGUEZ	3	0	1	0	1
47	MOCA	3	1	0	0	1
48	MOROVIS	2	1	0	1	0
49	NAGUABO	5	0	0	1	0
50	NARANJITO	1	1	0	0	0
51	OROCOVIS	1	1	0	1	0
52	PATILLAS	4	0	0	0	0
53	PEÑUELAS	4	0	0	0	1
54	PONCE	4	0	1	0	1
55	QUEBRADILLAS	2	0	0	0	0
56	RINCÓN	3	0	0	0	0
57	RÍO-GRANDE	1	0	1	0	0
58	SABANA-GRANDE	3	1	0	0	0
59	SALINAS	4	0	0	0	0
60	SAN-GERMÁN	3	1	0	1	0
61	SAN-JUAN/RIO-PIEDRAS	1	0	1	0	0
62	SAN-LORENZO	5	1	1	0	1
63	SAN-SEBASTIÁN	3	1	0	0	0
64	SANTA-ISABEL	4	0	0	1	0
65	TOA-ALTA	1	1	1	1	1
66	TOA-BAJA	1	0	1	0	1
67	TRUJILLO-ALTO	1	1	1	0	1
68	UTUADO	2	1	0	0	0
69	VEGA-ALTA	1	0	1	0	0
70	VEGA-BAJA	2	0	1	0	0
71	VILLALBA	4	1	0	1	0
72	YABUCOA	5	0	0	1	0
73	YAUCO	4	0	0	0	1

NOTE: the number "2" in the 3-way column indicates that the areal unit appeared in the sample twice.

APPENDIX D

Supplemental PR Milk Production Data Set

No.	Commonwealth of PR Dept. of Agriculture Figures 1964	1969	1974	1978	1982	1987	US Dept. of Agriculture Figures 1964	1969	1974	1978	1982
SAN JUAN REGION											
10	0	0	0	0	0	0	7550	41630	267000	*	*
11	2672093	1604655	102531	59878	35841	0	2922314	2138026	436000	536572	0
15	14033611	10262735	8280500	7625554	5573678	5000103	13317705	8541701	8002731	8214933	7403404
17	0	792002	1809013	2202096	2313870	1924085	9000	589360	2285110	2120430	1965864
21	264539	819848	7237	0	0	0	444527	388160	*	0	0
22	5696265	6624193	4658968	3596822	1560085	1460842	5366303	7022251	4773619	4117500	2218967
23	12006149	9195465	7617649	7760060	6979189	8292068	8755956	11634823	6253588	8738918	7584698
24	1150160	2350181	2262487	1660615	1838940	1873737	2232290	2306669	1497610	1705967	2722163
28	2610652	2775561	998994	349162	0	0	2901504	2506923	*	0	*
41	6241386	4691619	3856040	3102756	2484712	1643452	6664487	6862428	*	*	*
42	8551690	7017380	4839372	6541081	4993144	2959376	7301865	12619261	4641515	4706330	3663881
50	0	0	302921	495123	0	0	18590	442658	339480	*	0
51	0	0	1271284	4307398	3902081	3984938	92252	32940	*	5100810	*
57	3273996	2698469	1064259	1031370	89376	0	3309590	3648237	1680180	1068400	*
61	852592	0	0	0	0	0	1154442	718	0	0	0
65	7658083	12916817	9614295	10568260	9198407	7758118	7324650	12651144	10332300	13116449	6872900
66	7578411	9039653	8115627	7501947	5022776	3807065	8025577	10716794	8924721	4742634	3165081
67	2354925	2539162	2701469	2346560	0	0	2388112	2375006	2297300	2549350	1311750
69	3399819	4645413	4957906	4251061	3268073	2371432	4374838	4083319	5475860	2100163	4662996
ARECIBO REGION											
7	16937083	17286423	22884936	38203103	41604707	49935518	18708476	17167566	23747548	44698364	42011208
9	939454	1923720	6707175	7032163	5051707	5520844	625280	2838686	4437011	9396000	6252132
14	3498442	4762902	10578658	21996145	26365066	31295320	4708437	3571196	12665988	21714688	22087568
18	1353576	2188879	1475921	1822624	978042	1096727	1648322	1158836	1295750	1988301	*
30	28551031	44545030	59479609	75249565	68892249	69871825	30619976	40184120	61251096	70505928	68118912
38	0	0	0	429661	537296	188296	11590	69355	7255	353760	*
43	11348753	17181860	14921183	18922745	16680120	15001491	11051904	15986775	16406435	19155458	12154450
48	2310327	6045412	8443864	7202544	7742467	7453032	2615656	5845657	7766575	8194205	8147004
55	1029180	2481681	3343789	6532494	8143577	9419056	1321336	2637598	3770048	7664873	8674976

APPENDIX D (continued)

ARECIBO REGION											
68	23527	26441	474182	2945592	2583809	1677781	195610	270962	1229650	3540435	2738268
70	4301232	7885334	8165236	8235935	7578861	6216778	4459941	9794951	7291250	7546727	8927660
MAYAGUEZ REGION											
2	278923	0	0	0	0	0	309931	4140	7300	*	*
3	717523	1491909	1669646	1368574	2505215	3140789	848763	1619875	803400	787610	1934100
6	101374	0	0	584159	708788	417643	101937	0	0	*	*
12	4002427	3892377	3734611	5920545	3979471	2185403	6743712	5369830	6511819	4327010	2362072
25	531271	179337	0	0	0	0	524858	573400	*	*	0
31	195455	111211	0	0	0	0	205475	150168	3120	0	*
33	3632966	5011175	7289758	7854882	9636269	12095536	4716089	5201356	8795784	11059316	9156835
37	3849897	1956200	1137296	683703	507298	824164	6186601	4198887	1463410	2753139	*
39	0	0	0	0	281992	368173	3724	980	0	0	599848
44	0	0	0	0	0	0	1030	0	0	*	*
46	1130610	1039218	816562	1227122	713009	1281276	1005495	1125159	*	1424060	*
47	119255	302142	356029	577684	348507	621854	200650	582900	*	*	*
56	0	0	0	0	0	0	5918	0	0	0	0
58	732884	656405	267284	200813	220423	233982	575144	1146916	498210	*	525009
60	298423	224430	174984	491407	506155	503378	1054855	524990	108000	529160	*
63	396555	901372	694113	6981252	8848332	12936978	442693	1307507	1942760	7237823	11020401
PONCE REGION											
1	908627	1142647	848942	620860	176916	97279	1242242	977794	546438	1350000	*
8	64662	0	0	0	0	0	89185	*	2542	0	0
20	5267464	3867512	2860063	2121083	727575	830111	6416250	5673089	3383500	2073845	*
26	4012313	2758835	464146	664892	333980	267353	4353964	4850033	1392215	*	*
27	586467	68274	0	0	0	0	843190	37115	296	0	*
34	199410	1064447	932477	770818	0	0	338384	1203626	781860	782872	*
35	3001211	4119319	2763379	2415862	2140958	237283	3605426	4134317	2085250	3220646	1318170
52	0	0	0	0	0	0	132368	33463	270	0	*
53	1705323	1607139	887307	272120	0	0	2071033	1652362	950609	*	0
54	2487179	1469077	899810	0	0	0	3612609	1990616	2053278	*	0
59	4633845	3336205	2051691	1090473	0	0	5665700	*	*	*	*
64	323863	295727	153099	0	0	0	852231	*	*	0	0
71	0	0	0	0	0	0	2190	1379	100	7225	0
73	598096	786004	612769	1410978	304758	291937	870391	979082	623850	421820	*

APPENDIX D (continued)

CAGUAS REGION

4	1088860	1847578	2227461	2164103	1287068	159736	2004850	2086960	1283908	2845742	3151164
5	1066594	2297073	1886448	2940490	1482399	104001	1085500	1982400	2078500	3513060	2700000
13	16685750	18974610	18896886	19437033	12637575	11794647	15175599	22116022	19410784	19534736	10945389
16	957922	2331087	3889252	5770622	5340617	2542185	804680	1356000	3575130	5936594	5979697
19	2219135	5644346	4997297	5377585	5942051	6081887	3209062	7050726	3477860	5699949	4794619
29	7313575	10142017	9801715	10756693	9556050	7495314	7606994	12344297	8409906	11432838	10722916
32	766297	1067051	2560843	2800214	3044974	5527618	1654790	1165255	2682564	*	3860250
36	9554952	13201677	13929787	15150242	16451347	12641576	10060969	12323522	14083494	14865794	12325144
40	3281799	8690917	9157692	11553370	10989400	11515158	3527840	8708629	8026002	14808209	11363490
45	0	0	0	0	0	0	13290	30828	2376	*	0
49	8535251	14970646	15608767	16165177	15497338	17872680	7749925	13879823	14161049	15862336	14482284
62	2331379	4454020	5389758	6099982	5971860	4398981	2600692	5317006	6889868	5198315	6371337
72	2025864	3137574	2595787	4217771	2368430	1220452	1743698	3301757	3328658	3380363	2091690

NOTE 1: " No." refers to the municipio (areal unit) identification numbers appearing in Appendix A.
NOTE 2: missing data contained in the USDA data set, denoted by *, have been estimated prior to analyses undertaken in this text.

APPENDIX E

Puerto Rico Urban Population Components Data Set

No.	Name	R-U	1970 total	1970 urban	1980 total	1980 urban
SAN JUAN REGION						
10	BARRANQUITAS	0	20118	4508	21639	3618
11	BAYAMÓN/CATANO	1	182651	174011	222449	215996
15	CAROLINA	1	107643	96990	165954	159055
17	CEIBA	1	10312	2950	14944	9104
21	COMERÍO	0	18819	6297	18212	5736
22	COROZAL	0	24545	5211	28221	5889
23	DORADO	1	17388	4388	25511	19075
24	FAJARDO	1	23032	18249	32087	26928
28	GUAYNABO	1	67042	55310	80742	76378
41	LOÍZA/CANÓVANAS	1	39062	11272	52747	38623
42	LUQUILLO	0	10390	0	14895	4531
50	NARANJITO	0	19913	3283	23633	2849
51	OROCOVIS	0	20201	3684	19332	0
57	RÍO-GRANDE	1	22032	6999	34283	19253
61	SAN-JUAN/RÍO-PIEDRAS	1	463242	452749	434849	433901
65	TOA-ALTA	1	18964	3199	31910	19156
66	TOA-BAJA	1	46384	35874	78246	73141
67	TRUJILLO-ALTO	1	30669	18477	51389	47884
69	VEGA-ALTA	1	22810	8688	28696	20531
ARECIBO REGION						
7	ARECIBO	1	73468	35484	86766	52457
9	BARCELONETA/FLORIDA	0	20792	4515	26174	10845
14	CAMUY	0	19922	3892	24884	8234
18	CIALES	0	15595	4046	16211	3582
30	HATILLO	1	21913	2760	28958	18749
38	LARES	0	25263	4545	26743	5224
43	MANATÍ	1	30559	13483	36562	21727
48	MOROVIS	0	19059	2892	21142	2637
55	QUEBRADILLAS	0	15582	2840	19728	3770
68	UTUADO	0	35494	11573	34505	11113
70	VEGA-BAJA	1	35327	17089	47115	30796
MAYAGUEZ REGION						
2	AGUADA	0	25658	4590	31567	9244
3	AGUADILLA	1	51355	28538	54606	48613
6	AÑASCO	0	19416	4416	23274	5646
12	CABO-ROJO	0	26060	7181	34045	10292
25	GUANICA	1	14889	8979	18799	9628
31	HORMIGUEROS	1	10827	6531	14030	12441
33	ISABELA	0	30430	9515	37435	12087
37	LAJAS	0	16545	3391	21236	4275
39	LAS-MARÍAS	0	7841	0	8747	0
44	MARICAO	0	5991	0	6737	0

APPENDIX E (continued)

46	MAYAGUEZ	1	85857	69558	96193	85714
47	MOCA	0	22361	0	29185	5677
56	RINCON	0	9094	0	11788	0
58	SABANA-GRANDE	0	16343	5561	20207	7435
60	SAN-GERMÁN	0	27990	11613	32922	13054
63	SAN-SEBASTÍAN	0	30157	7169	35690	10619
PONCE REGION						
1	ADJUNTAS	0	18691	5319	18786	5239
8	ARROYO	0	13033	5429	17014	8435
20	COAMO	0	26468	12077	30822	12851
26	GUAYAMA	1	36249	23038	40183	27985
27	GUAYANILLA	0	18144	5189	21050	6163
34	JAYUYA	0	13588	3826	14722	3588
35	JUANA-DÍAZ	0	36270	13160	43505	20844
52	PATILLAS	0	17828	2543	17774	3172
53	PEÑUELAS	0	15973	3169	19116	4235
54	PONCE	1	158981	131660	189046	168868
59	SALINAS	0	21837	10007	26438	9238
64	SANTA-ISABEL	0	16056	4495	19854	6948
71	VILLALBA	0	18733	4134	20734	3469
73	YAUCO	0	35103	12922	37742	17204
CAGUAS REGION						
4	AGUAS-BUENAS	0	18600	3426	22429	3766
5	AIBONITO	0	20044	7582	22167	9331
13	CAGUAS	1	95661	64523	117959	105252
16	CAYEY	1	38432	21562	41099	23305
19	CIDRA	0	23892	6306	28365	9763
29	GURABO	1	18289	7611	23574	13010
32	HUMACAO	1	36023	12411	46134	26264
36	JUNCOS	1	21814	7985	25397	18455
40	LAS-PIEDRAS	0	18112	4636	22412	6046
45	MAUNABO	0	10792	0	11813	2987
49	NAGUABO	0	17996	4629	20617	4135
62	SAN-LORENZO	1	27755	7702	32428	18947
72	YABUCOA	0	30165	7850	31425	9358

APPENDIX F

Syracuse Pediatric Blood Lead Levels Data Set

census tract number	sample size	cases of elevated lead level	u-co-ordinate	v-co-ordinate	Hispanic	total population	census tract area	housing stock age <43	housing stock age 43-53	housing stock age >63	housing stock number of units
2	53	15	40.5774	76.9708	43	3548	1.0	99	174	1295	1649
3	11	3	40.6491	77.0276	13	1625	1.0	66	107	426	711
4	19	6	40.7176	76.9606	23	3879	2.0	229	317	1017	1673
5	49	20	40.5487	76.8640	24	1425	0.5	9	73	573	677
6	50	16	40.6124	76.9076	56	3136	1.0	47	124	1005	1669
7	16	1	40.6645	76.8993	18	1569	0.5	21	39	519	739
8	20	8	40.7238	76.8843	44	2578	1.0	57	161	870	1155
9	18	4	40.8301	76.8742	32	3613	1.0	295	292	893	1522
10	33	6	40.9585	76.9072	80	4099	1.0	788	292	935	2231
13	56	33	40.6160	76.7970	14	1351	0.5	134	102	453	673
14	70	24	40.6480	76.8397	73	2737	0.5	26	93	974	1265
15	47	16	40.7114	76.8038	41	2492	1.0	99	81	883	1171
16	24	8	40.7395	76.7653	50	3010	1.0	380	88	772	2011
17	47	16	40.8575	76.7750	117	5214	2.0	268	193	1634	2385
18	11	1	40.9599	76.8174	30	2844	1.0	248	162	813	1375
19	12	1	41.0640	76.8334	28	4460	2.0	375	271	1034	2044
20	33	17	40.2850	76.7449	43	2471	1.0	5	80	928	1144
21	28	12	40.4109	76.7239	37	1928	1.0	33	50	832	997
22	27	13	40.5148	76.7091	34	1298	1.0	0	17	510	616
23	67	30	40.6464	76.7507	68	2518	0.5	29	75	1117	1527
24	68	35	40.7456	76.7182	65	2052	0.5	54	41	813	1036
27	14	7	40.2954	76.6930	18	2098	1.0	91	61	809	1090
28	14	2	40.3370	76.6379	8	2058	0.5	48	68	746	907
29	22	12	40.4175	76.6609	4	1075	1.0	37	48	438	518
30	59	32	40.5205	76.6291	228	2284	1.0	345	117	413	1053
32	17	1	40.6612	76.6621	53	1300	1.0	8	0	38	979
33	33	7	40.7218	76.6694	8	920	0.5	14	0	52	508
34	30	20	40.7925	76.6665	17	1479	0.5	0	96	660	844
35	61	28	40.8718	76.6662	29	3131	1.0	31	102	731	1413
36	74	15	41.0101	76.7069	30	5077	3.0	441	339	556	2184

APPENDIX F (continued)

38	71	41	40.3751	76.5205	123	2617	1.0	120	164	591	1164
39	315	206	40.5130	76.5415	647	5842	1.0	228	143	1849	2547
40	79	52	40.5198	76.5837	392	2384	1.0	81	82	749	1052
41	10	3	40.6131	76.5444	8	383	0.5	11	13	79	142
42	46	13	40.6648	76.5638	57	1791	0.5	105	121	373	995
45	21	4	40.8887	76.5026	62	4060	1.0	175	232	1268	1736
46	22	4	41.0502	76.5823	52	5784	4.0	733	313	915	2262
49	19	7	40.3939	76.4224	26	1573	1.0	111	120	309	636
50	22	6	40.4665	76.4279	26	3122	1.0	83	80	892	1159
51	70	36	40.5421	76.3719	65	3058	1.0	158	240	692	1178
52	98	70	40.5500	76.4854	70	3064	1.0	177	166	704	1211
53	83	51	40.6466	76.4849	51	2794	1.0	25	89	798	1003
54	179	125	40.6618	76.4052	141	4053	1.0	40	255	1097	1504
55	29	3	40.7535	76.3700	77	4229	3.0	347	193	214	2072
57	48	17	40.5120	76.2970	77	2656	1.0	161	303	443	1122
58	95	64	40.6102	76.3219	43	3098	1.0	138	173	657	1146
59	84	43	40.6729	76.3135	120	2805	1.0	143	122	700	1032
60	21	4	40.5684	76.1287	25	3887	4.0	653	183	553	1701
61	64	14	40.7220	76.1303	116	9138	5.0	822	299	979	4385
129	22	4	40.1037	76.8313	52	2575	2	23	108	895	1208
130	12	1	40.0864	76.7644	70	4109	2	440	228	777	1895
143	19	3	41.2901	76.8293	24	3343	4	56	183	824	1489
146	21	2	41.3328	76.7217	81	5654	9	817	257	136	1904
162	17	4	40.6178	75.8480	10	2395	6	340	122	413	940

NOTE: We wish to thank Ms. Mary Burdick and Mr. Gary Urquhart of the Onondaga County Health Department for provision of the Lead Poisoning Control Program monitoring data included with these 1990 U.S. census data.

APPENDIX G

Metropolitan Toronto Origin-Destination Data Set

Origin	Destination Planning District (PD)															
	#1	#2	#3	#4	#5	#6	#7	#8	#9	#10	#11	#12	#13	#14	#15	#16
PD # 1	1804	644	406	746	211	648	73	315	69	143	229	139	281	94	83	207
PD # 2	637	911	266	120	33	70	41	195	39	64	33	16	41	12	7	19
PD # 3	417	263	1284	235	48	53	29	182	100	325	113	33	52	8	11	37
PD # 4	749	118	243	1319	211	186	14	73	26	95	217	77	104	22	21	77
PD # 5	222	28	50	213	624	109	7	29	13	52	115	133	196	19	28	108
PD # 6	648	71	53	194	109	1150	14	33	17	25	43	40	257	68	27	67
PD # 7	78	39	29	14	6	13	308	171	27	17	9	2	6	2	2	6
PD # 8	317	194	178	73	31	32	174	1550	208	88	32	14	27	4	5	16
PD # 9	72	40	96	28	14	18	27	205	711	151	27	11	18	4	5	17
PD #10	148	62	327	95	45	25	18	90	150	1178	164	44	45	9	9	52
PD #11	221	32	114	216	116	43	9	33	30	172	897	149	57	10	12	74
PD #12	135	15	33	73	137	38	3	13	10	45	150	421	79	10	18	160
PD #13	276	38	54	105	198	247	8	26	18	44	53	79	1700	204	201	405
PD #14	95	12	7	21	21	69	2	4	4	7	9	12	199	290	61	43
PD #15	83	6	12	19	29	30	2	5	6	10	14	17	193	61	479	98
PD #16	210	19	36	78	107	73	5	17	18	52	76	158	407	41	93	1298

Source: Data Management Group, University of Toronto/York University Joint Program in Transportation (1989), *Travel Survey Summary for the Greater Toronto Area*, p. 118

INDEX